ELEMENTS
OF
MECHANICS

ELEMENTS
OF
MECHANICS

P. F. Kelly

CRC Press
Taylor & Francis Group
Boca Raton London New York

CRC Press is an imprint of the
Taylor & Francis Group, an **informa** business

CRC Press
Taylor & Francis Group
6000 Broken Sound Parkway NW, Suite 300
Boca Raton, FL 33487-2742

Printed on acid-free paper
Version Date: 20150925

International Standard Book Number-13: 978-1-4822-0654-8 (Hardback)

Visit the Taylor & Francis Web site at
http://www.taylorandfrancis.com

and the CRC Press Web site at
http://www.crcpress.com

Maps:

Functions or transformations from a domain space, into a codomain which includes the range of the function. For example, the trajectory [sequence of positions] *of a moving particle can be represented by a map from a one-dimensional parameter space, "time," into the* [Euclidean] *space accommodating the particle.*

"... whilst the great ocean of truth lay all undiscovered [unmapped] *before me."*

Sir Isaac Newton

Contents

List of Examples

Preface

My original motivation for writing this series of books, *Elements of Mechanics*, *Properties of Materials*, and *Electricity and Magnetism*, was to provide students attending my classes with advance copies of notes for each lecture. Successive accretions of explanatory material [much of it generated in response to student questions] transformed this project into something more substantial, which could be read as

- A sole source for a sequence of introductory physics classes

- A student supplement to a standard textbook

- A review in preparation for graduate/professional/comprehensive examinations

The provenance of these volumes as notes for lecture is manifest throughout in style and content. Physics is, in this telling, an engaging endeavour rather than a spectator sport. Also, attempts are made to ensure that the reasons for studying physics:

LEARN	to acquire knowledge about nature,
CONTROL	to exercise dominion over nature,
CREATE	to experience the joys of invention and artistry,

are not lost entirely under the weight of detailed investigation of specific models for particular phenomena. Finally, while occasional mention is made of *au courant* topics, our primary concern is with the foundations of classical physics. Readers who wish to more fully appreciate the "modern" developments in physics are encouraged to work through these notes first.

Very early on, while each book in the trilogy was scarcely more than an inkling, they were dubbed $\mathcal{MAPS}$, $\mathcal{SPAM}$, and $\mathcal{AMPS}$. These names derived fuller significance by interpretation as acronyms obtained from permutations of the four Latin words

$$\textit{Mechanica} \qquad \textit{Ars} \qquad \textit{Physica} \qquad \textit{Scientia}.$$

This artifice is a nod toward the deeper mathematical structures hinted at in these notes. The words themselves have import too.

$\mathcal{M}\,\mathcal{A}$ The *Art of Mechanics* conveys the notion that the practice of physics is essentially creative. In another riff on this pair of words, the *Artes Mechanicae*, as traditionally understood, are practical skills[1] such as those employed by artisans in the production of useful goods and decorative artwork. Throughout these notes, we shall craft many mathematical models providing serviceable approximations to [pertinent aspects of] physical systems.

---

[1] In the past, *Artes Mechanicae* suffered invidious comparison with *Artes Liberales* or "Liberal Arts."

$\mathcal{PS}$ The *Science of Physics* is the ordering of knowledge[2] of the physical world. This ordering culminates in the discernment of fundamental symmetries and the formulation of conservation laws and theorems which characterise [some would dare to say "govern"] the behaviour of physical systems.

The three volumes in the series are partially sequenced: $\mathcal{MAPS}$ is propaedeutic, while $\mathcal{SPAM}$ and $\mathcal{AMPS}$ may be subsequently read in either order. There is also a gradual shift in writing style from a more discursive tone [with text and mathematical syntactic redundancy] in $\mathcal{MAPS}$ to a somewhat terser one in $\mathcal{SPAM}$ and $\mathcal{AMPS}$. An aim is to be welcoming to those new to the study of physics. Our intent is that upon completing this series students be prepared for specialised-subject upper-level undergraduate physics courses and textbooks.

A burden of thanks is owed to all of the students whom I have taught, and especially to those who greatly enjoyed it.

[This came as a surprise to some!]

Special credit is owed to those who have assisted me in the preparation of figures, especially Andy Geyer, and Carl and Jaspar von Buelow.

---

[2]This aspect of the courses hews more closely to traditional notions of liberal education.

Preface for Volume I

The acronym associated with this VOLUME is $\mathcal{MAPS}$, for multifarious reasons. The direct association with mathematical mappings has been alluded to in the first of the epigraphs. The latter epigraph is excerpted from a famous comment by Newton. Here, its purpose is to intimate that the formal study of [elementary Newtonian] classical mechanics provides one with a set of conceptual charts for organising glimpses of the *lay of the land* [of nature]. Furthermore, these rough charts help one *get one's bearings* in preparation for further explorations in VOLUMES II and III (and beyond). Another aphorism of Newton's, *i.e.*, that the scientists in every age stand on the shoulders of giants, is *apropos*. Throughout these pages, we shall make occasional pause to pay homage to those giants on whose scientific contributions we rely in our present-day understanding of the natural world.

There is yet another sense in which "maps" proves apposite. Notions of recursive approximation pervade both cartography and physics. Here, we shall consider an illustrative example. Throughout this VOLUME, and indeed the SERIES, we will be performing analogous analyses in a variety of physical contexts.

EXAMPLE [*Approximating Area and Predicting Perimeter*]

Prince Edward Island [PEI], the smallest and least populous Canadian province, is located at approximate latitude 46° 30′ N, and longitude 63° 00′ E.

> ASIDE: Charlottetown, PEI's capital, is called the *birthplace of Canada*, because the first of the discussions which culminated in the establishment of the Dominion of Canada (1867) took place there. However, PEI didn't enter into the Canadian Confederation until 1873!

On a globe that I have in my home, PEI appears as a tiny oblong blob. On a world map which adorns one of my walls, PEI is represented by a small seven-sided region. The map of Canada bedecking another wall includes a more faithful representation of the island's coastline. My road atlas contains a highway map which resolves many details pertinent to travel: primary and secondary roads and population centres. Mapping and satellite photo websites provide astonishingly detailed images of geographical features.

Oftentimes, it is only necessary to know that PEI exists, and then its description as a blob at the aforementioned approximate latitude and longitude suffices. For some other purposes, it may prove necessary to develop a model for the shape of the island. One's motivations for creating such a model, along with external circumstances and constraints, have a strong determining influence on the **accuracy** and **precision** of the ultimate construction. Considerations of accuracy and precision, in turn, help to guide one in formulating the model. This interplay is exhibited here as we attempt to estimate the area and the perimeter (coastline length) of PEI.

For instance, suppose that the goal is to estimate the land area to the nearest 10000 square kilometres,[3] and the perimeter to the nearest 10000 kilometres.[4] The blob and the septagon appearing on the globe and the world map enable one to roughly guess that the area and perimeter of PEI are

$$\text{Area} \sim 1 \times 10^4 \text{ km}^2 \qquad \text{and} \qquad \text{Perimeter} \sim 0 \times 10^4 \text{ km}.$$

---

[3]Ten thousand square kilometres is the area of a square with edge length 100 km, or [approx.] 62.5 miles. This corresponds to a region approximately three times the size of the state of Rhode Island or two-thirds the size of Connecticut.

[4]The metre's original definition was "one ten-millionth of the distance from the North Pole to the equator along the meridian passing through Paris." [It is reported that, while passing through Paris, the meridian stops for a *croissant et cafe au lait*.] Thus 10^4 km is one fourth of the circumference of the Earth.

One may feel underwhelmed by these very imprecise results. Nevertheless, we have accomplished a significant feat. Prior to this analysis, we had very little idea whatsoever about the size of PEI [except that it was smaller than the entire Earth].

IF these results are sufficient for our purposes, THEN we [usually] stop.

The qualifier "usually" appears here because one may always choose to further refine the model. In this particular case the result of ZERO for the perimeter was accurate, but not very satisfying. Even in the absence of a compelling need, one might wish to crank up the resolution in order to ascertain the relevant distance scale for the length of the coastline.

Consulting the map of PEI found in the road atlas, one can imagine "tiling" an area which approximates the shape of the island with five 1000 square kilometre tiles [squares with side 31.6 km, or 25 km × 40 km rectangles, *etc.*]. One can estimate the perimeter of the same shape to be ∼ 1 × 1000 km. Hence

$$\text{Area} \simeq 5 \times 10^3 \text{ km}^2 \qquad \text{and} \qquad \text{Perimeter} \simeq 1 \times 10^3 \text{ km}$$

are improved approximations for the area and coastline length of PEI. With enhanced resolution, one might[5] be able to further improve on the accuracy and the precision of these revised estimates.

We end this cartographic example by observing that **the resolution required of the model and its inputs is contingent upon the desired accuracy of the result.** The same is true for the modelling of physical phenomena.

ASIDE: The successes enjoyed by specific physical models in the description of particular phenomena must not be taken to mean that the models' constructs actually exist in nature. For example, despite the efficacy of the tiling model for a rough determination of PEI's geography, one would not expect the island to consist of five rectangular tiles. The best that we can say is that the models mimic some aspects of the presumed *reality*™.[6]

---

[5]The qualifier "might" appears for practical and theoretical reasons. Practical considerations (*e.g.*, lack of time or funding) may impede conversion of high-resolution geographical data into measured areas and perimeters. The theoretical reason is that the coastline may exhibit fractal behaviour. If this is the case, then the estimates for the perimeter will grow without bound as the resolution is increased!

[6]We whimsically employ the trademark symbol whenever we refer to physical reality as a clear and cautionary reminder that the assumption of its existence is metaphysical.

Part I

Mechanics

Chapter 1

Physics and Measurement

PHYSICS concerns itself with the fundamental processes which underlie all natural phenomena. Traditionally, there are two main branches of physics:

EXPERIMENT AND **THEORY**,

although, today, some suggest that computational physics constitutes a third branch.

EXPERIMENTAL PHYSICS   Experimental physicists attempt to quantify and measure specific natural phenomena.

THEORETICAL PHYSICS   Theorists attempt to derive [or otherwise explicate] physical laws adhered to by nature, and to make predictions of corroborative observable phenomena within specific models.

The ideas of physics find their most natural expression in the language of mathematics. The claim is often made that mathematics is, in the Platonic sense, pure, exact, and absolute. In contrast, physics involves an iterative sequence of approximate descriptions. Throughout these notes, we shall encounter various and sundry physical models which are [intended to be] approximations to the actual systems and processes of nature. No small part of the inestimable and ineffable beauty of the discipline of physics derives from the degree of success enjoyed by *simple* models.

This volume of lectures introduces and develops the subject of MECHANICS.

MECHANICS   The study of the motion of simple systems of particles.

PARTICLE   A particle is an object which experiences translational motion only. That is, there are no internal or relative motions, only so-called "bulk motion."

MECHANICS = KINEMATICS + DYNAMICS

KINEMATICS   Kinematics addresses the query *"How is it moving?"* and thus is essentially descriptive. [Chapters 1 to 8 deal primarily with kinematics.]

DYNAMICS   Dynamics takes up the question *"Why is it moving?"* and herein lies the greater part of the intellectual richness of physics.

The description of motion requires that quantities with dimensions of length, time, and [for later use] mass be defined and manipulated. In order to do this coherently and reproducibly, STANDARDS OF MEASURE must be introduced. In truth, any set of standards will do, but scientists have, as a community, adopted *le Système International d'Unités*, henceforth referred to as "SI," as the *universal* standard.

ASIDE: This pan-scientific consensus does not prevent individuals [and entire disciplines] from employing other schema of units, provided that they be prepared to convert to SI when communicating their results to the rest of us. For instance, astronomers and chemists, by tradition, generally use centimetre–gram–second [CGS] units.

The SI dimensional standards for length, time, and mass are given in the following table.

QUANTITY	$[Q]$	SI UNIT	ABBREVIATION
Length	$[L]$	metre	m
Time	$[T]$	second	s
Mass	$[M]$	kilogram	kg

The bracket notation "$[Q]$" is used consistently throughout these VOLUMES to mean "the SI unit associated with the dimensional quantity Q."

A feature of SI is that the basis units are naturally scalable [usually in powers of 10^3], which helps us to avoid errors and build intuition.

Prefix		Power		Prefix		Power
Yotta	Y	10^{24}		yocto	y	10^{-24}
Zetta	Z	10^{21}		zepto	z	10^{-21}
Exa	E	10^{18}		atto	a	10^{-18}
Peta	P	10^{15}		femto	f	10^{-15}
Tera	T	10^{12}		pico	p	10^{-12}
Giga	G	10^{9}		nano	n	10^{-9}
Mega	M	10^{6}		micro	μ	10^{-6}
kilo	k	10^{3}		milli	m	10^{-3}
hecto	h	10^{2}		centi	c	10^{-2}
deca	da	10^{1}		deci	d	10^{-1}

SI Prefixes and Their Associated Multipliers

From these elementary units of length, time, and mass, one can obtain new units by forming suitable combinations. An unsuitable combination would be

$$5.11\,\text{kg} + 1.25\,\text{m} + 3.64\,\text{s} \neq 10.00\,\text{kg} + \text{m} + \text{s},$$

as this is obviously nonsensical. On the other hand, a perfectly good and useful physical quantity is the mass per unit volume of a substance, *a.k.a.* the mass density:

$$\rho = \frac{\text{Mass}}{\text{Volume}}, \qquad \text{and hence} \qquad [\,\rho\,] = \frac{[\text{M}]}{[\text{L}]^3} = \frac{\text{kg}}{\text{m}^3}.$$

EXAMPLE [*Units of Density*]

Aluminium has mass density $\rho_{\text{Al}} = 2700\,\text{kg}/\text{m}^3$ [at standard temperature and pressure].

Q: A sample consists of two cubic centimetres of aluminium. What is its mass?

A: The mass of the sample is equal to its density multiplied by its volume. The volume, two cubic centimetres, is

$$2\,\text{cm}^3 = 2 \times (.01\,\text{m})^3 = 2 \times 10^{-6}\,\text{m}^3.$$

Hence,

$$M = 2700\,\frac{\text{kg}}{\text{m}^3} \times \left(2 \times 10^{-6}\,\text{m}^3\right) = 5.4 \times 10^{-3}\,\text{kg} = 5.4\,\text{g}.$$

ASIDE: Intriguing questions are:

Q: Why do different solid chemical elements [*e.g.*, aluminium, iron, carbon, *etc.*] have rather different densities?

A: The major contribution to density variance arises from elemental nuclear mass differences. The variation in atomic "size" [a fuzzy concept, to say the least, when quantum mechanics is considered] has only a minor effect.

Q: How is it possible for different phases of the same substance [*e.g.*, solid ice, liquid water, vaporous steam] to have different densities?

A: The empirical fact that the densities are different suggests that the spacing of atoms and molecules is phase-dependent.

Scientific Notation and Significant Figures

We presume that everyone is familiar with the standard form of representing numerical quantities, *viz.*, $X.abc \times 10^y$. In the case just displayed there are four significant figures. Implicit in this usage is the notion that the estimated **error** [of experimental and/or theoretical origin] is sufficiently small to render meaningful all of the quoted digits.

[In these notes we shall be lax and idiosyncratic in our use of scientific notation.]

Using Dimensional Analysis to Check Calculated Results

RULE:

IF the units are incommensurate with the dimension of the quantity,
THEN the answer is incorrect.

STANDARD DISCLAIMER: Getting consistent units in the solution to a problem does not guarantee that the result obtained is correct.

EXAMPLE [*Units of Velocity*]

IF one wishes to compute the velocity of a particle, THEN the dimensions of the result had better be $[\mathrm{L}/\mathrm{T}]$.

> ASIDE: There is a nearly[1] direct correspondence between the dimension of a quantity [general] and the SI unit [specific] associated with that dimension. For this reason, we will play *fast and loose* on occasion and elide the distinction between dimension and units.

EXAMPLE [*Dimensional Consistency*]

Suppose that a quantity, a, is known [from a definition, perhaps] to have SI units

$$[\,a\,] = \left[\frac{\mathrm{L}}{\mathrm{T}^2}\right] = \frac{\mathrm{m}}{\mathrm{s}^2}\,.$$

Further suppose that two other quantities, X and t, with dimensions of length and time, respectively, are mutually related through a. Unfortunately, we have forgotten the precise relation, but have narrowed it down to one of the following two forms:

$$X = \frac{1}{2}\,a\,t^{10} \qquad \text{OR} \qquad X = \frac{1}{10}\,a\,t^2\,.$$

Dimensional analysis enables us to choose correctly between these candidate formulae. The dimension of each left-hand-side, hereafter abbreviated LHS, is $[\,\mathrm{L}\,]$, by assumption.

FIRST FORMULA   The right-hand-side (*a.k.a.* RHS) of the first expression has dimension

$$\left[\frac{\mathrm{L}}{\mathrm{T}^2} \times \mathrm{T}^{10}\right] = \left[\mathrm{L} \times \mathrm{T}^8\right],$$

which cannot possibly be consistent. The first formula is certainly incorrect.

SECOND FORMULA   The RHS of the second formula has dimension

$$\left[\frac{\mathrm{L}}{\mathrm{T}^2} \times \mathrm{T}^2\right] = \left[\mathrm{L}\right],$$

and thus this expression constitutes the correct relation among the quantities a, X, and t in this particular instance.

---

[1] This caveat becomes clear once we realise that there are "fundamental" units (metres, seconds, kilograms, ..., plus four more) and "derived" units (*e.g.*, the unit of force, the newton, is equal to one kilogram metre per second-squared).

EXAMPLE [*Deriving the Form of Kepler's Third Law by Dimensional Consistency*]

Johannes Kepler (1571–1630) laboured mightily to discover the three laws of planetary motion[2] which bear his name. The third of these laws ["K3"] posits a relation

$$a^3 \propto \mathcal{T}^2 ,$$

between two parameters of planetary orbits:

a DISTANCE, a [the semi-major axis], and a TIME, $\mathcal{T}$ [the period].

Years later, Newton formulated the Law of Universal Gravitation [see Chapter 47 *et seq.*], characterised by a constant, G, with dimensions:

$$[\,G\,] = \frac{\mathrm{m}^3}{\mathrm{s}^2 \cdot \mathrm{kg}} .$$

Under the reasonable assumption that the mass of the sun, $M_\odot$, sets the scale for solar system gravitational effects [on account of its being so very much greater than those of the planets], naive dimensional analysis alone suggests that

$$a^3 = (\text{some constant}) \times G\,M_\odot \times \mathcal{T}^2 .$$

EXAMPLE [*Moving the Football*]

While packing toys in preparation for a household relocation, PK stumbled upon [literally!] his children's toy American-style football. This led him to ponder the following questions:

Qs: What is the radius of the sphere which has the same volume as the football?
Qc: What is the edge length of the cube which has the same volume as the football?

A: These are non-trivial questions!

The toy football is approximately ellipsoidal, with semi-major/minor axes $a \simeq 9\,\mathrm{cm}$, $b \simeq 6\,\mathrm{cm}$, and $c \simeq 6\,\mathrm{cm}$ respectively. [In truth, each of these estimated values is accurate to $\pm 0.5\,\mathrm{cm}$.] The volume formulae for an ellipsoid, a sphere of radius r, and a cube of edge length L are

$$V_e = \frac{4\,\pi}{3}\,a\,b\,c \quad , \qquad V_s = \frac{4\,\pi}{3}\,r^3 \quad , \qquad V_c = L^3 .$$

The scale of the system [that one can hold the toy football easily in one's hand] militates for the use of cubic centimetres to express the volume of the ellipsoid. Plugging in the measured values parameterising the size and shape of this particular football yields

$$V_e = \frac{4\,\pi}{3}\,9 \cdot 6 \cdot 6\ \mathrm{cm}^3 = 432\,\pi\ \mathrm{cm}^3 .$$

As: Setting the volume of the sphere equal to this amount and solving for the radius yields

$$\frac{4\,\pi}{3}\,r^3 = \frac{4\,\pi}{3}\,9 \cdot 6 \cdot 6\ \mathrm{cm}^3 \implies r^3 = 9 \cdot 6 \cdot 6\ \mathrm{cm}^3 \implies r = 6\,\sqrt[3]{\frac{3}{2}}\ \mathrm{cm} \simeq 6.87\,\mathrm{cm} \simeq 7\,\mathrm{cm} ,$$

the answer to the first question.[3] Interestingly, r turns out to be the **geometric mean** of the three semi-axes.

---

[2] Kepler's Laws are discussed in some detail in Chapter 50.
[3] When PK did the computation he also performed a simple error analysis which is discussed in the Addendum on the next page.

[Betcha never thought that the geometric mean would ever be useful, eh?]

Ac: In a similar manner, setting the volume of the cube equal to the volume of the ellipsoid and solving, we obtain

$$L = 6 \sqrt[3]{2\,\pi} \ \text{cm} \simeq 11.07 \ \text{cm} \simeq 11 \ \text{cm}$$

for the edge length of the cube.

ADDENDUM: Rudimentary Discussion of Error in the Football Example

Each of the measured ellipsoidal axes is believed to be accurate to $\pm 1/2$ cm. Therefore the fractional error in the volume is crudely estimated to be

$$\frac{\Delta V}{V} = \frac{\Delta a}{a} + \frac{\Delta b}{b} + \frac{\Delta c}{c} = \frac{0.5}{9} + \frac{0.5}{6} + \frac{0.5}{6} = \frac{2}{9}.$$

Thus, the calculated volume [in cubic centimetres] of the football, *viz.*, $432\,\pi$, is only accurate to about 22%. The computed values of the radius of the sphere and the edge-length of the cube possess fractional errors which are one-third that of the volume [owing to the taking of cube roots],

$$\frac{\Delta r}{r} \simeq \frac{\frac{\Delta V}{V}}{3} \simeq \frac{\Delta L}{L}.$$

The fractional errors in r and L are, therefore, approximately 2/27, or 7.4%, and the estimates for the radius and the edge-length are properly stated as

$$r = 6.9 \pm 0.5 \ \text{cm} \qquad \text{and} \qquad L = 11.1 \pm 0.8 \ \text{cm},$$

respectively.

Chapter 2

Kinematics in One Dimension

Q: Why restrict ourselves to one dimension when the world is clearly[1] three-dimensional?

A: Things are less complicated in one dimension [1-d], more clearly exposing the essential physics.

We'll begin with a collection of primitive definitions pertinent to the study of motion.

DISTANCE   Distance is a length traversed through some [explicitly or implicitly] specified time interval. Distance might be measured [employing an odometer for an automobile or bicycle, or a pedometer for a person walking or running], or may be inferred.

TIME INTERVAL   Mathematically speaking, time is any monotonic parameter employed to describe the motion of a particle or system. *The* time is read from a particular [calibrated] clock. In this sense, a time interval begins at some initial time, t_i, and ends at some final time, t_f, as measured on the clock [or on a pair of synchronised clocks]. The duration of the time interval is
$$\Delta t = t_f - t_i \, .$$

AVERAGE SPEED   The average speed of a particle is the total distance traversed by the particle through time interval Δt, divided by Δt. *I.e.,*

$$s_{\mathrm{av}} = \frac{\text{Distance traversed in time interval}}{\text{Duration of time interval}} \, .$$

The adjective AVERAGE is a reminder that this is a single representative value characterising the motion of the particle throughout the entire time interval under consideration.

DISPLACEMENT   Displacement is the difference in the absolute positions[2] of a particle through time interval Δt. Denoting the initial position of the particle [at time t_i corresponding to the start of the time interval] by x_i, and the final position of the particle [at t_f, the end of the time interval] by x_f,

$$\text{displacement} \; = \; \text{final position} \; - \; \text{initial position} \; = x_f - x_i \, .$$

Several aspects of displacement warrant particular attention here.

- Although distance is always positive indefinite, *i.e.*, greater than or equal to zero, displacement may be POSITIVE, ZERO, or NEGATIVE.

---

[1]Currently, there is much speculation that there exist additional spatial dimensions which happen to be hidden from our direct view.

[2]The notion of difference in position will become completely unambiguous when vectors are formally introduced in Chapter 5.

- Displacement is the NET distance travelled [colloquially: "as the crow flies"].

- Oftentimes, the goal is to ascertain the mathematical expression of a particle's **trajectory**, *i.e.*, its position as a function of time.

    IF the trajectory, $x(t)$, is known, THEN $\{x_f = x(t_f),\ x_i = x(t_i)\}$.

AVERAGE VELOCITY   The average velocity is the displacement through the time interval Δt, divided by Δt, *viz.*,

$$v_{\text{av}} = \frac{x_f - x_i}{t_f - t_i} = \frac{x(t_f) - x(t_i)}{t_f - t_i}\ .$$

EXAMPLE [*Displacement and Velocity vs. Distance and Speed*]

Let's apply these definitions of kinematic quantities in the context of three bouts of exercise on three successive days.

Day 1: PK jogged due north from the campus along a straight path for $10\,\text{km}$ in a total time of 1 hour. [We adopt the convention that northward is the "positive" direction.]

Distance = $10\,\text{km}$	$\Delta t = 1\,\text{h}$	Displacement = $+10\,\text{km}$
(average) speed = $10\,\text{km/h}$		(average) velocity = $+10\,\text{km/h}$

[Oooops! PK realised that he then had to, somehow, make his way back to the university.]

Day 2: PK ran $5\,\text{km}$ due north from campus, quickly turned around, and ran the $5\,\text{km}$ back (*i.e.*, south), all in 1 hour.

Distance = $10\,\text{km}$	$\Delta t = 1\,\text{h}$	Displacement = $0\,\text{km}$
(average) speed = $10\,\text{km/h}$		(average) velocity = $0\,\text{km/h}$

Day 3: For variety, PK chose to bicycle $5\,\text{km}$ due north (it took 20 minutes), to stop for a picnic (also 20 minutes), and to return in another 20 minute ride.

Distance = $10\,\text{km}$	$\Delta t = 1\,\text{h}$	Displacement = $0\,\text{km}$
(average) speed = $10\,\text{km/h}$		(average) velocity = $0\,\text{km/h}$

The differences between [and the limitations of] distance and displacement, and speed and velocity, are made manifest in these scenarios.

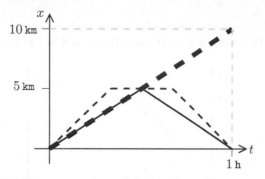

FIGURE 2.1 Rudimentary Spacetime Diagrams of PK Trajectories (The heavy-dashed, solid, and thin-dashed lines correspond to Days 1, 2, and 3, respectively.)

It is often helpful to draw "spacetime diagrams" as an aid to understanding the physics of a situation. Trajectories associated with each day's exercise considered in the example above, are represented in Figure 2.1.

A COMMENT, a CAVEAT, and a WARNING all bear mentioning here.

COMMENT: It is generally the case that the average speed will be greater than or equal to the magnitude of the average velocity.

CAVEAT: Do not be misled by the spacetime diagrams in Figure 2.1. That is, be cognizant of one's natural tendency to join **events** [points in spacetime] by unwarranted straight or smooth lines. The curves are intended to be suggestive, not authoritative, as illustrated by the higher-resolution sketch in Figure 2.2. Here, the grey line represents the average Day 1 behaviour, *i.e.*, 10 km in 1 h, while the black line shows that rather than maintaining a strictly constant pace, PK started slowly, hit a cruising pace which was faster than 10 km/h for the middle stretch, and cooled down at the end, running slowly. Determination of the more detailed trajectory required information gathered on much shorter time scales.

WARNING: Do not be again misled[3] by the apparent smoothness of the higher-resolution trajectory. Still higher temporal resolution may bring into view hitherto unseen aspects of PK's jog.

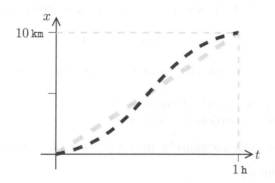

FIGURE 2.2 A More Detailed Portrayal of PK's First 10 km Run

---

[3] As the old saw goes: "Fool me once, shame on you; fool me twice, shame on me."

Archetypical behaviours of the average velocity are illustrated in Figure 2.3.

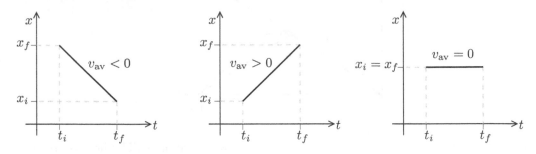

FIGURE 2.3   Archetypical Net Displacements and Average Velocities in One Dimension

Let's define another primitive kinematical quantity.

INSTANTANEOUS VELOCITY   The instantaneous velocity is the average velocity occurring through an infinitesimally small time interval, *viz.*, $\Delta t \to 0$. *I.e.*,

$$v = \lim_{\Delta t \to 0} v_{\text{av}} = \lim_{\Delta t \to 0} \frac{\Delta x}{\Delta t} = \frac{dx}{dt}\,.$$

The instantaneous velocity is the "default" velocity of interest and thus doesn't warrant a subscript distinguisher. On spacetime diagrams, $v(t)$ is the slope of the tangent line to the [putative] trajectory $x(t)$. As developed in calculus class, v may be approximated by values of the slopes of secant lines intersecting the trajectory in the vicinity of $\big(t, x(t)\big)$.

EXAMPLE [*Determining Velocity from a Known Trajectory*]

Suppose that the trajectory of a particular particle is [somehow] known to be

$$x(t) = b\,t^3\,, \quad \text{for some constant } b.$$

[Dimensional consistency demands that the units associated with b are $[b] = \mathbf{m/s^3}$.]

Q1: What is the velocity of the particle at time $t = 1$?
Q2: What is the velocity of the particle at time $t = 2$?

A1 & A2: We shall answer both questions by means of three different solution strategies.

METHOD ONE: [*Graphical*]
From a graph of $x(t)$, we can estimate the slopes of the tangents at the points $\big(1, b\big)$ and $\big(2, 8b\big)$. Two not-so-minor hitches may occur. We might find this process awkward to implement without specifying a particular value for b. Also, it is difficult to obtain high-precision results with this method.

METHOD TWO: [*Calculus Light*[4]]

Let's formally choose a temporal window including the time of interest, *viz.*,

$$t_i < t_0 < t_f \,.$$

The width of the window is dictated by the initial and final times:

$$t_f - t_i = \Delta t \,.$$

The corresponding displacement of the particle throughout the temporal window is

$$\Delta x = x_f - x_i = x(t_f) - x(t_i) \,.$$

Substituting $t_f = t_i + \Delta t$ into the given function for x yields

$$\Delta x = b \left(t_i + \Delta t\right)^3 - b \left(t_i\right)^3 = b \left[3\,t_i^2\,\Delta t + 3\,t_i\,(\Delta t)^2 + (\Delta t)^3\right] \,.$$

Hence,

$$v(t_0) = \lim_{\Delta t \to 0} \frac{\Delta x}{\Delta t} = b\left[3\,t_0^2\right] = 3\,b\,t_0^2 \,.$$

In the next-to-last equality, we recognised that "squeezing" the window necessarily entails $t_i \to t_0$. With this general result in hand, we can ascertain the velocities at the times $t = 1$ and $t = 2$.

A1: At $t = 1$, the velocity $v(1) = 3\,b$. | **A2:** At $t = 2$, the velocity $v(2) = 12\,b$.

METHOD THREE: [*Calculus Right*]

Let's just go ahead and take the derivative of the trajectory, *i.e.*, the position function,

$$v = \frac{dx}{dt} = \frac{d}{dt}\left(b\,t^3\right) = 3\,b\,t^2 \,.$$

Thus, the velocity is $3\,b$ at time $t = 1$ (**A1**), and $12\,b$ at time $t = 2$ (**A2**).

The velocity in this last example was not constant in time. The particle **accelerated** during the time interval from $t = 1$ to $t = 2$.

AVERAGE ACCELERATION   The average acceleration is the change in the velocity throughout the time interval Δt, divided by Δt,

$$a_{\mathrm{av}} = \frac{v_f - v_i}{t_f - t_i} \,.$$

As was the case for average velocity [deduced from spacetime graphs showing trajectories], there are three archetypes of average acceleration evident on graphs of [instantaneous] velocity *vs.* time.

---

[4]We cannot bear to write "Calculus Lite" (sic).

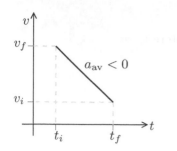
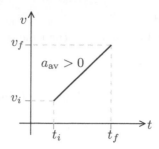
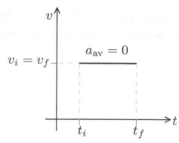

FIGURE 2.4 Archetypical Net Changes in Velocity in One Dimension

EXAMPLE [*Average Acceleration from Known Velocities*]

In the previous example, the average acceleration through the time interval from $t = 1$ to $t = 2$ is

$$a_{\text{av}} = \frac{v(2) - v(1)}{2 - 1} = \frac{12\,b - 3\,b}{2 - 1} = 9\,b.$$

Careful tracking of units reveals that $[\,a_{\text{av}}\,] = \text{m}/\text{s}^2$, as required.

Most often, the quantity of greatest physical import is the instantaneous acceleration.

INSTANTANEOUS ACCELERATION   Instantaneous acceleration is the average
acceleration through an infinitesimally small time interval, *viz.*, $\Delta t \to 0$. I.e.,

$$a = \lim_{\Delta t \to 0} a_{\text{av}} = \lim_{\Delta t \to 0} \frac{\Delta v}{\Delta t} = \frac{dv}{dt} = \frac{d}{dt}\left(\frac{dx}{dt}\right) = \frac{d^2 x}{dt^2}.$$

The instantaneous acceleration is regarded as the "default" acceleration. On spacetime
diagrams, $a(t)$ is the curvature of the trajectory $x(t)$ at time t. This is consistent with its
being the second time derivative of $x(t)$.

All of this is *peachy*, but presupposes that we somehow know the trajectory of the particle.

Q: Can knowledge of the acceleration lead to the velocity and then to the trajectory?
A: Yes! You betcha! We use integral calculus:

$$a = \frac{dv}{dt} \qquad \Longleftrightarrow \qquad \Delta v = \int_{t_i}^{t_f} a(t')\,dt'$$

$$v = \frac{dx}{dt} \qquad \Longleftrightarrow \qquad \Delta x = \int_{t_i}^{t_f} v(t')\,dt'$$

Chapter 3

More Kinematics in One Dimension

Let's revisit aspects of Chapters 1 and 2 in order to sharpen our diction in two respects and to examine the special case in which the instantaneous acceleration is constant throughout the time interval under consideration. First, we reiterate and amplify slightly our working definition of particle.

PARTICLE   A particle is an object which undergoes translational motion only. It moves through space in a *straightforward* manner. No consideration is given to any form of internal motion. Particles are often assumed to be point-like and/or structureless, but this need not be the case.

For example, when PK ran and bicycled in Chapter 2, he [along with his togs and equipment] was treated as a single particle. We chose to neglect the fact that, as PK ran and rode, his legs moved back-and-forth and up-and-down, his arms swung, his tongue lolled, *etc.*

> ASIDE: A kinesiologist might prefer a more refined model in which PK is comprised of a system of particles, each representing some part of his anatomy: trunk, head, knees, elbows, feet, hands, and so forth. Such a collection of particles is better able to accommodate a model description of the correlated internal motions associated with human locomotion.

Second, the AVERAGE appearing in average velocity, and average acceleration, is taken over time. To see why, let's employ some awfully wrong reasoning alongside correct reasoning and note the difference.

EXAMPLE [*Various Means of Average Taking*]

A car travels 5 km north at 10 km/h in 30 minutes. It then travels another 5 km north at 100 km/h in 3 minutes.

Q: What is the average velocity of the car during this trip? [North is the positive direction.]

As: The spatial average velocity of the car is

$$\widetilde{v}_{\text{av}} = \left(\frac{5 \,\text{km}}{10 \,\text{km}} \right) \times 10 \,\text{km/h} + \left(\frac{5 \,\text{km}}{10 \,\text{km}} \right) \times 100 \,\text{km/h} = 55 \,\text{km/h} \,.$$

At: The car's temporal average velocity is

$$v_{\text{av}} = \frac{\text{net displacement}}{\text{net time taken}} = \frac{10 \,\text{km}}{33 \,\text{minutes}} = \frac{600}{33} \,\text{km/h} \approx 18.2 \,\text{km/h} \,.$$

The two distinct averages yield sharply differing results. The temporal average better corresponds to our intuitive notion of what the average velocity should be, because it actually took the car slightly more than one-half hour to travel 10 km.

Third, we recall the last bit from Chapter 2:

$$a = \frac{dv}{dt} \quad \Longleftrightarrow \quad \Delta v = \int_{t_i}^{t_f} a(t')\,dt'$$

$$v = \frac{dx}{dt} \quad \Longleftrightarrow \quad \Delta x = \int_{t_i}^{t_f} v(t')\,dt'$$

These relations are correct and easily expressed, in principle, and yet oftentimes are hard to apply in practice. However, there exist a vast number of physical instances in which the acceleration is [to a reasonable approximation] **constant in time**. In such cases, we write

$$dv = \mathrm{a}\,dt\,,$$

where the roman typeface serves as a conspicuous reminder of the acceleration's constancy. Indefinite integration to obtain an expression for $v(t)$ must be done with care so as to ensure self-consistency. The integrals of each side of the equation, above, are

$$\mathrm{LHS} = \int dv = v + C_L\,, \quad \text{and} \quad \mathrm{RHS} = \int \mathrm{a}\,dt = \mathrm{a}\,t + C_R\,,$$

for constants of integration $C_{L,R}$. Equating the LHS with the RHS, and rearranging, yields an intermediate expression for the velocity as a function of time,

$$v(t) = \mathrm{a}\,t + C\,, \quad \text{for} \quad C = C_R - C_L\,.$$

Regarding this expression formally as a mathematical function of a single parameter, t, leads us to identify C as the **initial value** of the velocity [that assumed by the function when its parametric argument is set to zero]. Hence,

$$v(t) = \mathrm{a}\,t + v_0\,, \quad \text{where} \quad v_0 = v(0)\,.$$

ASIDE: The reason that we are making such a big deal out of this is that t, the time, is a parameter, and many things can be done to a mere parameter that should not meaningfully affect the physics of the situation. Let's illustrate this point with a facetious example.

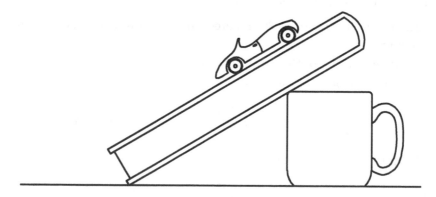

FIGURE 3.1   Toy Car on a Textbook Incline

Suppose that a wheeled toy car rolls down a constant-angle ramp made from a physics textbook and a coffee mug. While the car is on the ramp, it endures an acceleration which is, for all intents and purposes, constant. No problem, thus far, until I tell you that the time is determined by my wristwatch, which counts seconds elapsed from the previous midnight.[1] So, the value of the constant that I would put into the equation for velocity as a function of PK-wristwatch time would be that value which the velocity would have had at $t = 0$, had it actually been riding on the surface of an enormously large ramp from midnight until now. This interplay of 'constant' terms and parameterisation turns out to be a beneficent feature.

Performing definite integrations sidesteps the need for integration constants by direct calculation of particular finite differences. In this case,

$$\int_{v_i}^{v_f} dv = v_f - v_i \quad \text{and} \quad \int_{t_i}^{t_f} a\,dt = a\left(t_f - t_i\right),$$

where the initial, i, and final, f, times and particle velocities explicitly appear and thus obviate the need for the constant of integration. The two integrals inherit the equality of their integrands, and thus

$$v_f - v_i = a\left(t_f - t_i\right) \quad \text{or} \quad \Delta v = a\,\Delta t\,.$$

This is precisely what one expects when the acceleration is constant.

Having had so much fun computing the velocity, let's attempt a similar analysis for the position function. Substitution of the indefinite integral solution, $v(t)$, derived just above, into $dx = v\,dt$, yields

$$dx = (a\,t + v_0)\,dt\,.$$

Self-consistent indefinite integration of both sides of this equation produces a general result for the parameterised trajectory,

$$x(t) = \frac{1}{2}\,a\,t^2 + v_0\,t + x_0\,,$$

where x_0 is a new constant of integration vested with the interpretation that it is the position that the particle takes at $t = 0$ [under the assumption that the constant acceleration regime extends that far back into the past].

Performing a definite integration instead directly determines the displacement of the particle throughout the time interval,

$$x_f - x_i = \frac{1}{2}\,a\left(t_f^2 - t_i^2\right) + v_0\left(t_f - t_i\right)\,.$$

This expression does not reduce as did the velocity relation because of the intrinsic non-linearity [quadratic behaviour] of the position as a function of time.

ASIDE: Note that the determination of $x(t)$ employed the indefinite form of $v(t)$.

Q: Might one have used the velocity expressions obtained in the definite analysis?

A: Nope! Not without great care. [Think about this.]

---

[1] Don't quibble. My clock at home counts seconds from the Big Bang (*circa* 13.8 billion years ago, according to the 2013 estimate from the European Space Agency's Planck mission).

AN UNEXPECTED SURPRISE
(THE ORIGIN OF WHICH WILL BECOME CLEARER LATER ON)

IF the acceleration of the particle is truly constant in time, THEN it cannot depend on position either, since the particle is at different positions at different times!

[*Cool, eh?*]

Spatial constancy has consequences which we will now derive via two different methods.

METHOD ONE: [*Parametric Substitution*]
Time, t, acts simply as a parameter labelling correlated values of the velocity, $v(t)$, and position, $x(t)$, of the particle. Either of these equations may be inverted and combined with the other so as to eliminate the parameter. The simplest route is:

$$v = v_0 + a\,t \qquad \Longrightarrow \qquad t = \frac{v - v_0}{a},$$

which, when inserted into the expression for the position, yields

$$x - x_0 = v_0\,t + \frac{1}{2}\,a\,t^2 = v_0\left(\frac{v - v_0}{a}\right) + \frac{1}{2}\,a\left(\frac{v - v_0}{a}\right)^2.$$

Multiplying both sides of this last result by $2\,a$ cancels the denominators to produce

$$2\,a\left(x - x_0\right) = 2\,v_0\,v - 2\,v_0^2 + v^2 - 2\,v_0\,v + v_0^2 = v^2 - v_0^2,$$

an expression relating the change in the squared speed of the [constantly accelerating] particle to its net displacement. This relation may be further rearranged to isolate the squared speed of the particle at the final instant:

$$v^2 = v_0^2 + 2\,a\,\Delta x.$$

METHOD TWO: [*Chain Rule*]
Using the CHAIN RULE, one may rewrite the acceleration as

$$a = \frac{dv}{dt} = \frac{dv}{dx}\frac{dx}{dt} = \frac{dv}{dx}\,v = v\,\frac{dv}{dx}.$$

In the case at hand, the acceleration is constant, denoted by a, and one may assert that

$$a\,dx = v\,dv.$$

Consistent definite integration of both sides while maintaining the equality leads to

$$a\,x\Big|_{x_0}^{x} = \frac{1}{2}\,v^2\Big|_{v_0}^{v},$$

$$a\left(x - x_0\right) = \frac{1}{2}\left(v^2 - v_0^2\right).$$

Thus, the curious relation discovered above via simple algebraic manipulation,

$$v^2 = v_0^2 + 2\,a\,\Delta x,$$

is rederived here via elementary techniques of calculus.

The proper physical interpretation of this unexpected result, and an appreciation of its significance, must await the forthcoming analyses of mechanical work and energy in Chapter 23.

EXAMPLE [*Constant Acceleration: α Particle*]

> ASIDE: An alpha particle is a helium-4 nucleus consisting of two protons, two neutrons, and zero electrons [*i.e.*, a doubly ionised helium-4 atom]. Such particles are emitted by certain elemental isotopes undergoing radioactive decay. It so happened that α decay was the first radioactive process to be discovered (*circa* 1895).

An alpha particle, initially at rest, emerges from a two metre linear particle accelerator with speed 5×10^6 m/s [roughly 1.67 % the speed of light]. Inquiring minds want to know:

Qt: How long was the particle in the accelerator?

Qa: What was the average acceleration of the particle while it was in the accelerator?

In order to answer these questions, one must either know or make assumptions about the accelerating properties of the apparatus.

[HOW a linear accelerator works is a fitting topic for VOLUME III in this series.]

Here we eschew all the details and simply assume that the instantaneous acceleration of the α particle is constant while it remains inside the device.

METHOD ONE: [*First things first*]

The average velocity v_{av} is equal to the total displacement $\Delta x = 2$ m, divided by the time sought. On the other hand, since the acceleration is assumed to be constant, the velocity is a linear function, and hence

$$v_{av} = \frac{v_0 + v_{2m}}{2} = \frac{0 + 5 \times 10^6}{2} \text{ m/s} = 2.5 \times 10^6 \text{ m/s}.$$

At: Thus, the time of flight in the accelerator may be computed to be

$$\Delta t = \frac{\Delta x}{v_{av}} = \frac{2 \text{ m}}{2.5 \times 10^6 \text{ m/s}} = 8 \times 10^{-7} \text{ s} = 0.8 \, \mu s.$$

Aa: Now the answer to the second question is readily obtained:

$$a_{av} = \frac{\Delta v}{\Delta t} = \frac{5 \times 10^6 \text{ m/s}}{8 \times 10^{-7} \text{ s}} = 6.25 \times 10^{12} \text{ m/s}^2.$$

This enormously large value for the acceleration was not unexpected given the considerable speed with which the α particle leaves the relatively short apparatus.

METHOD TWO: [*Alphabetical order*]

Aa: The handy-dandy formula relating the change in speed-squared to the magnitude of the constant acceleration and the net displacement may be employed to determine the acceleration.

$$v^2 = v_0^2 + 2 \, a \, \Delta x \quad \Longrightarrow \quad a = \frac{v^2 - v_0^2}{2 \, \Delta x} = \frac{\left(5 \times 10^6\right)^2 - 0}{2 \times 2} \frac{(\text{m/s})^2}{\text{m}} = 6.25 \times 10^{12} \text{ m/s}^2.$$

At: With the constant acceleration in hand, one can compute the time of flight from the change in the velocity [**Q:** What's the average value of the constant acceleration?], *viz.*,

$$\Delta t = \frac{\Delta v}{a} = \frac{5 \times 10^6}{6.25 \times 10^{12}} \frac{\text{m/s}}{\text{m/s}^2} = 8 \times 10^{-7} \text{ s}.$$

EXAMPLE [*Constant Acceleration: PK and the Police Car*]

PK was driving his sports car at 54 km/h along a level and straight section of road for which the posted speed limit is 40 km/h. Just as PK passed a police car at rest on the side of the road, the police car began accelerating at 3 metres per second per second.

Qt: When does the police car pull alongside PK? *I.e.,* how much time elapses between the instant at which PK passes the police car and the instant at which it passes PK?

Qv: How fast is the police car going at the instant that it passes PK?

METHOD ONE: [*The viewpoint of an observer at rest*]

Let's analyse the motion from the point of view of an observer (call him "O") who remains at rest at the initial position of the police car. Set the origin of the space coordinate, $x = 0$, at that location, and the origin of the time coordinate, $t = 0$, at the instant at which the first passage occurs.

ASIDE: Here is a practical encounter with the most important aspect of the **art of solving problems**. One should choose spatial and temporal coordinates such that the motion of the system under consideration appears simple. A person may freely choose to do things the hard way, but simplicity is preferable.

O describes the constant-velocity motion of the sports car as

$$x_{\text{PK}} = v_{\text{PK}}\, t\,, \qquad \text{with} \qquad v_{\text{PK}} = 54\,\text{km/h} = 15\,\text{m/s}\,.$$

The police car is observed to accelerate from rest [at the origin] at a constant rate,

$$x_{\text{police}} = \frac{1}{2}\,\text{a}\,t^2\,, \qquad \text{and} \qquad v_{\text{police}} = \text{a}\,t\,, \qquad \text{with} \qquad \text{a} = 3\,\text{m/s}^2\,.$$

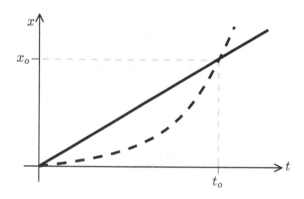

FIGURE 3.2   Position *vs.* Time for PK and Police Officer as Viewed by Observer O

Figure 3.2 shows a graph of the position *vs.* time functions for both the sports car (solid line) and the police car (dashed line). The trajectories intersect twice: at the origin, where PK passes the police car, and again at a later time, t_o, when the police car overtakes PK.

At: The answer to the first question is the time difference between the two crossings. The intersections of the trajectories occur when both particles are at the same position at the same time. This **constraint** that $x_{\text{PK}} = x_{\text{police}}$, at t_o, can be exploited to solve for t_o, *viz.*,

$$x_{o,\text{PK}} = v_{\text{PK}}\, t_o = \frac{1}{2}\,\text{a}\,t_o^2 = x_{o,\text{police}} \qquad \Longrightarrow \qquad 0 = \left(v_{\text{PK}} - \frac{1}{2}\,\text{a}\,t_o\right) t_o\,.$$

There are two solutions of this quadratic equation for the overtaking time. One of these is $t_o = 0$, which corresponds to the time of first passage,[2] while the other provides the time of second passage,

$$t_o = \frac{2\,v_{\text{PK}}}{a} = \frac{2 \times 15}{3} \, \frac{\text{m/s}}{\text{m/s}^2} = 10\,\text{s}\,.$$

Av: The speed of the police car at the instant of overtaking is

$$v_{\text{0police}} = a\,t_o = 3 \times 10 \, \frac{\text{m}}{\text{s}^2}\,\text{s} = 30\,\text{m/s} = 108\,\text{km/h}\,.$$

METHOD TWO: [*From a different perspective*]
The analysis in METHOD ONE, as fine as it is, requires the existence of O, who keeps track of the motions of both PK and the police car. Let's look at the situation from the perspective of another observer, NK, who is a passenger in the sports car. NK situates herself at the origin of her coordinates and chooses $t = 0$ to be the instant at which she passes the police car. NK believes that she is at rest and not accelerating, *ergo* her position function is constant [at the origin]. She sees the police car accelerate at a = $3\,\text{m/s}^2$, from an initial velocity of $-v_{\text{PK}} = 15\,\text{m/s}$ [backward], and an initial position at the origin. Hence, she describes the motion of the police car as

$$\widetilde{x}_{\text{police}} = -v_{\text{PK}}\,t + \frac{1}{2}\,a\,t^2\,,$$

as shown in Figure 3.3, and registers overtaking when $\widetilde{x}_{\text{police}} = 0$.

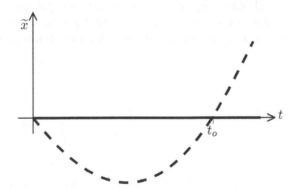

FIGURE 3.3 Position *vs.* Time for the Police Car as Viewed by NK

At: Solving for the times at which the police car is at $\widetilde{x}_{\text{police}} = 0$ yields both the first passage [extraneous] and second passage [significant] solutions

$$0 = \left(-v_{\text{PK}} + \frac{1}{2}\,a\,t_o\right)t_o \quad \Longrightarrow \quad t_o = 0 \quad \text{and} \quad t_o = \frac{2\,v_{\text{PK}}}{a} = 10\,\text{s}\,.$$

[We've skipped over some of the details, as they are presented above in METHOD ONE.]
Av: The velocity of the police car at the time of overtaking as observed by NK is

$$v_{\text{overtaking}} = -v_{\text{PK}} + a\,t_o = -15 + 3 \times 10 = +15\,\text{m/s} = 54\,\text{km/h}\,.$$

This stands to reason, since it is the correct **relative velocity** consistent with the result obtained above in METHOD ONE.

---

[2]This solution is often dismissively called extraneous or trivial.

The three object lessons of this example are:

1. There may be many ways to solve any particular problem.

2. Judicious choices of coordinates and parameterisations can aid in simplifying the computations needed to determine desired quantities.

    -   The way to become proficient at discerning the most advantageous choices of coordinates and origins is through practice, practice, practice, and more practice.

    -   The idea that one is free to do these sorts of computations from the points of view of various observers is the central tenet of the various efficacious Principles of Relativity. Today, we appreciate that inertial observers are governed by SPECIAL RELATIVITY elucidated by Einstein (1879–1955) in 1905. Prior to this, inertial observers were thought to obey the rules of GALILEAN RELATIVITY [so-called in honour of Galileo (1564–1642)]. In the **Newtonian Limit**, in which all velocities are small *vis-à-vis* the speed of light, Special Relativity reduces to Galilean Relativity. In these VOLUMES we'll focus exclusively on Galilean Relativity.

3. Different observers will obtain equivalent results for *bona fide* physical quantities [*e.g.,* the overtaking time in the above], and consistent results for quantities which are merely kinematical [*e.g.,* the velocity of the police car when it passes].

ADDENDUM: The Constant Acceleration Trajectory Reexamined

The above derivation [by direct integration] of the kinematical formula describing the trajectory of a particle undergoing constant acceleration,

$$x(t) = x_0 + v_0\, t + \frac{1}{2}\, a\, t^2\,,$$

may have seemed too *glib* or *facile*. Here, we shall reason our way to the same result, and thereby confirm its validity. A particular benefit of this alternative approach shall be a better appreciation of the significance[3] of the coefficient "1/2."

Let's start with the definition of acceleration as the time rate of change of velocity. IF the acceleration is constant, THEN the velocity varies linearly with time, as illustrated in Figure 3.4, AND the average velocity through any time interval extending from 0 to an unspecified time t is

$$v_{\text{av}} = v_{\text{midpoint}} = \frac{1}{2}\big(v_0 + v(t)\big) = \frac{1}{2}\big(v_0 + (v_0 + a\,t)\big) = \frac{1}{2}\big(2\,v_0 + a\,t\big) = v_0 + \frac{1}{2}\,a\,t\,.$$

This last result is reassuringly consistent with the fact that, for linear functions, the midpoint of the "rise" occurs at the midpoint of the "run."

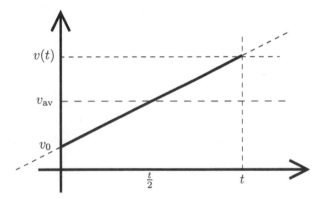

FIGURE 3.4 Velocity *vs.* Time under Constant Acceleration.

Also, the average velocity is defined by the total displacement occurring through the time interval,

$$v_{\text{av}} \equiv \frac{\Delta x}{\Delta t}\,.$$

The numerator on the RHS is the displacement from the initial position, $\Delta x = x(t) - x_0$. The denominator is the duration of the time interval $t - 0 = t$ [which we are free to leave unspecified].

Formally equating these two expressions for the average velocity yields

$$v_0 + \frac{1}{2}\,a\,t = \frac{x(t) - x_0}{t} \qquad \Longrightarrow \qquad x(t) = x_0 + v_0\,t + \frac{1}{2}\,a\,t^2\,,$$

which is precisely the relation that we set about to verify.

> ASIDE: When Newton was endowing physics with *language* and *grammar* by the invention of calculus, he provided "proofs" like the above to convince his friends of the veracity and utility of his then novel mathematical methodology.

---

[3] In Chapter 1, we argued for the consistency of a similar formula with coefficient equal to 1/10.

Chapter 4

Still More Kinematics in One Dimension

Recall that when the acceleration is constant in time (and space!) one obtains:

$$v(t) = a t + v_0 , \qquad x(t) = \frac{1}{2} a t^2 + v_0 t + x_0 , \quad \text{and} \quad v^2 = v_0^2 + 2 a \Delta x .$$

Let's consider instances in which the acceleration of a particle is not constant in time.

....

EXAMPLE [*Velocity from a Known Trajectory (Reprised)*]

Recall the example from Chapter 2 in which the explicit form of the trajectory of a particle was quoted: $x(t) = b t^3$, for b, a constant.

Q: How did we know this?

A: Perhaps it had been ascertained that the acceleration of the particle increased linearly with time. A felicitous choice of time parameterisation enabled a single simple expression to describe the acceleration in a manner consistent with the initial data:

$$a(t) = 6 b t , \qquad \text{and} \qquad v_0 = v(t) \Big|_{t=0} = 0 , \qquad x_0 = x(t) \Big|_{t=0} = 0 .$$

Integrating the acceleration once to get the velocity function [and employing the boundary condition to fix the constant of integration] yields

$$v(t) = 3 b t^2 .$$

Integrating once more to obtain the position function leads to

$$x(t) = b t^3 ,$$

illustrating the inherent self-consistency of our analysis.

....

EXAMPLE [*Kinematics of Simple Harmonic Oscillation*]

Let's preview SIMPLE HARMONIC OSCILLATION[1] [SHO] for which

$$x(t) = A \sin (\omega t) ,$$

where the amplitude A, with units $[\,\mathtt{m}\,]$, and the angular frequency ω, with units $[\,\mathtt{radians/s}\,]$, are both constant in time. Taking the time derivative of the position function generates the velocity function

$$v(t) = \omega A \cos (\omega t) .$$

As the radian is a dimensionless quantity, the units associated with ωA are equivalent to $[\,\mathtt{m/s}\,]$, maintaining dimensional consistency.

---

[1] This topic will be touched upon in Chapter 15 and considered in depth in VOLUME II.

Taking the time derivative of the velocity function produces the acceleration function

$$a(t) = -\omega^2 A \sin(\omega t) ,$$

and again the units work out consistently. The essential feature of SHO is the fact that

$$a(t) \propto -x(t) !$$

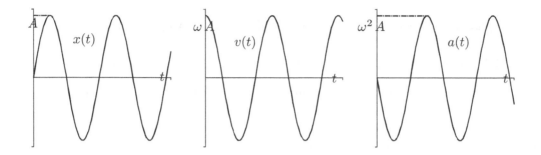

FIGURE 4.1   Position, Velocity, and Acceleration *vs.* Time for SHO

ASIDE: From calculus, we know that the general solution of

$$\frac{d^2 x(t)}{dt^2} = -\omega^2 x(t) \qquad \text{is} \qquad x(t) = A \sin(\omega t + \phi)$$

$$= A \cos\left(\omega t + \widetilde{\phi}\right)$$

$$= A_s \sin(\omega t) + A_c \cos(\omega t) .$$

Note: each form of the solution possesses two constants of integration.

EXAMPLE [*Kinematics of ZENOish Motion*]

Let's investigate ZENOish[2] motion. Suppose that

$$x(t) = \frac{V_0}{k}\left(1 - e^{-kt}\right),$$

where k and V_0 are positive constants. The units of k are necessarily $[\,\mathrm{s}^{-1}\,]$ to render the exponent dimensionless, and hence the units of V_0 must[3] be $[\,\mathrm{m/s}\,]$. This trajectory possesses a horizontal asymptote, *cf.* Figure 4.2, and the formal limit,

$$\lim_{t \to \infty} x(t) = \frac{V_0}{k} .$$

One is forced to conclude that an infinite amount of time is required for the particle to traverse the finite distance from $x(t)\big|_{t=0} = 0$ to $x = V_0/k$, notwithstanding the fact that, as proven just below, the particle is moving forward at all times after $t = 0\,\mathrm{s}$!

----

[2]Zeno of Elea (*circa* 450 BC) was a Greek philospher of the Parmenidean School. He developed several paradoxes in support of his view that, contrary to one's senses, all motion is illusory.
[3]This provides a rationale for the designation "V." We'll see in a moment why the "naught" subscript is appended.

ASIDE: This feature of the motion inspired the appellation "ZENOish."

Taking the time derivative of the trajectory yields the velocity function:

$$v(t) = V_0\, e^{-k\,t}, \quad \text{and thus} \quad v(t)\big|_{t=0} = V_0\,,$$

providing *a posteriori* justification for including the naught in the initial parameterisation of the trajectory. A graph of the velocity function appears in Figure 4.2. Were it not the case that the velocity [constrained to remain positive, *i.e.,* forward] tends strongly to zero as t grows large, it would be impossible for the displacement through infinite elapsed time to remain finite.

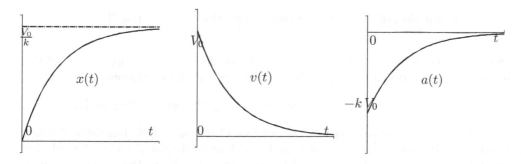

FIGURE 4.2 Position, Velocity, and Acceleration *vs.* Time for ZENOish Motion

Continuing the kinematical analysis, the time derivative of the velocity function generates the acceleration function:

$$a(t) = \frac{dv}{dt} = \frac{d^2 x}{dt^2} = -k\, V_0\, e^{-k\,t}\,,$$

which is also illustrated in Figure 4.2. The acceleration is always negative, as the velocity is always decreasing. The essential feature of ZENOish motion is that the acceleration is proportional to the velocity and oppositely directed. That is, the particle slows down in direct proportion to how fast it is moving:

$$a(t) \propto -v(t)\,.$$

We shall re-encounter this situation when we consider linear drag forces in Chapter 20.

ASIDE: **Q:** What would happen if the acceleration were proportional to the [positive] velocity?

EXAMPLE [*Acceleration Regimes I*]

Consider an [otherwise constant] acceleration which switches on and off, thus lasting only for a finite time interval.

[Mathematically speaking, such an acceleration function is piecewise constant.]

For definiteness, let us suppose that a particle experiences the acceleration [UNITS $= \mathrm{m/s^2}$]

$$a(t) = \begin{cases} 0\,, & -\infty < t < 0 \quad \text{Regime I} \\ -2\,, & 0 \le t \le 3 \quad \text{Regime II} \\ 0\,, & 3 < t < \infty \quad \text{Regime III} \end{cases}.$$

At time $t = 0$, it is located at $x_0 = +10\,\text{m}$ and moving with velocity $v_0 = +4\,\text{m/s}$.

There are three distinct **dynamical regimes** to consider here. Of these three, the first and third happen to be somewhat trivial.

I The trajectory, $x(t)$, of the particle for all times preceding $t = 0$ can be obtained by realising that the particle does not accelerate in Regime I and that the quoted initial state is equivalent to final data for this regime. Thus [with units for VELOCITY, POSITION, and TIME understood to be m/s, m, and s, respectively],

$$v(t) = 4 \qquad \text{and} \qquad x(t) = 10 + 4\,t\,, \qquad \text{throughout Regime I.}$$

[One might call this a "retrodiction," rather than a "prediction."]

II The trajectory of the particle in Regime II can be determined by application of the constant acceleration formulae incorporating the specified initial conditions. That is,

$$v(t) = 4 - 2\,t \qquad \text{and} \qquad x(t) = 10 + 4\,t - t^2\,, \qquad \text{throughout Regime II.}$$

These formulae describe the velocity and the position of the particle as functions of times corresponding to Regime II. In particular, at $t = 3$ [just as Regime II ends], the velocity of the particle is $v(3) = -2$, while its position is $x(3) = 13$. From these formulae, it is manifestly clear that the particle has turned around, and that it is to the right of its initial position at the start of Regime III.

III The subsequent ($t > 3$) trajectory of the particle [in Regime III] is easily described in any of the three following manners.

METHOD ONE: [*In words*]

The particle moves at constant velocity $-2\,\text{m/s}$ *from its starting point of* $+13\,\text{m}$.

METHOD TWO: [*In an equivalent formula*]

Let's introduce a new time parameter, $\widetilde{t}$, taking the value $\widetilde{t} = 0$ at the start of Regime III, which occurs at time $t = 3$. Thus

$$v(t) = -2 \qquad \text{and} \qquad x(t) = 13 - 2\,\widetilde{t}(t) = 13 - 2(t - 3)\,, \qquad \text{throughout Regime III.}$$

Analysis in terms of $\widetilde{t}$ is hardly radical. Instead, it amounts to no more than the **split timing** of intervals that one encounters in sports training.

METHOD THREE: [*In a conventionally simplified manner*]

The above expression for the position as a function of time can be simplified to read

$$x(t) = 19 - 2\,t\,, \quad \text{throughout Regime III.}$$

This formula promotes the fiction[4] that the particle was at $x = 19$ at time $t = 0$. This is precisely the position that some other particle, experiencing only the acceleration characteristic of Regime III, must have had at time $t = 0$ in order for its position and velocity to coincide with those of the actual particle at $t = 3$, the beginning of Regime III.

To summarise, we have obtained a complete description of the position and the velocity of the particle subjected to this particular off–on–off variable acceleration.

---

[4] *Cf.* the discussion of the toy car, textbook, and coffee mug in Chapter 3.

EXAMPLE [*Acceleration Regimes II*]

For our final example, suppose that the acceleration varies in such a way that the motion of the particle is best understood in terms of a sequence of non-trivial regimes. Consider the case of a particle, initially at rest at the origin, subjected to an acceleration which

0 → 2   increases linearly from 0 to $2\,\text{m/s}^2$ in the time interval from $t = 0$ to $t = 2\,\text{s}$,

2 → 3   remains constant, at $2\,\text{m/s}^2$, between $t = 2\,\text{s}$ and $t = 3\,\text{s}$, and

3 → 4   decreases linearly back to zero in the following one-second interval,

as illustrated in Figure 4.3.

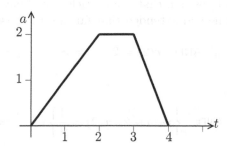

FIGURE 4.3   Variable Acceleration *vs.* Time Requiring Dynamical Regimes

Our goal is to compute the velocity and position of the particle at $t = 4\,\text{s}$.

It is not possible to achieve this goal in a single computational step. Instead, the analysis must proceed methodically through the regimes corresponding to the piecewise domains of the acceleration function.

[SI units are implicit throughout the remainder of the analysis.]

$$a(t) = \begin{cases} t, & t \in [0,2] & \text{Regime I} \\ 2, & t \in [2,3] & \text{Regime II} \\ 8 - 2t, & t \in [3,4] & \text{Regime III} \end{cases}$$

Separate, yet linked, analyses must be conducted in each regime.

I   The acceleration function $a(t) = t$. Integrate once to get the velocity function,

$$v(t) = \frac{1}{2}t^2 + v_0, \quad \text{AND} \quad v_0 = 0 \quad \Longrightarrow \quad v(t) = \frac{1}{2}t^2.$$

Integrating the velocity yields the position function,

$$x(t) = \frac{1}{6}t^3 + x_0, \quad \text{AND} \quad x_0 = 0 \quad \Longrightarrow \quad x(t) = \frac{1}{6}t^3.$$

Hence, in Regime I, the trajectory and the motion of the particle are determined completely by the above solutions. However, Regime I comes to an end at $t = 2$.

II   The acceleration is constant, and thus one can immediately write

$$v(t) = 2\,t + v_0 \qquad \text{and} \qquad x(t) = t^2 + v_0\,t + x_0\,,$$

where v_0 and x_0, ... Whoops! The initial conditions specified in the problem are for Regime I, not Regime II. There are two ways to proceed. In each we must first calculate the exit data for Regime I because these constitute the initial data for Regime II. At the end of Regime I, the velocity and position are

$$v_I = v(2) = 2 \qquad \text{and} \qquad x_I = x(2) = \frac{4}{3}\,.$$

METHOD ONE: [*With a single time*]

Let's compute the velocity and position that the particle would have had at time $t = 0$, if the constant acceleration regime had extended that far into the past. That is,

$$v(t) = 2\,t + v_0 \quad \text{AND} \quad v(2) = 2 \qquad \Longrightarrow \qquad v_0 = -2\,.$$

Similarly,

$$x(t) = t^2 + v_0\,t + x_0 \quad \text{AND} \quad \left\{ v_0 = -2,\ x(2) = \frac{4}{3} \right\} \qquad \Longrightarrow \qquad x_0 = \frac{4}{3}\,.$$

The particle's trajectory in the time interval from $t = 2$ to 3 is given by

$$v(t) = 2\,t - 2\,, \qquad \text{and} \qquad x(t) = t^2 - 2\,t + \frac{4}{3}\,.$$

METHOD TWO: [*With split timing*]

Nothing prevents us from making use of the "split timing" feature on our wristwatches, so LET $\tilde{t} = t - 2$, such that $\tilde{t} = 0$ at the start of Regime II. The constant acceleration kinematical formulae are easily adapted to the present case:

$$v(\tilde{t}) = 2\tilde{t} + v_I = 2\tilde{t} + 2 \qquad \text{and} \qquad x(\tilde{t}) = \tilde{t}^{\,2} + v_I\tilde{t} + x_I = \tilde{t}^{\,2} + 2\tilde{t} + \frac{4}{3}\,.$$

> ASIDE: As a consistency check, one can substitute $\tilde{t} = t - 2$ into the METHOD TWO result and obtain the METHOD ONE expressions.
>
> Note that it is accidental that the values of x_0 in the single time and split time parameterisations happen to be equal.

Regime II ends at $t = 3$ [equivalently $\tilde{t} = 1$] with the onset of Regime III.

III   Once again, the analysis must be anchored by consideration of the exit conditions from the previous regime:

$$v_{II} = v(3) = 4 \qquad \text{and} \qquad x_{II} = x(3) = 4\frac{1}{3}\,.$$

METHOD ONE: [*Still one time*]

In this regime, the acceleration is given by $a(t) = 8 - 2t$, which implies

$$v(t) = -t^2 + 8t + C \quad \text{and} \quad x(t) = -\frac{1}{3}t^3 + 4t^2 + Ct + D,$$

where C and D are constants of integration determined by matching to the initial data for Regime III. *I.e.,* at time $t = 3$,

$$v(3) = -3^2 + 8 \times 3 + C \quad \text{and} \quad x(3) = -\frac{1}{3}3^3 + 4 \times 3^2 + 3C + D$$
$$4 = 15 + C \qquad\qquad 4\tfrac{1}{3} = 27 + 3C + D$$

Solving yields $C = -11$ and $D = 10\frac{1}{3}$, and hence

$$v(t) = -t^2 + 8t - 11 \quad \text{and} \quad x(t) = -\frac{1}{3}t^3 + 4t^2 - 11t + 10\frac{1}{3},$$

describe the velocity and trajectory of the particle while it is in Regime III.

METHOD TWO: [*Split again*]

LET $\hat{t} = t - 3$ be employed in Regime III. The acceleration is

$$a(\hat{t}) = 2(1 - \hat{t}),$$

as is easily confirmed by noting that $a = 2$ at $\hat{t} = 0$, and $a = 0$ at $\hat{t} = 1$. Integrating the acceleration function once,

$$v(\hat{t}) = -\hat{t}^2 + 2\hat{t} + v_{II} = -\hat{t}^2 + 2\hat{t} + 4,$$

and then again,

$$x(\hat{t}) = -\frac{1}{3}\hat{t}^3 + \hat{t}^2 + 4\hat{t} + x_{II} = -\frac{1}{3}\hat{t}^3 + \hat{t}^2 + 4\hat{t} + 4\frac{1}{3}.$$

Finally, after working systematically through regimes I, II, and III, we have determined that at time $t = 4\,\text{s}$, the particle has velocity

$$v(t)\Big|_{t=4} = v(\hat{t})\Big|_{\hat{t}=1} = 5\,\text{m/s},$$

and position

$$x(t)\Big|_{t=4} = x(\hat{t})\Big|_{\hat{t}=1} = 9\,\text{m}.$$

Chapter 5

Vectors

In this chapter, we'll endure a brief introduction to vector algebra.[1]

SCALAR QUANTITIES   Scalar quantities are entirely specified by a single number, along with the appropriate units.

- The temperature at a particular location is a real-valued scalar function of time. The units can be Celsius or kelvin or even [*gasp*] degrees Fahrenheit [a non-SI unit].

- The number of people in a certain room is a non-negative integer-valued function of time. "Person" is a suitable unit in this context.

- The area of a hockey rink is a real-valued quantity with units m^2 (in Canada and Europe) or ft^2 (in the USA and Canada[2]).

SCALARS possess all of the familiar mathematical properties of real numbers and real-valued functions.

N-VECTOR QUANTITIES   Vectors are completely specified by a collection of N numbers, along with the appropriate units. [The technical definition is more strongly constraining than this.] There are two complementary modes of description for vectors.

$\longrightarrow$ GEOMETRIC: Quoting the MAGNITUDE (1 number) and the DIRECTION (specified by $N - 1$ numbers[3]).

$\longrightarrow$ COMPONENT: Quoting the vector's N Cartesian[4] components. These are conveniently and customarily labelled by subscripts, *i.e.*, the x-component of the vector $\vec{A}$ is A_x. Often, the vector will be represented using N-tuple notation, *viz.*, $\vec{A} = (A_x, A_y, A_z, \dots)$. [The vector is the collection of its components].

---

[1]Our presentation of vector properties will neither be rigorous nor mathematically self-contained. Instead, we rely on the reader's intuitive understanding of vectors to expedite this treatment. Detailed and careful investigation of the structure of vector spaces is a fitting subject for a mathematics course. Fortunately, physicists seldom encounter the sorts of pathologies which would invalidate naive application of vector techniques.

[2]Canada is *officially bilingual,* eh?

[3]For an instance of direction-specifying numbers, think of latitude and longitude determining a ray from the centre of the Earth.

[4]Other valid systems of coordinates may be used. The default system for these notes is Cartesian.

VECTOR PROPERTIES

EQUALITY: Two vectors, $\vec{A}$ and $\vec{B}$ are equal, $\vec{A} = \vec{B}$, IFF [meaning IF AND ONLY IF] their magnitudes are equal AND have the same units, AND their directions are the same. Note that $\vec{A}$ and $\vec{B}$ do not have to start and end at the same locations, *i.e.*, overlap completely, to be equal.

[We are allowed to carefully slide vectors around in the spaces in which they reside.]

Also, $\vec{A} = \vec{B}$ IFF all of the corresponding components are equal, *viz.*,

$$\vec{A} = \vec{B} \qquad \Longleftrightarrow \qquad A_x = B_x, \quad A_y = B_y, \quad A_z = B_z, \ldots .$$

ADDITIVE CLOSURE: The addition of two vectors, resident in the same vector space, results in a third vector also inhabiting the common vector space,

$$\vec{D} = \vec{A} + \vec{B}.$$

The idea, as shown in Figure 5.1, is that $\vec{B}$ is slid until its tail coincides with the head of $\vec{A}$. The sum is the unique vector from the tail of $\vec{A}$ to the head of $\vec{B}$.

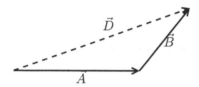

FIGURE 5.1   Addition of $\vec{B}$ to $\vec{A}$ Results in $\vec{D}$

In terms of components, we have

$$\vec{D} = \vec{A} + \vec{B} \qquad \Longleftrightarrow \qquad D_x = A_x + B_x, \quad D_y = A_y + B_y, \quad D_z = A_z + B_z, \ldots .$$

Figure 5.2 illustrates that vector addition is both COMMUTATIVE

$$\vec{A} + \vec{B} = \vec{B} + \vec{A},$$

and ASSOCIATIVE

$$\vec{A} + \vec{B} + \vec{C} = \left(\vec{A} + \vec{B} \right) + \vec{C} = \vec{A} + \left(\vec{B} + \vec{C} \right).$$

SCALAR MULTIPLICATION: Given a vector, $\vec{A}$, and a scalar, s, a new vector, $\vec{B} = s\,\vec{A}$, is produced by scalar multiplication. The idea of scalar multiplication arises when one generalises the notion of adding a vector to itself some number of times.

IF the scalar is dimensional, THEN $\vec{A}$ and $\vec{B}$ lie in different vector spaces.

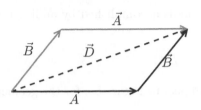 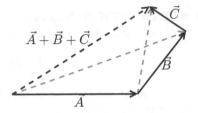

FIGURE 5.2   Commutativity and Associativity of Vector Addition

Geometrically, $\vec{A}$ and $\vec{B}$ point in the "same" direction [we say this *cautiously* when they lie in different spaces] with the magnitude of $\vec{B}$ equal to s times the magnitude of $\vec{A}$.

In terms of components,

$$\vec{B} = s\,\vec{A} \qquad \Longleftrightarrow \qquad B_x = s\,A_x\,, \quad B_y = s\,A_y\,, \quad B_z = s\,A_z\,, \ \ldots\,.$$

Scalar multiplication is DISTRIBUTIVE over vector addition, which means that:

$$\forall\,\{\vec{A},\vec{B}\}\ \text{vectors in the same space, AND}\ \forall\,\{a,b\}\ \text{scalars with the same units}\,,$$
$$\left(a+b\right)\left(\vec{A}+\vec{B}\right) = a\left(\vec{A}+\vec{B}\right) + b\left(\vec{A}+\vec{B}\right) = \left(a+b\right)\vec{A} + \left(a+b\right)\vec{B}\,.$$

ZERO VECTOR: There exists a unique vector, the ZERO vector, with the property that when it is added to any other vector in the space, the result is equal to the other vector.

$$\vec{A} + \vec{0} = \vec{A}\,, \ \text{for all vectors}\ \vec{A}.$$

Note that $\vec{0}$ necessarily has zero magnitude and **arbitrary** direction.

> ASIDE: **Q:** Is this essential and intrinsic arbitrariness necessary?
> **A:** Consider two distinct, finite vectors and imagine shrinking[5] them while preserving their directions. In the limit of infinite shrinkage, both original vectors converge to the same ZERO vector. Thus one is obligated to state that the direction associated with $\vec{0}$ is arbitrary.

In terms of components, the ZERO vector is

$$\vec{0} = (0\,,0\,,0\,, \ldots)\,.$$

ADDITIVE INVERSE: For each vector, $\vec{A}$, in the space, there exists a unique additive inverse, denoted by $-\vec{A}$, such that

$$\vec{A} + \left(-\vec{A}\right) = \vec{0}\,.$$

The inverse has the same magnitude as $\vec{A}$, but it is oppositely directed.

In terms of components,

$$\vec{A} = (A_x\,, A_y\,, A_z\,, \ldots) \qquad \Longleftrightarrow \qquad -\vec{A} = (-A_x\,, -A_y\,, -A_z\,, \ldots)\,.$$

This notation makes clear the fact that $-\vec{A} = (-1)\,\vec{A}$.

---

[5] Iterated scalar multiplication by a shrink factor, s, with $0 < s < 1$, generates a sequence of parallel vectors possessing ever-smaller magnitudes.

VECTOR SUBTRACTION: Subtraction of vectors is accomplished by adding the additive inverse of the vector to be subtracted, *viz.*,

$$\vec{D} = \vec{A} - \vec{B} = \vec{A} + (-\vec{B}).$$

A picture can save us about 10^4 words, and so we appeal to Figure 5.3 for the geometrical understanding of vector subtraction.

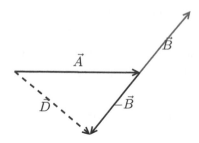

FIGURE 5.3   Subtraction of $\vec{B}$ from $\vec{A}$ Results in $\vec{D}$

In terms of components, vector subtraction is precisely as one might expect,

$$\vec{D} = \vec{A} - \vec{B} \qquad \Longleftrightarrow \qquad D_x = A_x - B_x, \quad D_y = A_y - B_y, \quad D_z = A_z - B_z, \ldots.$$

This completes the short list of vector algebra tools necessary to facilitate the quantitative study of kinematics in more than one dimension. **While these properties may seem arcane, they provide justification for the practical use of vectors in physics.**

The above properties are quite primitive in the mathematical sense of the term. A somewhat higher-level construction is the DOT PRODUCT of two vectors.

DOT PRODUCT   The dot product takes two vectors and produces a scalar. From the geometric viewpoint, the dot product is expressed in terms of the magnitude of each of the vectors and the interior angle, $0 \le \theta \le \pi$, lying between them [in the plane defined by the two vectors] when placed tail-to-tail,[6] as in Figure 5.4.

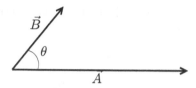

FIGURE 5.4   Vectors $\vec{A}$ and $\vec{B}$ Placed Tail-to-Tail

Computation of the dot product from geometric data is via

$$\vec{A} \cdot \vec{B} = \left|\vec{A}\right| \left|\vec{B}\right| \cos(\theta),$$

---

[6]Note the liberties taken when the vectors lie in different spaces.

where $|\vec{X}|$ denotes the magnitude of the vector $\vec{X}$ and θ is the interior angle noted in Figure 5.4. It turns out that the dot product is a **symmetric bilinear operator**:

$$\vec{A} \cdot \vec{B} \equiv \vec{B} \cdot \vec{A}, \quad \text{and} \quad (\vec{A} + \vec{B}) \cdot \vec{C} = \vec{A} \cdot \vec{C} + \vec{B} \cdot \vec{C}.$$

The interpretation of the dot product is subtle and profound. A moment's thought confirms that $\vec{A} \cdot \vec{B}$ is equal to the projection of $\vec{B}$ onto the direction of $\vec{A}$, multiplied by the magnitude of $\vec{A}$. It is also equal to the projection of $\vec{A}$ in the direction of $\vec{B}$ times the magnitude of $\vec{B}$, as shown in Figure 5.5.

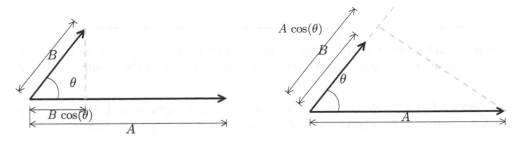

FIGURE 5.5 The Dot Product of Vectors is Symmetric

Knowledge of the Cartesian components of the two vectors enables economical formulation of the dot product as

$$\vec{A} \cdot \vec{B} = A_x B_x + A_y B_y + A_z B_z + \ldots .$$

The symmetry and linearity arise from the ordinary arithmetic commutativity and associativity of the components.

Q: What can one *do* with the dot product?
A: See the discussions immediately below. Better still, *stay tuned* for its appearance in routine kinematical computations, and especially when dynamics is reformulated in terms of energy in Chapters 21 and beyond.

The dot product of a vector with itself yields the square of the magnitude of the vector because the angle between any non-zero vector and itself is necessarily zero.

$$\vec{A} \cdot \vec{A} = |\vec{A}|^2 \cos(0) = |\vec{A}|^2.$$

In terms of components,

$$\vec{A} \cdot \vec{A} = A_x^2 + A_y^2 + A_z^2 + \ldots .$$

This expression for the length–squared of the vector conforms exactly to that obtained via the Pythagorean Theorem.

With a robust means of calculating the magnitudes of vectors in hand, one is led to construct UNIT VECTORS. A unit vector has magnitude precisely equal to 1, and thus is a means of signifying a "pure" direction. This feature makes it useful for both generating and comparing vectors.

[In Chapter 2, PK went running "due north," and "due south." These are unit vectors.]

We shall denote the unit vector in the direction of $\vec{A}$ by $\hat{A}$, and [uniquely] determine it from $\vec{A}$ via scalar multiplication by the reciprocal of its magnitude, *viz.*,

$$\hat{A} = \frac{\vec{A}}{|\vec{A}|} .$$

From a linearly-independent set of vectors spanning a vector space, one can construct[7] an orthonormal set of basis vectors. **Orthonormal** connotes both **ortho**gonality and **nor**malised.

ORTHOGONALITY   IF $\vec{A} \cdot \vec{B} = 0$, AND neither vector is equal to $\vec{0}$, THEN the vectors $\vec{A}$ and $\vec{B}$ are orthogonal. [This is a generalisation of the notion of "perpendicular".]

NORMALISED   A vector is normalised, IF it has length 1 [in appropriate units].

The canonical example of an orthonormal set of basis vectors is $\{\hat{\imath}, \hat{\jmath}, \hat{k}, \ldots\}$, whose elements correspond to [or as is often said, "generate"] the Cartesian x-, y-, z-, $\ldots$-axes:

$$\hat{\imath} \cdot \hat{\imath} = 1 \qquad \hat{\imath} \cdot \hat{\jmath} = 0 \qquad \hat{\imath} \cdot \hat{k} = 0 \qquad \ldots$$
$$\hat{\jmath} \cdot \hat{\imath} = 0 \qquad \hat{\jmath} \cdot \hat{\jmath} = 1 \qquad \hat{\jmath} \cdot \hat{k} = 0 \qquad \ldots$$
$$\hat{k} \cdot \hat{\imath} = 0 \qquad \hat{k} \cdot \hat{\jmath} = 0 \qquad \hat{k} \cdot \hat{k} = 1 \qquad \ldots$$
$$\vdots \qquad\qquad \vdots \qquad\qquad \vdots \qquad\qquad \ddots$$

These relations extend to other orthonormal bases.

It is now time to confess that the ideas of dot product and orthonormal Cartesian basis vectors have been implicitly employed from the start of this chapter! For instance, the component representation of any vector is necessarily with respect to a basis set, and

$$A_x = \vec{A} \cdot \hat{\imath}, \qquad A_y = \vec{A} \cdot \hat{\jmath}, \qquad A_z = \vec{A} \cdot \hat{k}, \qquad \ldots .$$

With the properties of the orthonormal basis vectors and the components established, the component form of the expression for the dot product may be derived as follows. First, write the vectors $\vec{A}$ and $\vec{B}$ as sums of their constituent components:

$$\vec{A} = A_x\,\hat{\imath} + A_y\,\hat{\jmath} + \ldots, \qquad \text{and} \qquad \vec{B} = B_x\,\hat{\imath} + B_y\,\hat{\jmath} + \ldots .$$

Second, the bilinearity of the dot product enables systematic expansion in terms of dot products of the basis vectors:

$$\begin{aligned}
\vec{A} \cdot \vec{B} &= (A_x\,\hat{\imath} + A_y\,\hat{\jmath} + \ldots) \cdot (B_x\,\hat{\imath} + B_y\,\hat{\jmath} + \ldots) \\
&= A_x\,B_x\,\hat{\imath} \cdot \hat{\imath} + A_x\,B_y\,\hat{\imath} \cdot \hat{\jmath} + \ldots \\
&\quad + A_y\,B_x\,\hat{\jmath} \cdot \hat{\imath} + A_y\,B_y\,\hat{\jmath} \cdot \hat{\jmath} + \ldots \\
&\quad + \ldots \\
&= A_x\,B_x + A_y\,B_y + \ldots .
\end{aligned}$$

---

[7]Precisely how one does this in general need not concern us now. Suffice it to say that there is an algorithm, the Gram–Schmidt procedure, which can be employed.

It is the orthonormality property of the Cartesian basis set of vectors which ensures that only the "diagonal" terms [among the set of all possible products of components of $\vec{A}$ with components of $\vec{B}$] arise in the expression for the dot product.

The geometric formulation requires the existence of a dot product to unambiguously define magnitudes and angles. *E.g.*, for vectors $\vec{A}$ and $\vec{B}$, and θ, the angle between them,

$$|\vec{A}| = \sqrt{\vec{A} \cdot \vec{A}}, \qquad |\vec{B}| = \sqrt{\vec{B} \cdot \vec{B}}, \qquad \text{and} \quad \cos(\theta) = \frac{\vec{A} \cdot \vec{B}}{|\vec{A}||\vec{B}|} = \frac{\vec{A} \cdot \vec{B}}{\sqrt{\vec{A} \cdot \vec{A}} \, \sqrt{\vec{B} \cdot \vec{B}}} \, .$$

There is one more type of vector product to consider.

CROSS PRODUCT   The cross product takes two [3-d] vectors as input and produces a unique vector as output. Geometrically, we write

$$\vec{A} \times \vec{B} = |\vec{A}| \, |\vec{B}| \, \sin(\theta) \, [\,\text{Right Hand Rule}\,] \, ,$$

where, once again, θ is the interior angle between the two vectors $\vec{A}$ and $\vec{B}$ when they are placed tail-to-tail.

MAGNITUDE   A feature of fundamental importance is that the magnitude of the resultant vector is the area of the parallelogram generated by the two input vectors. Hence, the magnitude of the cross product is maximised when $\vec{A}$ and $\vec{B}$ are perpendicular, *i.e.*, $\theta = \pm\pi/2$; and ZERO when $\vec{A}$ and $\vec{B}$ are parallel, $\theta = 0$, or antiparallel, $\theta = \pi$.

DIRECTION   The resultant is directed **perpendicular to the plane formed from $\vec{A}$ and $\vec{B}$**, with its **orientation** assigned by means of the RIGHT HAND RULE [RHR[8]].

> ASIDE: If the two input vectors are (anti-)parallel, then the perpendicular direction is not uniquely specified. Consistency is preserved because the magnitude is zero and thus the cross product yields the ZERO vector.

RHR   There are a variety of practical ways to express the RHR, and all require the use of one's right hand. PK's preferred method is to

- arrange the vectors so that they are tail-to-tail,

- align his hand with the first vector in the cross product,

    [The heel of PK's hand lies at the tail, and his fingers point toward the head.]

- and curl/rotate/sweep his metacarpals[9] and digits onto the second vector,

    [This hand motion occurs parallel to the plane fixed by the two vectors.]

- and PK's right thumb provides the orientation of the cross product.

An alternate method is to make a crude 3-d set of axes with one's thumb, index, and middle fingers. Align the thumb with the first vector and rotate until the index finger is along the second vector, and *presto* the middle finger provides the orientation.

ASIDE: **Q:** What if a person applies a LEFT HAND RULE instead?

> **A:** Such a *gauche* or *sinister* person will obtain exactly the opposite orientation for the result of the cross product!

---

[8]There happen to be three Right Hand Rules employed in various vector contexts. We shall encounter a second form in Chapters 34 and 40, and a third in VOLUME III.

[9]PK occasionally *starts off on the wrong foot* and attempts this move with his metatarsals.

MAGNITUDE   The instantaneous speed of the particle at time t is the magnitude of its velocity at that instant. *I.e.,* speed $= |\vec{v}(t)|$.

DIRECTION   The velocity vector at time t, $\vec{v}(t)$, is tangent to the trajectory of the particle at that instant.

In a similar manner, expressions for the average and instantaneous acceleration may be formulated. The quantity of greatest import is the instantaneous acceleration.

AVERAGE ACCELERATION   The average acceleration is the change in the instantaneous vector velocity throughout the time interval Δt, divided by the duration of the interval. *I.e.,*

$$\vec{a}_{\text{av}} = \frac{\Delta \vec{v}}{\Delta t} = \frac{\vec{v}(t_f) - \vec{v}(t_i)}{t_f - t_i} .$$

INSTANTANEOUS ACCELERATION   The instantaneous acceleration is the average acceleration through an infinitesimal time interval,

$$\vec{a} = \lim_{\Delta t \to 0} \vec{a}_{\text{av}} = \lim_{\Delta t \to 0} \frac{\Delta \vec{v}}{\Delta t} = \frac{d\vec{v}(t)}{dt} .$$

In terms of Cartesian coordinates,[3]

$$\vec{a} = (a_x, a_y, a_z, \ldots) = \left(\frac{dv_x}{dt}, \frac{dv_y}{dt}, \frac{dv_z}{dt}, \ldots \right)$$
$$= \frac{d\vec{v}(t)}{dt} = \left(\frac{d^2 r_x}{dt^2}, \frac{d^2 r_y}{dt^2}, \frac{d^2 r_z}{dt^2}, \ldots \right) = \frac{d^2\vec{r}(t)}{dt^2} .$$

These differential relations are integrable [at least in principle],

$$\vec{a} = \frac{d\vec{v}}{dt} \qquad \Longleftrightarrow \qquad \Delta \vec{v} = \int_{t_i}^{t_f} \vec{a}(t')\, dt' ,$$

$$\vec{v} = \frac{d\vec{r}}{dt} \qquad \Longleftrightarrow \qquad \Delta \vec{r} = \int_{t_i}^{t_f} \vec{v}(t')\, dt' .$$

---

[3]We reaffirm that the analogous non-Cartesian coordinate expressions are usually more complicated.

There is an important subtlety to accelerated motion once we consider spaces of more than one dimension. Two distinct **types** of acceleration may occur.

Type 1   The speed changes while the direction remains constant. This situation has been encountered previously in Chapter 2. [There, the accelerations could only be of this type, because we restricted our analysis to motions occurring in one-dimensional space]. In this instance the vector acceleration is (anti-)parallel to the vector velocity.

Type 2   The speed is constant while the direction of the motion changes. This is novel and reflects the particle's freedom to follow a curved trajectory. In this instance the vector acceleration is perpendicular to the vector velocity.

Any actual acceleration is expected to be a mixture of these two types. However, these are the two important limiting cases. Any acceleration at a particular instant may always be uniquely decomposed into a component parallel to the velocity and part(s) perpendicular to the velocity.

> ASIDE: The distillation into these types is experientially militated, as well. For example, when one is driving a sports car on a winding road, pressing the gas and brake pedals effectuates the first type of acceleration, while turning the steering wheel produces accelerations of the second type. This particular instance will be further elaborated upon in Chapter 8.

As in Chapter 3, let's pay careful attention to the case in which the acceleration happens to be constant in time. For clarity and ease of notation in this analysis, we'll restrict ourselves to two dimensions when we write explicit formulae involving components.

Suppose that the acceleration is constant, *i.e.*,

$$\vec{a} = \text{constant vector} = \frac{d\vec{v}}{dt}.$$

Choosing Cartesian coordinates x and y, generated by the basis unit vectors $\hat{\imath}$ and $\hat{\jmath}$, we can express the acceleration in terms of components:

$$\vec{a} = (a_x, a_y) = a_x\,\hat{\imath} + a_y\,\hat{\jmath}.$$

Constancy of the vector acceleration implies that its components, a_x and a_y, are each constant. Furthermore, the velocity can be decomposed with respect to the same basis:[4]

$$\vec{v} = (v_x, v_y) = v_x\,\hat{\imath} + v_y\,\hat{\jmath}.$$

We can [formally] take the time derivative of the velocity

$$\frac{d\vec{v}}{dt} = \frac{d}{dt}\left[v_x\,\hat{\imath} + v_y\,\hat{\jmath}\right] = \frac{dv_x}{dt}\,\hat{\imath} + \frac{dv_y}{dt}\,\hat{\jmath},$$

under the assumption that the unit vectors, $\{\hat{\imath}, \hat{\jmath}\}$, do not vary [explicitly or implicitly] with time. From the definition of the acceleration as the time rate of change of velocity,

$$a_x\,\hat{\imath} + a_y\,\hat{\jmath} = \frac{dv_x}{dt}\,\hat{\imath} + \frac{dv_y}{dt}\,\hat{\jmath},$$

---

[4]Well, not actually the same basis, because the acceleration, velocity, and position vectors belong to different vector spaces. What we really mean is that the respective bases for the various spaces are chosen in a mutually consistent manner. For simplicity we call this "the same" basis.

and because the basis is orthonormal, we can establish equalities among the respective components, *i.e.*,

$$a_x = \frac{dv_x}{dt}, \quad a_y = \frac{dv_y}{dt} \quad \implies \quad dv_x = a_x\, dt, \quad dv_y = a_y\, dt.$$

The x and y motions have decoupled and these two equations for the components of the velocity can be integrated, yielding

$$v_x = a_x\, t + v_{0x} \quad \text{and} \quad v_y = a_y\, t + v_{0y},$$

for constants of integration [*a.k.a.* initial data] v_{x0} and v_{y0}. We can formally reconstitute the expressions for the components of the velocity into a single vector formula

$$\vec{v}(t) = \vec{a}\, t + \vec{v}_0,$$

making the vector result look just like the natural generalisation of the 1-d formula for the velocity when the acceleration is constant.

By extension of the analysis above, $\vec{r}(t)$ is straightforwardly obtained:

$$\vec{r} = (r_x,\, r_y) = r_x\, \hat{i} + r_y\, \hat{j}, \quad \text{and} \quad \frac{d\vec{r}}{dt} = \frac{d}{dt}\left[r_x\, \hat{i} + r_y\, \hat{j} \right] = \frac{dr_x}{dt}\, \hat{i} + \frac{dr_y}{dt}\, \hat{j}.$$

From the definition of the velocity as the time rate of change of position, it follows that

$$v_x\, \hat{i} + v_y\, \hat{j} = \frac{dr_x}{dt}\, \hat{i} + \frac{dr_y}{dt}\, \hat{j},$$

and because the basis is orthonormal, the respective components are equal. *I.e.*,

$$v_x = \frac{dr_x}{dt}, \quad v_y = \frac{dr_y}{dt} \quad \implies \quad \begin{cases} dr_x = v_x\, dt = \left[a_x\, t + v_{0x} \right] dt \\ dr_y = v_y\, dt = \left[a_y\, t + v_{0y} \right] dt \end{cases}.$$

Again, the x and y motions have decoupled and each part can be simply integrated:

$$r_x = \frac{1}{2}\, a_x\, t^2 + v_{0x}\, t + r_{0x}, \quad \text{and} \quad r_y = \frac{1}{2}\, a_y\, t^2 + v_{0y}\, t + r_{0y},$$

for two additional constants of integration r_{0x} and r_{0y}. The decoupled expressions for the components of the position may be recombined into a single vector formula,

$$\vec{r}(t) = \frac{1}{2}\, \vec{a}\, t^2 + \vec{v}_0\, t + \vec{r}_0,$$

making the vector result look just like the natural generalisation of the one-d formula for the position when the acceleration is constant.

Wow! All of this has been quite amazing.

One might expect to next derive a formula analogous to the "third" kinematic relation. However, we cannot repeat the algebraic steps in our simple derivation because we are unable to consistently solve for the time by dividing the vector velocity by the vector acceleration. That is,

$$\vec{v} = \vec{a}\, t + \vec{v}_0 \quad \text{does not allow us to conclude that} \quad \text{``} t = \frac{\vec{v} - \vec{v}_0}{\vec{a}}. \text{''}$$

The simplest intuitive way to see the abject pathology here is to note that the very idea of dividing by a vector is *dodgy*. *I.e.*, one can divide by any scalar quantity, but how does one divide by up, or north, or to the left?

Chapter 7

Projectile Motion

To recapitulate: motion under the influence of a constant acceleration, $\vec{a}$, expressed in terms of fixed Cartesian coordinates, is described by

$$\vec{v}(t) = \vec{a}\,t + \vec{v}_0\,, \quad \text{and} \quad \vec{r}(t) = \frac{1}{2}\,\vec{a}\,t^2 + \vec{v}_0\,t + \vec{r}_0\,.$$

Now, let's consider an application of these formulae: PROJECTILE MOTION.

ASSUMPTIONS

- In the standard case, the particle is projected over a flat, level surface with no obstructions. Also, its launching and landing occur at the same height.

- The particle remains near the Earth's surface and travels a relatively short distance so as to ensure that the local gravitational field is effectively constant. A rigorous introduction to fields, in the context of Newton's LAW OF UNIVERSAL GRAVITATION, will commence in Chapter 47. For our present purposes, it is sufficient to say that the gravitational field gives rise to an **acceleration due to gravity**, which amounts to a phenomenological property of material bodies. Experiment[1] has determined that

$$\vec{g} = g\,[\text{down}] = 9.81\,\text{m/s}^2\,[\text{down}] \approx 10\,\text{m/s}^2\,[\text{down}]\,,$$

  and this represents the [typical] magnitude and direction of the acceleration due to gravity on or near the Earth's surface.

  The two restrictions imposed on the motion of the particle warrant more detailed examination.

  <small>Near Surface</small> The magnitude of the local acceleration due to gravity diminishes as one ascends above or burrows beneath the surface of the Earth, as will be seen in Chapter 47. Insisting that changes in the magnitude of the local field experienced by the particle throughout its trajectory be insignificant[2] restricts the extent to which the particle may rise and fall in the course of its motion.

---

[1]Not unlike Galileo's apocryphal dropping of various objects from the Leaning Tower of Pisa.

[2]**Q:** What constitutes a significant difference in the field? **A:** To answer this question, one must consider the natural scales inherent in the physical system, the putative accuracy of any physical measuring devices employed in an experimental analysis of the particle, and the relative sizes of inaccuracies introduced by other approximations.

Short Range   The acceleration vector is directed locally downward. Suppose that a football[3] is punted from New York to New Delhi. It is evident [by inspection of a globe] that the downward direction in New Delhi differs from that in New York. In order that changes in the direction of the local field experienced by the particle along its trajectory be insignificant, the range must perforce be "short enough."

- The effects of air resistance and wind are neglected. In most instances, this approximation turns out to be the largest source of error/deviation.

- An often unmentioned approximation is that the motion of the Earth is ignored.

  ASIDE: The **Coriolis effect**, arising from the rotation of the Earth on its axis, influences the trajectories of all freely moving objects.

SKETCH

Figure 7.1 illustrates the essential quantities in our investigation of projectile motion.

FIGURE 7.1   Projectile Motion (The dashed curve represents the model trajectory in which air resistance is neglected. The dotted curve includes the effects of air resistance.)

SETTING UP THE ANALYSIS

By assumption, the constant acceleration kinematical formulae [reiterated at the start of this chapter] apply. The task at hand is to tailor them to these particular circumstances. Conventions for coordinates and the setting of the clock are chosen to facilitate the analysis.

○   Coordinates

We shall employ Cartesian coordinates, (x, y), in our efforts to determine the motion of the particle. Let x denote the horizontal direction in which the projectile moves[4] with downrange assigned to be the positive sense. Let y be directed vertically with increasing values going upward.

$\vec{a}$ The acceleration due to gravity has constant magnitude and acts exclusively in the downward vertical direction, *viz.*, $\vec{a} = \vec{g} = (0, -g)$.

$\vec{v}$ The components of the velocity are $\vec{v} = (v_x, v_y)$.

$\vec{r}$ The components of the position are $\vec{r} = (r_x, r_y)$.

---

[3]Whether this football is [approximately] spherical or ellipsoidal has no bearing on the argument.
[4]The direction is unique, as there is no wind to blow the projectile sideways.

o   Clock

Let's set our watch to read $t = 0$ at the instant when the particle is launched. In this case, the $\vec{v}_0$ appearing in the kinematical formula is the actual initial velocity of the projectile,

$$\vec{v}_0 = (v_{0x} \, , \, v_{0y}) \, .$$

One can re-express the initial velocity in terms of magnitude and direction, *viz.*,

$$\vec{v}_0 = (v_{0x} \, , \, v_{0y}) = (v_0 \cos(\theta_0) \, , \, v_0 \sin(\theta_0)) = v_0 \, (\cos(\theta_0) \, , \, \sin(\theta_0)) = v_0 \, \widehat{v_0} \, .$$

Here v_0 is the initial speed[5] with which the particle moves, while θ_0 denotes the launch angle [measured above the horizontal as indicated in Figure 7.1], and $\widehat{v_0} = (\cos(\theta_0) \, , \, \sin(\theta_0))$ is the unit vector parallel to the initial direction of the projectile's motion.

o   Origin

The origin of the spatial coordinates, $(0 \, , \, 0)$, may be chosen to coincide with the point from which the projectile is launched. Therefore,

$$\vec{r}_0 = (r_{0x} \, , \, r_{0y}) = (0 \, , \, 0) \, .$$

ASIDE: Our choices of clock and coordinate origins are mutually consistent. The requirement of consistency is not at all onerous. This may be seen by recalling the manner in which the constants appearing in the kinematical formulae compensated for split timing in the examples treated in Chapters 3 and 4.

APPLYING THE KINEMATICAL FORMULAE

The vector equations describing the trajectory of the projectile become

$$\vec{v} = \vec{a}\,t + \vec{v}_0 \quad \Longleftrightarrow \quad \begin{cases} v_x(t) = a_x\,t + v_{0x} = v_0 \cos(\theta_0) \\ v_y(t) = a_y\,t + v_{0y} = -g\,t + v_0 \sin(\theta_0) \end{cases} , \quad \text{and}$$

$$\vec{r} = \frac{1}{2}\,\vec{a}\,t^2 + \vec{v}_0\,t + \vec{r}_0 \quad \Longleftrightarrow \quad \begin{cases} r_x(t) = \frac{1}{2}\,a_x\,t^2 + v_{0x}\,t + r_{0x} = v_0 \cos(\theta_0)\,t \\ r_y(t) = \frac{1}{2}\,a_y\,t^2 + v_{0y}\,t + r_{0y} = -\frac{1}{2}\,g\,t^2 + v_0 \sin(\theta_0)\,t \end{cases} .$$

We emphasise that the time t is merely a parameter whose specification determines the position of the particle along its trajectory. We have exercised our freedom to choose inessential things [*i.e.,* coordinate directions and origins, clock units and that instant at which $t = 0$] in order to make these relations appear *simple*.

Q: Well, are they simple?

A: Let's examine them more closely.

The x-component of the velocity is constant, $v_x = v_0 \cos(\theta_0)$, as a direct consequence of the fact that $a_x = 0$. The y-component of the velocity starts at $v_0 \sin(\theta_0)$ and decreases linearly with time.

ASIDE: The acceleration due to gravity is generally neither parallel nor anti-parallel to the velocity, except in cases in which the projectile is fired straight up [or down], and then the resultant motion is strictly one-dimensional.

---

[5] Recall that speed is the magnitude of the velocity and is thus a scalar quantity.

LAUNCH	$t = 0$	$\theta = \theta_0$
MAXIMUM HEIGHT	$t_{\max} = \dfrac{v_0 \sin(\theta_0)}{g}$	$\theta = 0$
LANDING	$t_R = \dfrac{2\,v_0 \sin(\theta_0)}{g}$	$\theta = -\theta_0$

Another reason for subjecting ourselves to these last two examples is to show very explicitly that projectile motion does not fall into either of the two pure archetypes for accelerated motion. Neither the angle nor the speed of the projectile is constant in standard projectile motion.

Q: What kinds of non-standard cases exist?

A: Interesting scenarios include: firing the projectile from a rampart above a level field [as the enemy approaches on foot] or from a cliff above the sea [as the foe approaches by ship]; ski-jumping over sloped terrain; requiring that the projectile pass over a net or under a crossbar; ..... The possibilities are endless!

Chapter 8

Circular Motion

Recall that in an archetypical form of accelerated motion the particle's speed remains constant while the direction of its velocity vector changes. The simplest instance of this is UNIFORM CIRCULAR MOTION, in which a particle traverses a circular path[1] of radius R, at constant speed.

Suppose that at time t, the particle is located at the point $\vec{r}(t)$, and its velocity is $\vec{v}(t)$.

[Tautologically, the velocity is tangent to the circular trajectory of the particle.]

ASIDE: If the origin of coordinates is placed at the centre of the circular path, then the velocity vector is perpendicular to the position vector at all times as they rotate together around the circle.

At a slightly later time, $t + \Delta t$, the particle will have moved to a new position. The relative displacement of the particle during the time interval, Δt, may be characterised by the relative angle $\Delta\theta$ [quantified in radians]. The velocity vector at the later time, $\vec{v}(t + \Delta t)$, is tangent to the circular path at the new location of the particle, as exhibited in Figure 8.1.

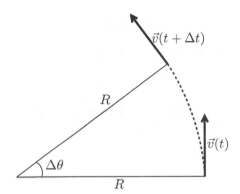

FIGURE 8.1 Motion of a Particle along a Segment of Circular Path

The change in the velocity of the particle through the time interval is

$$\Delta\vec{v} = \vec{v}(t + \Delta t) - \vec{v}(t).$$

Several [well-defined and well-motivated] steps take us from this velocity difference to an expression for the instantaneous acceleration of the particle.

- In the concurrent limits $\{\Delta t \to 0 \text{ AND } \Delta\theta \to 0\}$, $\Delta\vec{v}$ is directed toward the centre of the circular path. Hence, the instantaneous acceleration of the particle points directly toward the centre of the circular path. The adjective **centripetal**, meaning "centre-seeking," is used to denote this type of acceleration.

---

[1]More precisely, the particle need only be moving along a segment of circular arc.

General Analysis of 2-D Motion in Polar Coordinates

Let's embark upon a general analysis of single-particle motion in terms of polar coordinates.

[By so doing, we shall also more fully appreciate Cartesian coordinates].

The trajectory of a particle moving in 2-d may be described in terms of its Cartesian coordinate component functions $(x(t), y(t))$. As per Chapters 6 and 7, the basis vectors, $\hat{\imath}$ and $\hat{\jmath}$, are assumed to possess no explicit or implicit time dependence. That is, the Cartesian basis is fixed.

On the other hand, the description of the particle's motion may be framed in terms of its distance from a fixed point [the origin], and the angle between the ray from the origin to the particle and some fixed reference direction. Without loss of generality [WLOG], we shall insist that the Cartesian and polar coordinate systems share a common origin as illustrated in Figure 8.4.

>   ASIDE: This assumption was made in the above discussion pertaining to interconversion between Cartesian and polar coordinates. If it so happens that the two coordinate systems do not share a common origin, then one may transform the existing Cartesian coordinates to a new set [with the same basis vectors] whose origin is coincident with that of the polar system, by means of a [global, finite] translation. Only the values of the [Cartesian] position component functions will change, in the expected manner, leaving the velocity and acceleration unaffected.

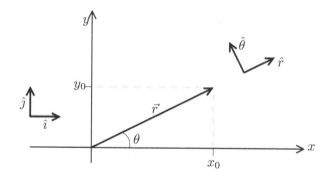

FIGURE 8.4   The Position of a Particle in Cartesian and Polar Coordinates

There always[3] exist unit vectors $\hat{r}$ and $\hat{\theta}$ which characterise the directions of increasing r and increasing θ [at the instantaneous location of the particle] respectively. Thus, these basis vectors are not fixed; rather, they vary with the position of the particle. One might say that they inherit a dependence on time due to the motion of the particle.

Both $\{\hat{\imath}, \hat{\jmath}\}$ and $\{\hat{r}, \hat{\theta}\}$ are valid orthonormal bases for the description of the instantaneous position of the particle. Since these basis sets can be used to describe any 2-d vectors, they can be employed to provide expressions for each other, *viz.*,

$$\hat{r} = \cos(\theta)\,\hat{\imath} + \sin(\theta)\,\hat{\jmath} \qquad\qquad \hat{\imath} = \cos(\theta)\,\hat{r} - \sin(\theta)\,\hat{\theta}$$
$$\hat{\theta} = -\sin(\theta)\,\hat{\imath} + \cos(\theta)\,\hat{\jmath} \qquad\qquad \hat{\jmath} = \sin(\theta)\,\hat{r} + \cos(\theta)\,\hat{\theta}\ ,$$

where the context for the choices of trigonometric functions and signs appears in Figure 8.4.

---

[3]The sole exception to the "always" occurs when the particle is at the origin, because there, and only there, the "theta direction" is ambiguous.

ASIDE: Those familiar with linear algebra will recognise that the coefficients in the formulae relating the basis vectors are the matrix components associated with rigid rotations in 2-d:

$$\begin{bmatrix} \tilde{x} \\ \tilde{y} \end{bmatrix} = \begin{bmatrix} \cos(\theta) & \sin(\theta) \\ -\sin(\theta) & \cos(\theta) \end{bmatrix} \begin{bmatrix} x \\ y \end{bmatrix}.$$

In Chapter 6 we derived the velocity and acceleration from the explicit formulae describing the particle's position as a function of time [with respect to a Cartesian basis]:

$$\vec{r}(t) = r_x(t)\,\hat{\imath} + r_y(t)\,\hat{\jmath},$$

$$\vec{v}(t) = \frac{d\vec{r}}{dt} = \frac{dr_x}{dt}\,\hat{\imath} + \frac{dr_y}{dt}\,\hat{\jmath},$$

$$\vec{a}(t) = \frac{d\vec{v}}{dt} = \frac{d^2\vec{r}}{dt^2} = \frac{d^2 r_x}{dt^2}\,\hat{\imath} + \frac{d^2 r_y}{dt^2}\,\hat{\jmath}.$$

In polar coordinates, the position of the particle is expressed as

$$\vec{r}(t) = r(t)\,\hat{r} = (\text{magnitude})\,[\text{direction}].$$

The velocity is the time rate of change of the position, and thus

$$\vec{v}(t) = \frac{d\vec{r}}{dt} = \frac{dr}{dt}\,\hat{r} + r\,\frac{d\hat{r}}{dt}.$$

The first term accounts for the change in radial distance to the origin, while the second arises from the change in the radial direction at the location of the particle.

Q: How does one determine the time rate of change of the radial direction?

A: Just watch us! Expressing $\hat{r}$ in terms of the [fixed] Cartesian basis enables straightforward computation of the rate at which it is changing.

$$\begin{aligned} \frac{d\hat{r}}{dt} &= \frac{d}{dt}\Big[\cos(\theta)\,\hat{\imath} + \sin(\theta)\,\hat{\jmath}\Big] \\ &= -\sin(\theta)\,\frac{d\theta}{dt}\,\hat{\imath} + \cos(\theta)\,\frac{d\theta}{dt}\,\hat{\jmath} \\ &= \omega\Big[-\sin(\theta)\,\hat{\imath} + \cos(\theta)\,\hat{\jmath}\Big] \\ &= \omega\,\hat{\theta}. \end{aligned}$$

Q: Wow! Can this result be true?

A: You betcha! The vector $\hat{r}$ is changing in time, but it is [and remains] a unit vector and therefore has constant length. Since it cannot extend [i.e., change in its own direction], it can only be changing in the "other direction(s)." There is only one such possible direction in 2-d. Thus, the radial unit vector can only twist around in the $\hat{\theta}$ direction.

Analogous behaviour must hold for the direction of increasing angle: it can only twist in the radial direction. Let's verify this by direct computation.

$$\begin{aligned} \frac{d\hat{\theta}}{dt} &= \frac{d}{dt}\Big[-\sin(\theta)\,\hat{\imath} + \cos(\theta)\,\hat{\jmath}\Big] \\ &= -\cos(\theta)\,\frac{d\theta}{dt}\,\hat{\imath} - \sin(\theta)\,\frac{d\theta}{dt}\,\hat{\jmath} \\ &= -\omega\Big[\cos(\theta)\,\hat{\imath} + \sin(\theta)\,\hat{\jmath}\Big] \\ &= -\omega\,\hat{r}. \end{aligned}$$

Chapter 9

Dynamics and Newton's First Law

Up until this point, we have been exclusively concerned with How things move [*i.e.*, position, velocity, acceleration, *etc.* of a particle as functions of time], without much consideration of Why[1] they are moving.

Let's begin to remedy this *lacuna* with a phenomenological definition of force.

FORCE   A particle [of constant mass] accelerates when it is acted upon by a net [external] force.

Some necessary [weaselly] legalese has been inserted as fine print.

CONSTANT MASS   Constancy of mass excludes cases in which the particle is losing mass, like a rocket burning its fuel, or gaining mass, like a raindrop accreting water vapour.

EXTERNAL   This qualifier is somewhat moot, since we are now considering instances of the motion of single particles, but will become essential once we encounter systems of particles in Chapter 29 *et seq.*

Our first [trivial] inference is that **force is a vector quantity**, since it is closely associated with acceleration. Laying the groundwork for further investigation of force begins with a discussion of Newton's FIRST LAW OF MOTION.

Newton's First Law [N1]   *a.k.a.* The Law of Inertia

There exist a variety of different [and yet equivalent] formulations of N1. We'll start with the following expression.

N1   A particle [of constant mass] will continue in a state of uniform motion unless it is acted upon by a NET [external] force.

In a **state of uniform motion** the particle moves with constant[2] velocity. This includes the special case $\vec{v} = \vec{0}$, which has given rise to the colloquial version of the First Law.

N1   An object at rest remains at rest unless acted upon by a NET [external] force.

Yet another version of N1 finds expression in terms of INERTIAL REFERENCE FRAMES.

N1   An object in a state of uniform motion [including "at rest"], as observed in an inertial reference frame [IRF], will remain in this state in the absence of any NET [external] force exerted upon the object.

---

[1]Some would contend that "Why" is too freighted a word to use here. If you subscribe to these sentiments, then substitute "a somewhat deeper sense of How," for this usage of "Why."

[2]A constant velocity is unvarying both in magnitude and direction.

In this last formulation, Newton's First Law of Motion hardly appears to involve "motion" at all. In fact, it is a profound statement about the invariance of physical laws under inessential changes in the choice of coordinates. Let's investigate how this works.

- An inertial reference frame, hereafter IRF, is a coordinate system, along with a clock, in which N1 is observed to hold.

- Two IRFs which are moving uniformly with respect to one another constitute equally valid venues for the analysis of motion and the accurate determination of underlying dynamics.

The second point amounts to an expression of the PRINCIPLE OF GALILEAN RELATIVITY [*a.k.a. the democracy*[3] *of inertial observers*]. While observers in different IRFs may very well disagree on the kinematical description of the motion of a given particle [positions and velocities], they will always agree on the dynamical aspects [accelerations and forces].

This is heady stuff! Let's consider a few examples to refine our understanding.

EXAMPLE [*PK and the Police Car (Reprised)*]

Our first example reprises the sports car and police car chase example found in Chapter 3. Recall that the kinematic analysis was performed [twice] from the viewpoint of an observer "O," who was standing by the side of the road, and also from the perspective of "NK," who was riding along with PK. The two descriptions of the cars' trajectories are different, as evident in the contrast between Figures 3.2 and 3.3.

ASIDE: In the statement of the problem, we were simply told that the acceleration of the police car was the same for both observers. This did not seem at all unreasonable. Still, we might have been pleasantly surprised that each computed exactly the same time interval between the two passings.

It happens that both O and NK are inertial observers, and their coordinates and clocks each constitute a valid IRF for the analysis of the motion. The non-equivalent elements in their respective descriptions are inessential [the values assigned to constants of integration], while truly dynamical properties [acceleration and the computed time interval] are exactly the same.

EXAMPLE [*Riding on the Inertial Express*]

For our second example, imagine[4] that two of the *Steinein* brothers, *Albert* and *Bertal*, henceforth "A" and "B," are riding on a passenger railcar which is moving at constant velocity $\vec{V}$ along a straight smooth horizontal section of track. Between the brothers is an approximately frictionless table[5] across which they will slide a particle.

---

[3] Although this appears to be a misuse of the political term "democracy," in that the Laws of Physics are not determined by majority vote, it has become standard usage and provides a convenient shorthand way of expressing the idea that all inertial observers are equally able to determine the universally valid Laws of Physics. This stands in contrast to some *gnostic* version of nature whose laws are only apprehendable by a select group of "preferred observers."

[4] What follows is a specific instance of a *Gedanken experiment* [*a.k.a.* thought experiment] performed entirely in our minds. Einstein, for one, employed *Gedanken experiments* to wonderful effect. The ironclad rule for these musings is that every aspect must, always, be realisable in practice, in principle, to any desired level of approximation. Thus, a frictionless horizontal plane is allowed, while travelling back in time to chat with your forebears is not! A general rule for *Gedanken experiments* is that they are best carried out while sitting in one's most comfortable chair with a nice cup of tea or coffee near at hand.

[5] 'Approximately frictionless' means that friction is a sub-dominant contribution to the overall margin of experimental error.

A imparts a constant initial velocity, v_0 [directed toward B], to the particle.

Let's introduce two more inertial observers: "R," who rides along in the railcar,[6] and "T," who stands trackside. Now let's explicitly consider the two cases of PARALLEL and PERPENDICULAR motions.

- PARALLEL

  Suppose that A and B are facing the front and rear of the train, respectively, as in Figure 9.1. The velocity of the particle as observed by R is $\vec{v}_0$, while T sees the particle moving with velocity $\vec{V} + \vec{v}_0$. Clearly, R and T do not agree on the magnitude of the velocity of the particle [and may even argue about the direction too]. However, each may determine that the particle is not accelerating and thus, in accord with N1, is not being acted upon by a net force.

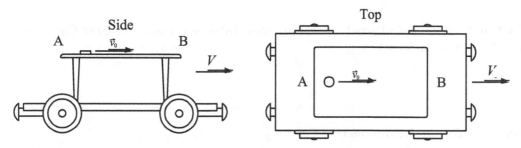

FIGURE 9.1 Motion of a Particle on a Frictionless Table on a Train Moving at Constant Velocity (Parallel Case)

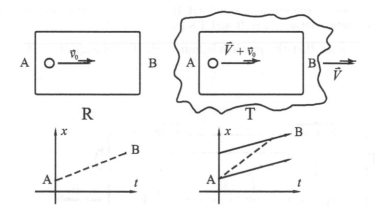

FIGURE 9.2 Two Views of the Motion of a Particle on a Frictionless Table on a Train Moving at Constant Velocity (Parallel Case)

- PERPENDICULAR

  Suppose that A and B are seated as in Figure 9.3. The velocity of the particle as observed

---

[6]The snarky among us might venture that R stands for "redundant," since A, B, and R are all mutually at rest. We include R so to not conflate the rôles of inertial observer and participant in the present example.

by R is $\vec{v}_0$, while T sees the particle moving with velocity $\vec{V} + \vec{v}_0$. In this instance, R and T disagree about **both** the magnitude and the direction of the velocity of the particle. However, they each may determine that the particle is not accelerating and, thus, by N1, is not being acted upon by a net force.

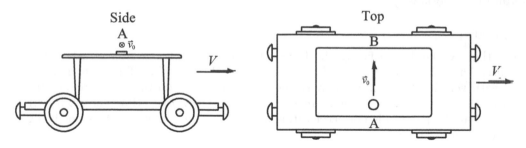

FIGURE 9.3   Motion of a Particle on a Frictionless Table on a Train Moving at Constant Velocity (Perpendicular Case)

EXAMPLE [*Riding on the Non-Inertial Express*]

As the third example, suppose that the moving railcar is undergoing a constant acceleration $\vec{a} = a$ [forward].   From observations of the trajectory of the railcar, T is able to infer the magnitude of the acceleration: a. Similarly, R can determine the magnitude of $\vec{a}$ by careful observation of the trajectories of fixed trackside objects, which he perceives to be accelerating in the [backward] direction.

ASIDE: The acceleration of the railcar means that R is NOT an inertial observer, and this vitiates the democratic equivalence of R and T's points of view!

Suppose that at the instant that the particle is launched, the velocity of the train is $\vec{V}$.

- PARALLEL

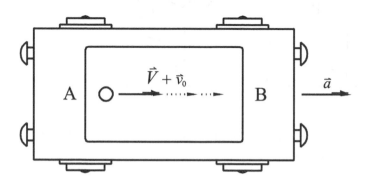

FIGURE 9.4   Motion of a Particle on a Frictionless Table on an Accelerating Train (Parallel Case)

Suppose that A and B are facing the front and rear of the train, as in Figure 9.4. With some care, T is able to *track* the motions of the particle and the train separately. From the observation that the particle is moving with constant velocity, $\vec{V} + \vec{v}_0$, T is able to conclude that no net force acts on the particle. By composition of the inertial motion of the particle and the accelerated motion of the railcar, T is able to [accurately and completely] ascertain the APPARENT motion of the particle observed by R.

    o IF $\vec{v}_0$ and $\vec{a}$ lie in the same direction, THEN the particle appears to slow down as it proceeds, because the table [affixed to the railcar] gains speed on the particle.

        [Circumstances exist in which the particle slows, stops, and then returns to A!]

    o IF $\vec{v}$ and $\vec{a}$ are oppositely directed, THEN the particle appears to speed up on its journey across the table.

Indeed, R does perceive the velocity of the particle to be changing after it is launched. His careful measurements reveal that the acceleration of the particle has magnitude "a" in the [backwards] direction. *Ergo*, in light of N1, he might surmise that there is a force acting on the particle so as to produce this rearward acceleration.

    ASIDE: Further experimentation determines that the acceleration does not depend on the mass or composition of the particle! That is, the acceleration experienced by the particle is not unlike the constant gravitational acceleration encountered in our analysis of projectile motion in Chapter 7. Einstein had such a realisation (*circa* 1907) *en route* to his formulation of the General Theory of Relativity.

And yet, postulating some novel physical interaction to account for this force is an enormous step to take. In addition, T's observations militate against the introduction of a new force to explain R's observations.

It is more prudent then, in this instance, to [provisionally] conclude that N1 does not apply for R,[7] because the railcar is not an IRF. It is customary practice to account for R's observations [when necessary] by invoking a **pseudoforce** [*a.k.a.* fictitious force] which acts in R's non-inertial reference frame.

• PERPENDICULAR

Suppose that A and B are seated as in Figure 9.5. Careful inspection by T reveals that the particle is moving with constant velocity $\vec{V} + \vec{v}_0$, while the table is accelerating in the [forward] direction. The combination of these motions leads T to conclude that the particle trajectory will appear to curve toward the rear of the train as it moves across the table. This is precisely what R observes. The urge to describe these observations using the formulae developed for projectile motion should be tempered by the realisation that the "force" responsible for the acceleration of the particle is fictitious.

To summarise: IF the railcar is accelerating, THEN it no longer constitutes an IRF, AND N1 fails to hold for particles observed by R. The inertial observer standing trackside, T, is deemed able to accurately observe and analyse the underlying physics.

    ASIDE: The weasel word "deemed" appears in recognition of the facts that the Earth is rotating on its axis and revolving around the Sun; the Solar System is revolving about the centre of the Milky Way galaxy; and the galaxy is hurtling toward the Virgo supercluster. Fortunately, however, an observer moving uniformly on the surface of the Earth is effectively inertial for all but the most sensitive experiments.

---

[7] *A priori*, either [or both] of the observers could reside in a non-inertial frame. The fact that T sees inertial behaviour where it might reasonably be expected, while R does not, greatly increases the odds that R resides in a non-inertial reference frame.

EXAMPLE [*Correct and Incorrect Applications of Newton's Third Law*]

CORRECT REASONING

When PK goes skydiving and is in *freefall*,[6]

the Earth "pulls down" on PK with a force of about 850 newtons.

ASIDE: **Q:** How do we know what the force on PK is?

A: At this early stage in our development of dynamics, we must appeal to N2, and the phenomenology of projectile motion, to determine the magnitude of this downward force. *To wit*, PK's mass is $m \sim 85\,\text{kg}$, and on exit from the airplane he experiences an acceleration, $\vec{a} = \vec{g} \sim 10\,\text{m/s}^2$ [downward] [neglecting air resistance]. *Ergo*, the magnitude of the net [gravitational] force acting on PK is about $850\,\text{N}$.

Meanwhile, **PK "pulls up" on the Earth with a force of 850 newtons, too.**

The force that the Earth exerts on PK profoundly affects his motion,[7] as it causes him to accelerate at roughly 10 metres/second per second. On the other hand, the force which PK exerts on the Earth produces a negligible effect on its motion, $a_\oplus \simeq 1.25 \times 10^{-22}\,\text{m/s}^2$, owing to the enormous degree of inertia ($m_\oplus \simeq 6 \times 10^{24}\,\text{kg}$) possessed by the Earth.

[It is very unlikely that such a tiny acceleration could ever be detected.]

INCORRECT REASONING

A person who does not properly apprehend N3 may be led to reason *falsely* thus:

"IF forces of interaction always occur in action–reaction pairs,
THEN it is impossible to exert a NET force on anything!"

It is evident that this line of argument is seriously flawed, as one would never be convinced by it to go skydiving without a parachute. Its logical flaw abides in the failure to precisely note the distinction between the body which is **producing** the force and the particle which is being **acted upon**. The *actor* and the *actee* are different particles, even when N3 exposes the symmetry of the interaction between them. It is not appropriate to add together forces acting on different particles,[8] and thus action–reaction forces do not cancel.

---

[6] *Wheeeeeeee!*

[7] It's fair to say that it *effects* his motion, too.

[8] We will have occasion to qualify this statement somewhat when considering systems of particles in Chapter 30 and thenceforth.

Chapter 11

Solving Dynamics Problems Using Newton's Laws

Our overarching goal, as physicists, is to develop models which can reproduce and explain the observed regularities and behaviours of physical systems. Newton's Laws of Motion provide us with two axioms which codify observed[1] regularities:

N1 Net [external] force is required to change the state of motion of an object.

N3 Forces of interaction necessarily occur pairwise.

along with a quantitative definition of inertia relating cause [force] to effect [acceleration]:

N2 $m\,\vec{a} = \vec{F}_{\text{NET ext'l}}$.

Axioms and definitions are well and good, but ...

Q: How might we achieve our overarching goal?

A: Apply $m\,\vec{a} = \vec{F}_{\text{NET ext'l}}$ to any particular physical situation so as to derive or construct the **equation(s) of motion** which hold for the particle(s) under study.

The first step in application of N2 is the identification of those objects [particles] participating in the dynamics of the physical system.

> ASIDE: We presume here that there will always be some reliable means by which the participants may be distinguished.
>
> [This tenet is relaxed in statistical mechanics]
>
> A common practice is to label particles by their salient dynamical properties: mass, charge, surface colour, composition, *etc.* Furthermore, any unique specification of $\mathcal{N}$ particles is equivalent to labelling by the set of positive integers, $\{1, 2, \ldots \mathcal{N}\}$, which we shall employ in the generic case.

Second, the forces which act upon each participating object must be enumerated. Forces acting on the nth particle, $n \in [1, 2, \ldots, \mathcal{N}]$, may depend on:

- The time, t

- Its position, $\vec{r}_n(t)$, and the positions of the other particles, $\{\vec{r}_m\}|_{m \neq n}$

- Its velocity, $\vec{v}_n(t)$, and the velocities of other particles, $\{\vec{v}_m(t)\}|_{m \neq n}$

- The particle's intrinsic attributes: mass (m_n), electric charge (q_n), *etc.*

---

[1]One must tread carefully here. The ancient Greeks—Aristotle in particular—would argue that these axioms are far from self-evident. In Aristotle's reckoning, the only way in which a particle can be induced to remain in a state of uniform motion, other than at rest, is through constant application of external force. This fundamental discrepancy between Aristotle and Newton/Galileo has been resolved in Newton's favour. In defence of the Ancients, however, one must appreciate that the technologies to overcome the effects of friction and other dissipative forces were little-advanced in the ancient world.

2. The block is the sole dynamical constituent. It is subject to three distinct forces.

Weight ○   The weight of the block is $\vec{W} = m\,g$ [vertically down].

Normal ○   The normal force on the block is $\vec{N} = N$ [vertically up]. [It acts perpendicular to the surface of contact between the block and the plane.]

Applied ○   The applied force is $\vec{F}_A = F_A$ [to the right].

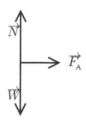

FIGURE 11.2   FBD for the Block

3. Coordinates

Each force enumerated above acts either horizontally or vertically, while the motion of the block takes place along a horizontal straight line. So, let's choose Cartesian coordinates, (x, y), where y is vertical [positive increasing upwards], while x is horizontal and aligned with the applied force [positive increasing to the right]. The precise location of the origin need not concern us now, but its specification should take into account the initial data for the block.

4. $m\,\vec{a} = \vec{F}_{\text{NET}}$

Vertical: $m\,a_y = F_{\text{NET}y} = N - m\,g.$
As the block remains on the horizontal surface, there can be no acceleration in the vertical direction, *i.e.*, $a_y = 0$. Hence, $F_{\text{NET}y} = N - m\,g = 0$, and so,

$$N = m\,g,$$

the magnitude of the normal force must be equal to the weight of the block.

Horizontal: $m\,a_x = F_{\text{NET}x} = F_A.$
Thus, the acceleration in the x-direction is

$$a_x = \frac{F_A}{m} = \text{constant.}$$

The block undergoes constant acceleration along the horizontal plane!

5. The constant acceleration kinematical formulae describe the motion of the block.

[The y-direction is trivial; all of the excitement takes place in the x-direction.]

Therefore, the completely general solution for the trajectory of the block is:

$$v_x = \frac{F_A}{m} t + v_{0x} \qquad\qquad v_y = 0$$

$$r_x = \frac{F_A}{2m} t^2 + v_{0x}\, t + r_{0x} \qquad\qquad r_y = r_{0y}$$

,

where the values of the integration constants v_{0x} and (r_{0x}, r_{0y}) are not yet specified, as they depend on inessential details such as the placement of the origin and the instant at which $t = 0$ on the clock. *Artful* choices conform with the initial data and lead to spare and simple formulae for the trajectory.

6. The result obtained seems quite reasonable: no motion off the plane and constant acceleration in the direction of the constant applied force.

EXAMPLE [*Variation on a Block on a Horizontal Frictionless Plane*]

A block of mass m is pushed along a frictionless horizontal surface with constant applied force $\vec{F_A} = F_A$ [down and to the right at angle θ below the horizontal]. Again, suppose that the block follows a straight-line path.

We proceed in steps as directed by the recipe.

1. Sketch

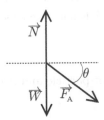

FIGURE 11.3 Another Block of Mass m Pushed along a Frictionless Plane

2. The block is the sole dynamical constituent. It is acted upon by three forces.

Weight ○ $\vec{W} = m\,g$ [vertically down].

Normal ○ $\vec{N} = N$ [vertically up].

Applied ○ The applied force is $\vec{F_A} = F_A$ [down and right at angle θ].

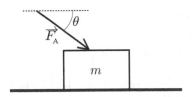

FIGURE 11.4 FBD for the Block

3. Coordinates: Cartesian coordinates, (x, y) [(horizontal , vertical)], will do nicely.

   The weight and normal forces act vertically, while the applied force has non-zero projection onto both the horizontal and vertical axes. Meanwhile, the block moves along the frictionless plane in a horizontal straight line.

4. $m\,\vec{a} = \vec{F}_{\text{NET}}$

   Vertical: $m\,a_y = F_{\text{NET}y} = N - m\,g - F_A \sin(\theta)$.

   There is no acceleration in the vertical direction since the block remains on the horizontal surface, and thus $a_y = 0$. Therefore

   $$0 = F_{\text{NET}y} = N - m\,g - F_A \sin(\theta), \qquad \Longrightarrow \qquad N = m\,g + F_A \sin(\theta).$$

   The normal force cancels both the weight of the block and the downward component of the applied force.

   Horizontal: $m\,a_x = F_{\text{NET}x} = F_A \cos(\theta)$.

   Thus, the acceleration in the x-direction is

   $$a_x = \frac{F_A \cos(\theta)}{m} = \text{constant.}$$

   The block undergoes constant horizontal acceleration.

5. The constant acceleration kinematical formulae apply. Hence,

   $$v_x = \frac{F_A \cos(\theta)}{m}\,t + v_{0x} \qquad\qquad\qquad v_y = 0$$

   $$r_x = \frac{F_A \cos(\theta)}{2m}\,t^2 + v_{0x}\,t + r_{0x} \qquad\qquad r_y = r_{0y}$$

   is the general solution for the trajectory of the block, with the kinematical constants remaining to be specified.

6. No motion off the plane and constant acceleration in the x-direction is eminently reasonable, *encore!*

EXAMPLE [*Another Variation on a Block on a Horizontal Frictionless Plane*]

A block of mass m is pulled along a frictionless horizontal surface with constant applied force $\vec{F}_A = F_A$ [at angle θ above the horizontal, to the right]. Suppose yet again that the block moves only in the direction of the horizontal component of the applied force.

   ASIDE: We assume that the applied force is not large enough to lift the block off the plane.

1. Sketch—see Figure 11.5.

2. The block is the only constituent particle and it experiences the actions of three separate forces.

Weight ∘ $\vec{W} = m\,g$ [vertically down].

Normal ∘ $\vec{N} = N$ [vertically up].

Applied ∘   The applied force is $\vec{F}_A = F_A$ [at θ to the horizontal, up and right].

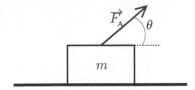

FIGURE 11.5   Yet Another Block of Mass m Pulled along a Frictionless Plane

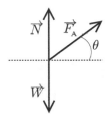

FIGURE 11.6   FBD for the Block

3. Coordinates: Cartesian coordinates, *viz.*, $(x\,,\,y)$ [(horizontal , vertical)], suffice.

4. $m\,\vec{a} = \vec{F}_{\text{NET}}$

Vertical: $m\,a_y = F_{\text{NET}y} = N - m\,g + F_{\text{A}}\sin(\theta).$
   Once again, the block remains on the horizontal plane, so $a_y = 0$. This, in turn, requires that

$$F_{\text{NET}y} = N - m\,g + F_{\text{A}}\sin(\theta) = 0\,, \qquad \Longrightarrow \qquad N = m\,g - F_{\text{A}}\sin(\theta)\,.$$

   The magnitude of the normal force is equal to the block's weight, less the amount by which the applied force acts upward on the block.

Horizontal: $m\,a_x = F_{\text{NET}x} = F_{\text{A}}\cos(\theta).$
   Thus, the acceleration in the x-direction is

$$a_x = \frac{F_{\text{A}}\cos(\theta)}{m} = \text{constant.}$$

   This is precisely the same result for the acceleration as was obtained in the previous example, when the applied force acted at θ below the horizontal!

5. The constant acceleration kinematical formulae apply. Therefore, the block's trajectory is *exactly* the same as was found in the previous example.

6. *Cool, eh?*

Chapter 12

Ropes and Pulleys

EXAMPLE [*A Two-Block and Pulley System*]

Two blocks, labelled 1 and 2, of mass M_1 and M_2, respectively, are joined by a piece of ideal rope which passes over an ideal pulley, as shown in the sketch below. The first block slides along a frictionless horizontal plane, while the second drops vertically downward. We shall assume that drag forces are negligible and that the acceleration due to gravity, g, is constant.

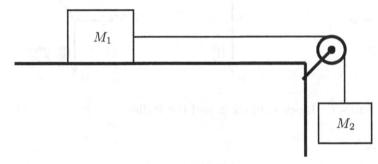

FIGURE 12.1 A Two-Block and Pulley System

Q: What is the acceleration of this system of blocks?

ASIDE: In any multi-constituent system, the particles' accelerations are correlated. In this particular case, the blocks' accelerations have the same magnitude, while their respective directions differ. The sought-for answer is this common magnitude.

A: To find the accelerations of the blocks and the system, let's follow the recipe.

1. Sketch—see Figure 12.1 above.

2. There are three dynamical constituents: Block 1, Block 2, and the Pulley. Each of these is acted upon by various forces.

   Block 1

   Weight 1 $\circ$ $\vec{W_1} = M_1\,g$ [vertically down].
   Normal 1 $\circ$ $\vec{N_1} = N_1$ [vertically up].
   Tension 1 $\circ$ The tension in the rope, $\vec{T_1} = T_1$ [horizontally rightward].

   Block 2

   Weight 2 $\circ$ $\vec{W_2} = M_2\,g$ [vertically down].
   Tension 2 $\circ$ The tension in the rope, $\vec{T_2} = T_2$ [vertically up].

Pulley

Weight $\mathbb{P}$ ∘ We are assuming that the pulley is ideal, and hence inertialess. Thus, its mass, $M_{\mathbb{P}}$, and weight, $\vec{W}_{\mathbb{P}} = M_{\mathbb{P}}\,g$ [vertically down], are both zero.

Tension 1′ ∘ The tension in the rope segment lying between the Pulley and Block 1, $\vec{T}_1' = T_1'$ [horizontally leftward].

Tension 2′ ∘ The tension in the rope segment lying between the Pulley and Block 2, $\vec{T}_2' = T_2'$ [vertically up].

Normal $\mathbb{P}$ ∘ There is a support force, $\vec{P}$, holding the pulley in place.

Friction $\mathbb{P}$ ∘ We're assuming that friction and drag forces are negligible.

In light of the above enumeration of the forces on each dynamical constituent, the three FBDs are illustrated below.

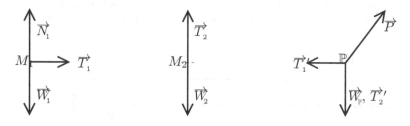

FIGURE 12.2   FBDs for Block 1, Block 2, and the Pulley

3. Coordinates

Each of the forces listed above acts in either the horizontal or vertical direction, with the possible exception of $\vec{P}$. Block 1 slides horizontally to the right, Block 2 falls vertically downward, and the Pulley is fixed in place. Exercising our freedom to choose different coordinate sets, (x_i , y_i), for each constituent, we arrive at the choices displayed in the following table.

Constituent	x-coordinate	y-coordinate
Block 1	horizontal [increasing RIGHT]	vertical [increasing UP]
Block 2	vertical [increasing DOWN]	extraneous
Pulley	horizontal [increasing RIGHT]	vertical [increasing UP]

4. $m\,\vec{a} = \vec{F}_{\text{NET}}$

Block 1

Vertical: $M_1\,a_{1y} = F_{\text{NET}1y} = N_1 - M_1\,g.$

The sliding block remains on the horizontal surface, and so the vertical component of its velocity remains fixed at zero throughout its motion. The constancy of the vertical component of the velocity means that there must be an absence of net acceleration in this direction, *i.e.*, $M_1\,a_{1y} = 0$. Thus,

$$0 = F_{\text{NET}1y} = M_1\,g - N_1 \qquad \Longrightarrow \qquad N_1 = M_1\,g.$$

[The magnitude of the normal force is equal to the weight of the block.]

Horizontal: $M_1 a_{1x} = F_{\text{NET}1x} = T_1$.

Thus,
$$M_1 a_{1x} = T_1.$$

ASIDE: Since T_1 has no compelling reason to change with time, one might suspect that Block 1 endures constant acceleration.

CAVEAT: If the rope has inertia, then one expects the tension to vary.

Block 2

Vertical: $M_2 a_{2x} = F_{\text{NET}2x} = W_2 - T_2$.

The tension force and the weight act in opposite directions and thus partially cancel, leading to
$$M_2 a_{2x} = M_2 g - T_2.$$

The signs occur the way they do on account of the orientation of the x-axis.

ASIDE: As T_2 has no compelling reason to change with time, one might suspect that the acceleration of Block 2 is also[1] constant.

Horizontal: $M_2 a_{2y} = 0 = F_{\text{NET}2y}$.

No forces act on Block 2 in the horizontal direction. Thus, this component of the net force vanishes, and hence so too does the horizontal component of its acceleration. Block 2 falls straight downward.

Pulley

The pulley is fixed in place. Its acceleration is zero, and the net force acting upon it vanishes.

Vertical: $0 = P_y - T_2' - W_{\mathbb{P}} = P_y - T_2'$.

The upward component of the pulley support force counters[2] the force of tension T_2' acting down.

Horizontal: $0 = P_x - T_1'$.

The rightward component of the pulley support force cancels the tension T_1' acting to the left.

5. To this point, the analysis has been adamantly *reductionist* insofar as each constituent has been considered in isolation. Now we shall link together the parts of the system by imposition of the constraints *enjoined* by the ideal rope and pulley.

Ideal Rope   An ideal rope transmits tension without any diminution in its magnitude:
$$T_1 = T_1', \qquad \text{AND} \qquad T_2 = T_2'.$$

Also, the acceleration of all points along the rope is the same,
$$a_{1x} = a_{2x} = \underline{\text{the}} \text{ characteristic acceleration of the system } = a.$$

---

[1] These suspicions about the accelerations of the blocks and the tension forces they experience are mutually consistent.

[2] If the pulley were not massless, then the support force would counter its weight, too. Our repeated mention of the weight of the putatively ideal pulley is so as to not prejudice the modelling of rotational inertia in Chapter 37 *et seq.*

Ideal Pulley An ideal pulley changes the direction of the tension in the associated rope without changing its magnitude,

$$T_1' = T_2' = \underline{\text{the}} \text{ characteristic tension in the rope } = T.$$

Rewriting our dynamical equations in light of these linking relations yields

$$\left.\begin{cases} M_1\, a = T \\ M_2\, a = M_2\, g - T \end{cases}\right\}, \ \textit{i.e.,} \text{ two equations with two unknowns: } \{a, T\}.$$

The particular choices of coordinates made earlier have had the favourable effect of diminishing the algebraic complexity of the set of equations. Solving these equations [by whatever means[3] one might prefer] results in

$$a = \frac{M_2}{M_1 + M_2}\, g \quad \text{and} \quad T = \frac{M_1 M_2}{M_1 + M_2}\, g\,.$$

The acceleration of the system and the tension throughout the rope are both constant in time.

> ASIDE: If our aim were to determine the trajectory of one or both of the blocks, then we would insert the acceleration that was determined just above, along with the requisite initial data, into the constant acceleration kinematical formulae.

6. The comments scattered throughout the analysis indicate that the results are in accord with our expectations. The consistency of the solutions can be verified by paying particular attention to two physically relevant limits.

    ◇ IF $M_1 \gg M_2$, **THEN** the system will have a small acceleration.
    Our intuition suggests that the relatively small net external force [magnitude equal to $M_2\, g$] will have only a small influence on this system [with total inertia, $M_{\text{Total}} = M_1 + M_2$]. In fact, the general formula for the acceleration reveals[4]

    $$\lim_{M_1 \gg M_2} a = \lim_{M_1 \gg M_2} \frac{M_2}{M_1 + M_2}\, g \to 0^+,$$

    confirming our expectations.

    ◇ IF $M_1 \ll M_2$, **THEN** the system will accelerate rapidly.
    In this case, the relatively large net force [magnitude $M_2\, g$] ought to produce a strong response in this system. Examination of the acceleration formula reveals that in this limit,

    $$\lim_{M_1 \ll M_2} a = \lim_{M_1 \ll M_2} \frac{M_2}{M_1 + M_2}\, g \to g^-,$$

    which again conforms to our expectations.

---

[3]Suitable techniques include guesswork, Gauss–Jordan reduction, matrix methods, or what have you.
[4]The superscripted $+(-)$ indicates that the limiting value is approached from above (below).

EXAMPLE [*Another Two-Block and Pulley System*]

Two blocks, labelled 1 and 2, with masses M_1 and M_2, respectively, are joined by a piece of rope, which passes over a pulley as shown in Figure 12.3.

[The idealisations in the previous example apply here too.]

Q: What is the acceleration of this system of blocks?
A: Let's follow the recipe to find out.

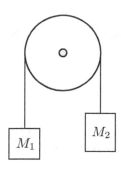

FIGURE 12.3   Another Two-Block and Pulley System

1. Sketch—see Figure 12.3.

2. There are three dynamical constituents: Block 1, Block 2, and the Pulley, $\mathbb{P}$.

Block 1	**Block 2**
Weight 1 ∘ $\vec{W_1} = M_1\, g\ [\downarrow]$.	Weight 2 ∘ $\vec{W_2} = M_2\, g\ [\downarrow]$.
Tension 1 ∘ $\vec{T_1} = T_1\ [\uparrow]$.	Tension 2 ∘ $\vec{T_2} = T_2\ [\uparrow]$.

Pulley

Weight $\mathbb{P}$ ∘   The weight of the pulley is assumed to vanish, $\vec{W_p} = \vec{0}$.

Tension 1' ∘ $\vec{T_1'} = T_1'\ [\text{vertically down}] = T_1'\ [\downarrow]$.

Tension 2' ∘ $\vec{T_2'} = T_2'\ [\text{vertically down}] = T_2'\ [\downarrow]$.

Normal $\mathbb{P}$ ∘   The support force, $\vec{P}$, holds the axle in place.

Friction $\mathbb{P}$ ∘   Friction and drag forces are assumed to be negligible.

The above enumeration of the forces on each dynamical constituent is restated in the three separate FBDs illustrated in Figure 12.4.

3. Coordinates

The relevant direction is vertical. Let y be upward positive for Block 1 and the Pulley, and downward positive for Block 2.

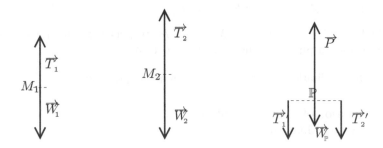

FIGURE 12.4 FBDs for Block 1, Block 2, and the Pulley

4. $m\,\vec{a} = \vec{F}_{\text{NET}}$

Block 1	Block 2
$M_1\,a_{1y} = F_{\text{NET}1y} = T_1 - M_1\,g$	$M_2\,a_{2y} = F_{\text{NET}2y} = M_2\,g - T_2$

Pulley

The pulley is fixed in place, and thus it does not accelerate. The support force precisely cancels the combined effects of the forces of tension T_1' and T_2' [along with the vanishing weight of the pulley].

5. The idealised properties of the rope and the pulley serve to link the dynamical constituents.

Ideal Rope Ideal rope is a perfect transmitter of the tension force,

$$T_1 = T_1' \qquad \text{AND} \qquad T_2 = T_2'.$$

That the ideal rope does not stretch implies that

$$a_{1y} = a_{2y} = a\,,$$

where a is the acceleration characteristic of the system.

> ASIDE: Do note that $a_{1y} = a_{2y}$, rather than $a_{1y} = -a_{2y}$, on account of the choices we made in orienting the blocks' y-coordinate axes.

Ideal Pulley An ideal pulley changes the direction of a tension force without changing its magnitude. Hence,

$$T_1' = T_2' = T\,,$$

where T is the magnitude of the tension throughout the entire ideal rope.

Rewriting the dynamical equations in light of these linking relations produces

$$\left\{ \begin{array}{l} M_1\,a = T - M_1\,g \\ M_2\,a = M_2\,g - T \end{array} \right\}, \quad \text{two equations for two unknowns: } \{a, T\}.$$

The solution of this set of equations is

$$a = \frac{M_2 - M_1}{M_1 + M_2}\,g\,, \qquad \text{and} \qquad T = \frac{2\,M_1 M_2}{M_1 + M_2}\,g\,.$$

Again, both the acceleration and the tension throughout the string are constant in time. The trajectory of each block is given by [suitably adapted] constant-acceleration kinematical formulae.

6. Let's argue for the consistency of this solution.

⋄ IF $M_1 \ll M_2$, *i.e.*, the masses of the blocks are quite disparate, THEN the system will have a relatively large rate of acceleration. [Little cancellation of the respective weight forces occurs, and $M_{\text{Total}} = M_1 + M_2$ is only slightly larger than M_2.] Taking the limit of the acceleration formula bears this out:

$$\lim_{M_1 \ll M_2} a = \lim_{M_1 \ll M_2} \frac{M_2 - M_1}{M_1 + M_2} g \to g^-.$$

The tension in the rope *gives rise* to the large upward acceleration of Block 1 [nearly g] in addition to cancelling its weight, so it is not surprising that

$$\lim_{M_1 \ll M_2} T = \lim_{M_1 \ll M_2} \frac{2\,M_1\,M_2}{M_1 + M_2} g \to 2\,M_1\,g.$$

⋄ IF $M_1 \simeq M_2$, *i.e.*, the two blocks have comparable degrees of inertia, THEN the system will accelerate slowly. [The weights of the blocks nearly cancel.] In this limit, the acceleration of the system tends toward

$$\lim_{M_1 \simeq M_2} a = \lim_{M_1 \simeq M_2} \frac{M_2 - M_1}{M_1 + M_2} g \to 0.$$

Also in this limit, the tension in the rope becomes

$$\lim_{M_1 \simeq M_2} T = \lim_{M_1 \simeq M_2} \frac{2\,M_1\,M_2}{M_1 + M_2} g \simeq M_1\,g \simeq M_2\,g,$$

since the blocks are suspended by the rope as the system accelerates slowly.

An intriguing realisation is that the above-described systems are indeed accurately thought of in combined terms: *i.e.*, that the "total inertia" of the system is a combination of the inertia contributions of the constituents, and that the system as a whole responds to the net external force acting upon it.

Q: Why does the tension not contribute to the net external force?

A: The tension is internal and is thus incapable of causing acceleration of the system.

Q: Can this be true? The motions of the blocks are profoundly affected by the tension.

A: The internal tension cannot give rise to bulk motion of the entire system.

[Go ahead and try to lift yourself off the ground by pulling upward on your shoelaces!]

It is certainly the case that internal forces may affect the motions of parts of the system, but when all is said and done—and N3 is rigorously satisfied—the internal forces have zero net effect on the entire system under consideration.

Q: Why does the tension assume its respective values [(one-half of/one times) the harmonic mean of the weights of the blocks] in the two examples?

A: The tension is reactive in that it assumes the magnitude necessary to ensure that the accelerations of the blocks are entirely consistent with the overall acceleration of the system. Furthermore, there also exists a symmetry in these two cases: the tension is the same if we switch the roles [by switching the labels] of Blocks 1 and 2.

EXAMPLE [*Statics Problem: The Suspended Block*]

Three pieces of ideal rope are knotted together at a **junction** [taken to occupy a single point]. The free end of one rope is tied to a block of mass M, while the other two loose ends are affixed to a horizontal ceiling. All ropes are straight[5] and the two that attach to the ceiling form [acute] interior angles α and β as shown in Figure 12.5. We assume that the system is **static**; it is not changing with time.

[A non-static system can be "at rest" for an instant only; a static system is continually at rest.]

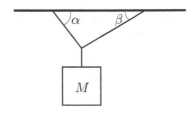

FIGURE 12.5   A Hanging Block System

Our goal is to predict the tensions in the ropes given the particular conditions specified by the mass of the block and the two interior angles.

> ASIDE: The IDEA is that these estimates of the tensions may then be compared with manufacturer's product information so as to determine in advance the grade of material (*i.e.,* string, chain, fishing line, binder twine, steel cable, or carbon fibre) needed [and hence the feasibility of using a particular material] to implement the proposed design.

Q: What are the tensions in the ropes?
A: To find these out, let's follow the recipe.

1. Sketch—see Figure 12.5.

2. There are two dynamical constituents: the block, and [surprisingly, as it has no inertia] the junction.

Block

Weight ∘ $\vec{W} = M\,g$ [vertically down].
Tension 1 ∘ $\vec{T_1} = T_1$ [vertically up].

Junction

Tension 1 ∘ $\vec{T_1}' = T_1'$ [vertically down].
Tension 2 ∘ $\vec{T_2} = T_2$ [up and right at angle β above the horizontal].
Tension 3 ∘ $\vec{T_3} = T_3$ [up and left at angle α above the horizontal].

This enumeration of the forces on each dynamical constituent enables the drawing of the two FBDs in Figure 12.6.

---

[5]This condition ensures that all tensions are non-zero, avoiding a pathological case.

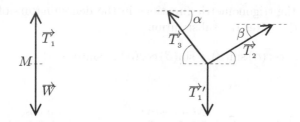

FIGURE 12.6 FBDs for the Hanging Mass Problem

3. Coordinates

Let's employ (horizontal , vertical) coordinates (x , y) for the analysis.

4. $m\,\vec{a} = \vec{F}_{\text{NET}}$

Block

Vertical: $M\,a_y = F_{\text{NET}y} = T_1 - M\,g$.

Since the block remains at rest, there must be no acceleration [in any direction]. This requires that the magnitude of the tension in rope 1 be equal to the weight of the block.

$$0 = F_{\text{NET}y} = T_1 - M\,g \qquad \Longrightarrow \qquad T_1 = M\,g\,.$$

Horizontal: *Irrelevant.* No forces act horizontally on the block.

Junction

The tension forces from the three ropes must completely cancel in order for the system to remain static.

Vertical: $0 = F_{\text{NET}y} = -T_1' + T_2 \sin(\beta) + T_3 \sin(\alpha)$.

Horizontal: $0 = F_{\text{NET}x} = +T_2 \cos(\beta) - T_3 \cos(\alpha)$.

5. Ideal rope constraints link together the constituents.

Ideal Rope   Ideal rope is a perfect transmitter of force, hence

$$T_1 = T_1'\,.$$

Rewriting the dynamical equations in light of this linking relation yields

$$\left.\begin{cases} 0 = M\,g - T_1 \\ 0 = -T_1 + T_2 \sin(\beta) + T_3 \sin(\alpha) \\ 0 = T_2 \cos(\beta) - T_3 \cos(\alpha)\,. \end{cases}\right\}\;,\quad \text{three equations for unknowns } \{T_1, T_2, T_3\}\,.$$

Solving this set of equations, by one's favourite method, yields:

$$T_1 = M\,g\,,$$

$$T_2 = \frac{M\,g\,\cos(\alpha)}{\cos(\alpha)\sin(\beta) + \sin(\alpha)\cos(\beta)}\,,\quad \text{and}$$

$$T_3 = \frac{M\,g\,\cos(\beta)}{\cos(\alpha)\sin(\beta) + \sin(\alpha)\cos(\beta)}\,.$$

The structure of the trigonometric functions in the denominators of the expressions for T_2 and T_3 invites the substitution

$$\cos(\alpha)\sin(\beta) + \sin(\beta)\cos(\alpha) \equiv \sin(\alpha + \beta)$$

leading[6] to

$$T_1 = Mg, \qquad T_2 = \frac{Mg\cos(\alpha)}{\sin(\alpha + \beta)}, \qquad T_3 = \frac{Mg\cos(\beta)}{\sin(\alpha + \beta)}.$$

The upshot of all of this is that knowledge of M, and the angles α and β, enables computation of the expected tensions in the three ideal ropes.

6. It is heartening to see the symmetry of the solutions under interchange of labels: $(2, \beta) \Longleftrightarrow (3, \alpha)$. This is vital, since how we choose to identify the ropes should not affect the physics. Let's also explore two particular physically relevant limits.

    ⋄   IF $\alpha = \beta$, THEN the tensions in the two upper ropes are equal,

$$T_2 = T_3 = T = \frac{Mg\cos(\alpha)}{2\sin(\alpha)\cos(\alpha)} = \frac{Mg}{2\sin(\alpha)}.$$

    This result conforms with our intuitions, for IF $\alpha = \beta \rightarrow \pi/2$, THEN $T \rightarrow Mg/2$, whereas IF $\alpha = \beta \rightarrow 0$, THEN $T \rightarrow \infty$.

    ⋄   IF $a \neq \beta$ AND $\beta \rightarrow \pi/2$, THEN

$$T_2 \rightarrow \frac{Mg\cos(\alpha)}{\cos(\alpha) \times (1) + \sin(\alpha) \times (0)} \simeq Mg,$$

$$T_3 \rightarrow \frac{Mg \times 0}{\cos(\alpha) \times (1) + \sin(\alpha) \times (0)} \simeq 0,$$

    indicating that the second rope does almost all of the work in holding the block, as per our expectations.

---

[6] The temptation to invoke the theorem that the sum of the interior angles in a plane triangle is equal to π **radians**, *a.k.a.* 180°, shall be resisted.

Chapter 13

Blocks in Trains and in Contact

EXAMPLE [*The Train of Blocks*]

Three blocks, with masses M_1, M_2, and M_3, respectively, are joined by ideal rope connectors to make a train, as in Figure 13.1. An external agent applies a horizontal constant force $\vec{F}_A$ to an ideal rope attached to the first car in the train. The plane on which the blocks slide is frictionless.

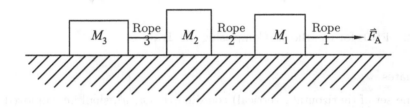

FIGURE 13.1 A Train of Three Blocks Being Pulled by an External Force

Q: What is the acceleration of the train?
A: To find this out, let's follow the recipe.

1. Sketch—see Figure 13.1 above.

2. All three blocks participate in the dynamics. From "caboose" (3) to "engine" (1), the forces on each are enumerated below.

   Block 3

   Weight 3 ∘ $\vec{W}_3 = M_3\, g$ [vertically down].

   Normal 3 ∘ $\vec{N}_3 = N_3$ [vertically up].

   Tension 3R ∘   The tension in Rope 3, acting on Block 3, is $\vec{T}_{3R} = T_{3R}$ [to the right].

   Block 2

   Weight 2 ∘ $\vec{W}_2 = M_2\, g$ [vertically down].

   Normal 2 ∘ $\vec{N}_2 = N_2$ [vertically up].

   Tension 3L ∘   The tension in Rope 3, acting on Block 2, is $\vec{T}_{3L} = T_{3L}$ [to the left].

   Tension 2R ∘   The tension in Rope 2, acting on Block 2, is $\vec{T}_{2R} = T_{2R}$ [to the right].

Block 1

Weight 1 ∘ $\vec{W_1} = M_1\,g$ [vertically down].

Normal 1 ∘ $\vec{N_1} = N_1$ [vertically up].

Tension 2L ∘   The tension in Rope 2, acting on Block 1, is $\vec{T}_{2L} = T_{2L}$ [to the left].

Applied 1 ∘   The applied force, $\vec{F_A} = F_A$ [to the right], is transmitted via Rope 1.

The forces on each block appear in the FBDs in Figure 13.2.

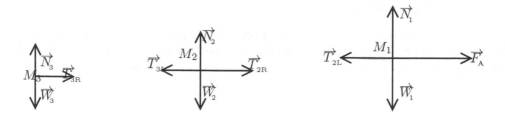

FIGURE 13.2   FBDs for the Blocks Comprising the Train

3. Coordinates

   The same set of (horizontal , vertical) coordinates, $(x\,,y)$, shall be employed for all three constituent blocks.

4. $m\,\vec{a} = \vec{F}_{\text{NET}}$

   ### Block 3

Vertical: $M_3\,a_{3y} = F_{\text{NET}3y} = N_3 - M_3\,g.$
   Since Block 3 rides along the horizontal surface, the vertical component of its acceleration is zero. Thus, the y-component of the net force acting on the block vanishes, necessitating that $N_3 = M_3\,g$.

Horizontal: $M_3\,a_{3x} = F_{\text{NET}3x} = T_{3R}.$

   ### Block 2

Vertical: $M_2\,a_{2y} = F_{\text{NET}2y} = N_2 - M_2\,g.$
   As was the case for Block 3, $N_2 = M_2\,g$ cancels the weight of Block 2.

Horizontal: $M_2\,a_{2x} = F_{\text{NET}2x} = T_{2R} - T_{3L}.$

   ### Block 1

Vertical: $M_1\,a_{1y} = F_{\text{NET}1y} = N_1 - M_1\,g.$
   Ditto for Block 1, leading to $N_1 = M_1\,g$.

Horizontal: $M_1\,a_{1x} = F_{\text{NET}1x} = F_A - T_{2L}.$

5. The linkages among the equations of motion for the blocks are provided by the ideal ropes.

Ideal Rope Both the tension at, and the acceleration of, all points throughout the ideal rope take on dynamically determined constant values. Hence,

$$T_{3R} = T_{3L} = T_3 , \qquad \text{AND} \qquad T_{2R} = T_{2L} = T_2 ,$$
$$\text{WHILE} \qquad a_{3x} = a_{2x} = a_{1x} = a .$$

The acceleration, a, is characteristic of the system as a whole.

After linking, we have three dynamical equations:

$$\left\{ \begin{array}{l} M_3\, a = T_3 \\ M_2\, a = T_2 - T_3 \\ M_1\, a = F_{\!A} - T_2 \end{array} \right\} , \text{ involving the unknowns } \{a, T_2, T_3\} .$$

The solution of this set of equations is

$$a = \frac{F_{\!A}}{M_1 + M_2 + M_3} , \qquad \begin{aligned} T_2 &= \frac{M_2 + M_3}{M_1 + M_2 + M_3}\, F_{\!A} \\ T_3 &= \frac{M_3}{M_1 + M_2 + M_3}\, F_{\!A} \end{aligned} .$$

6. The following considerations attest to the soundness of these results.

   ◇ **3** The tension in Rope 3 is precisely that which will cause Block 3 to accelerate at a.

   ◇ **32** The tension in Rope 2 ensures that the Block 2–Block 3 subsystem accelerates at a.

   ◇ **321** The acceleration of the whole system is equal to the net [external] force, divided by the total inertia of the system.

The train of blocks on the frictionless plane is dynamically indistinguishable from a stack of the same blocks!

[This will not necessarily remain the case when friction acts.]

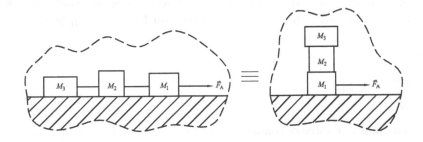

FIGURE 13.3 A Train on a Frictionless Plane is Dynamically Equivalent to a Stack

EXAMPLE [*Forces of Contact and Newton's Third Law*]

Two blocks, labelled 1 and 2, of mass M_1 and M_2, respectively, are placed in contact with one another on a frictionless horizontal plane. Together they are pushed with a horizontal applied force, F_A, as shown in Figure 13.4. We assume that drag forces are negligible, that g is constant, and that the blocks do not deform.

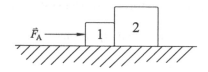

FIGURE 13.4   A Two Blocks in Contact System

Q: What is the acceleration of the system of blocks?

A: Let's embark on a brief aside [predicting the result] before following the recipe.

> ASIDE: In light of the previous examples, one might surmise that—under the assumptions listed above—the system will behave as though it were a particle of mass $M_{\text{Total}} = M_1 + M_2$. The net external [applied] force would then cause the system of blocks to accelerate with
>
> $$\vec{a} = \frac{F_A}{M_1 + M_2}\ [\rightarrow].$$
>
> Knowing the result in advance is generally a good thing. We'll follow the recipe through anyway and shall serendipitously encounter yet another *cool* aspect of nature.

1. Sketch—see Figure 13.4.

2. There are two blocks: "1" and "2." Each is acted upon by various forces.

    Block 1

    Weight 1 ○ $\vec{W_1} = M_1\,g\ [\downarrow]$.

    Normal 1 ○ $\vec{N_1} = N_1\ [\uparrow]$.

    Applied ○ $\vec{F_A} = F_A\ [\rightarrow]$.

    Contact 21 ○   A NORMAL FORCE OF CONTACT occurs between the two blocks since their sides are touching. That force exerted by 2 on 1 is $\vec{C_{21}} = C_{21}\ [\leftarrow]$.

    Block 2

    Weight 2 ○ $\vec{W_2} = M_2\,g\ [\downarrow]$.

    Normal 2 ○ $\vec{N_2} = N_2\ [\uparrow]$.

    Contact 12 ○   The [normal] force of contact exerted by 1 on 2 is $\vec{C_{12}} = C_{12}\ [\rightarrow]$.

    With the above enumeration of the forces on each dynamical constituent, we can draw the FBDs illustrated in Figure 13.5.

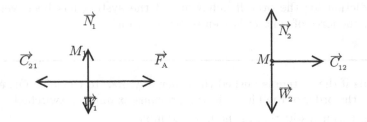

FIGURE 13.5 FBDs for Blocks 1 and 2 in Contact

3. Coordinates

    Each of the forces listed above acts solely in either the horizontal or vertical direction, while the blocks move horizontally to the right. A single set of coordinates (x, y), (horizontal , vertical), suffices for both blocks.

    > ASIDE: We refrain from specifying the location of the origin as it is not necessary to do so.

4. $m\vec{a} = \vec{F}_{\text{NET}}$

    Block 1

    Vertical: $M_1 a_{1y} = F_{\text{NET}1y} = N_1 - W_1.$
    That Block 1 remains on the horizontal surface requires that $N_1 = M_1\, g.$
    Horizontal: $M_1 a_{1x} = F_{\text{NET}1x} = F_A - C_{21}.$

    Block 2

    Vertical: $M_2 a_{2x} = F_{\text{NET}2x} = N_2 - W_2.$
    The horizontal trajectory of Block 2 requires that $N_2 = M_2\, g.$
    Horizontal: $M_2 a_{2x} = F_{\text{NET}2x} = C_{12}.$

5. The forces of contact are constrained by N3 to be equal and opposite, and this fact serves to link the individual blocks into the two block system, *viz.,*

$$C_{12} = C_{21} = C\,.$$

Furthermore, IF the blocks remain in contact AND they do not deform, THEN each must experience exactly the same acceleration:

$$a_{1x} = a_{2x} = a\,, \ \underline{\text{the}} \text{ acceleration of the system.}$$

Rewriting the dynamical equations in light of these linking relations yields,

$$\left. \begin{array}{l} M_1\, a = F_A - C \\ M_2\, a = C \end{array} \right\}, \ \text{two equations involving unknowns } \{a, C\}\,.$$

Solving these equations results in

$$a = \frac{F_A}{M_1 + M_2} \quad \text{and} \quad C = \frac{M_2}{M_1 + M_2}\, F_A\,.$$

> ASIDE: Yay! The acceleration is precisely in accord with our surmise.

6. Our prediction for the overall behaviour of the system has been verified. In addition, the force of contact is seen to be constant.

Q: What happens if the system is pushed the other way, *i.e.,* to the left? Or, equivalently: What happens if the order of the blocks in the previous example is switched?

A: Let's perform this investigation in the next example.

....

EXAMPLE [*Forces of Contact and Newton's Third Law (Encore)*]

The two blocks, labelled 1 and 2, from the previous example are reordered on the horizontal frictionless surface and given another push with an identical copy of the earlier applied force. Everything is just as it was in the previous example (except for the rearrangement of the blocks) as displayed in Figure 13.6.

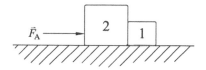

FIGURE 13.6   Another Two Blocks in Contact System

Qa: What is the acceleration of the system of blocks in this instance?

Aa: Again, let's guess the result prior to following the recipe.

> ASIDE: The difference between this case and the previous example amounts to exchanging the labels, $1 \leftrightarrow 2$. Thus, the acceleration, $\vec{a} = F_A/(M_1 + M_2)$ [rightward], is unchanged.

> [The really interesting question concerns the force of contact.]

Qc: What is the value of the force of contact?

> Is it the same as in the previous example?
>
> OR
>
> Is it obtained by the $1 \leftrightarrow 2$ switcheroo in the formulae from the previous example?

Ac: Let's follow the recipe to determine the answers to all of the questions.

1. Sketch—see Figure 13.6.

2. The two blocks experience sundry forces.

Block 2	**Block 1**
Weight ∘ $\vec{W_2} = M_2\, g$ [↓].	Weight ∘ $\vec{W_1} = M_1\, g$ [↓].
Normal ∘ $\vec{N_2} = N_2$ [↑].	Normal ∘ $\vec{N_1} = N_1$ [↑].
Applied ∘ $\vec{F_A} = F_A$ [→].	Contact ∘ $\vec{C}_{21} = \tilde{C}_{21}$ [→].
Contact ∘ $\vec{C}_{12} = \tilde{C}_{12}$ [←].	

In consideration of the enumeration of the forces on each block, the FBDs are presented in Figure 13.7.

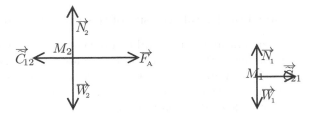

FIGURE 13.7 FBDs for Blocks 2 and 1 in Contact

3. Coordinates

   Let's choose the same set (x, y), (horizontal, vertical), as previously.

4. $m\vec{a} = \vec{F}_{\text{NET}}$

Block 2	**Block 1**
$M_2\,a_{2x} = F_{\text{NET}2x} = F_A - \tilde{C}_{12}$	$M_1\,a_{1x} = F_{\text{NET}1x} = \tilde{C}_{21}$

5. The forces of contact are constrained to satisfy N3, and so

$$\tilde{C}_{12} = \tilde{C}_{21} = \tilde{C}\,.$$

Furthermore, under our assumptions, the blocks have the same acceleration,

$$a_{1x} = a_{2x} = a\,.$$

Rewriting our dynamical equations in light of these linkages yields

$$\left\{ \begin{array}{l} M_2\,a = F_A - \tilde{C} \\ M_1\,a = \tilde{C} \end{array} \right\}\,, \quad \text{two equations for two unknowns } \{a, \tilde{C}\}\,.$$

Solving these equations yields

$$a = \frac{F_A}{M_1 + M_2} \quad \text{and} \quad \tilde{C} = \frac{M_1}{M_1 + M_2}\,F_A\,.$$

6. Our questions are answered and the acceleration prediction has been verified.

   The acceleration is exactly as we predicted, and the force of contact is that obtained from the previous result by the switcheroo $1 \leftrightarrow 2$.

   Q: Is the fact that the contact force depends crucially on the order of the blocks in accord with our intuitive grasp of nature?

   A1: Here's a *Gedanken* experiment to which we can all relate.

   Suppose that your belongings are packed into boxes in preparation for a move. Two of the boxes rest on a tile floor. One contains heavy books and paper, the

FIGURE 14.2   FBD for the Block and Decomposition of the Weight Force

3. Coordinates

    Let x denote the position along the incline [↘ positive], and let y represent the position perpendicular to the surface of the incline [↗ positive]. These coordinates are well-adapted to the motion of the block.

    > ASIDE: There is, at this juncture, no compelling need to specify the position of the origin.

    [The plane on which the block slides may be fixed by the coordinate condition, $y = 0$.]

4. $m\vec{a} = \vec{F}_{\mathrm{NET}}$

    Block

    With recourse to elementary trigonometry, one can determine that the components of the weight parallel to, and perpendicular to, the plane are

    $$W_{\parallel} = M\,g\,\sin(\theta) \qquad \text{and} \qquad W_{\perp} = -M\,g\,\cos(\theta)\,.$$

    The parallel, $\parallel$, component is positive because it is directed "down" along the incline, while the perpendicular, $\perp$, component is negative as it is directed into, or "below," the surface.

    IN PLANE: $M\,a_x = F_{\mathrm{NET}x} = W_{\parallel} = M\,g\,\sin(\theta).$

    OFF PLANE: $M\,a_y = F_{\mathrm{NET}y} = N + W_{\perp} = N - M\,g\,\cos(\theta).$
        Since the block rides on the plane, $a_y = 0$. Therefore,

    $$N = M\,g\,\cos(\theta)\,,$$

    i.e., the normal force cancels $W_{\perp}$.

5. According to the IN PLANE equation of motion, the acceleration of the block along the inclined surface is
    $$a_x = g\,\sin(\theta)\,.$$

    This acceleration is constant [assuming constancy of g and θ], and hence the constant acceleration kinematical formulae describe the trajectory of the block.

6. To test the veracity of this result, let's look at two particular limiting cases.

⬦ IF $\theta \to 0$ (*i.e.*, the plane is, in fact, horizontal), THEN the acceleration,

$$a_x\big|_{\theta \to 0} = g \sin(0) = 0 \,.$$

The horizontal frictionless plane constitutes an IRF!

⬦ IF $\theta \to \frac{\pi}{2}$ (*i.e.*, the plane is a vertical cliff), THEN the acceleration,

$$a_x\big|_{\theta \to \frac{\pi}{2}} = g \sin\left(\frac{\pi}{2}\right) = g \,.$$

In this situation, the block is in freefall.

Variations on the theme of holding a block at rest on the surface of an inclined plane with a rope, or some other means,[1] comprise a large class of standard dynamics problems.

Just for grins, consider the following rather non-standard problem.

Q: How might one keep a block at a fixed position on an inclined plane, without directly exerting an external force on the block?

At first blush, this would seem to be impossible. The key to resolving this conundrum is to be found amongst the weasel words buried in the question. The block need only be at rest *with respect to the incline*, which allows for accelerated motion of the entire block–incline system as long as there is no relative motion of the constituents.

A: We'll investigate three distinct solutions in three separate examples to follow.

EXAMPLE [*Fixing a Block on a Frictionless Incline I*]

IF the block–incline system is in freefall, THEN both constituents are accelerating at $\vec{g}$, AND IF they have zero relative velocity to begin with, THEN they will remain relatively at rest at later times.

> ASIDE: This is not unlike an astronaut's experience of "weightlessness" while orbiting the Earth [a type of FREEFALLING]. Astronauts observe that, when they let go of an item, it just seems to hang there *in much the same way that bricks don't*.[2]

EXAMPLE [*Fixing a Block on a Frictionless Incline II*]

IF the block–incline system is itself riding on a larger frictionless plane which is inclined at the same angle, THEN the system will accelerate down the larger plane at a constant rate, AND provided the block is initially at rest on the smaller plane, it will remain at relative rest.

Cool, eh?

---

[1] Are you *inclined* to *spring* ahead to the next chapter to encounter one of these other means?

[2] This is a brazen attempt to win the Award for the Most Gratuitous Reference to *The Hitchhiker's Guide to the Galaxy* in a Screenplay or Lecture Notes.

EXAMPLE [*Fixing a Block on a Frictionless Incline III*]

Suppose that the block–incline system is itself riding on a horizontal frictionless plane. IF an applied force, $\vec{F_A} = F_A\,[\rightarrow]$, say, acts on the block–incline system, THEN for a particular choice of the magnitude, F_{A0}, the block remains at constant position on the surface of the incline.

> ASIDE: The particular value, F_{A0}, is a CRITICAL VALUE of the applied force, in that qualitative characteristics and behaviours of the physical system are determined by whether the applied force is greater than, or less than, F_{A0}.

> > *A priori*, the value of the critical force, F_{A0}, is unknown.
> > In general, there may be more than one critical value for a given parameter.

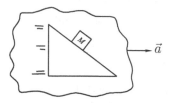

FIGURE 14.3   An Accelerated Block–Incline System

Q: How do we know that the applied force must be precisely tuned to a critical value?
A: Consider the following.

- TOO SMALL   Several possible behaviours ensue. [There exists another critical value.]

    ◇   IF $F_A = 0$, THEN the block will slide down the frictionless inclined plane, AND the plane will recoil[3] to the left! The recoil of the incline has the effect of presenting a steeper apparent angle of incline to the block, and hence $a_\parallel > g\,\sin(\theta)$.

    ◇   There exists a particular applied force, F_{Ab}, which braces the inclined plane, completely eliminating the recoil.

    <   IF $F_A > 0$ AND $F_A < F_{Ab}$, THEN the leftward recoil acceleration of the incline is diminished, as is the acceleration of the block along the incline.

    =   IF $F_A = F_{Ab}$, THEN the block moves down the incline in exactly the manner illustrated in the example presented at the beginning of this chapter, *i.e.*, $a_\parallel = g\,\sin(\theta)$.

    >   IF $F_A > F_{Ab}$ (but not too large), THEN the incline will accelerate to the right, whilst the block continues to slide down and to the right. The forward acceleration of the incline has the effect of presenting a shallower apparent slope to the block, and consequently $a_\parallel < g\,\sin(\theta)$.

---

[3] Patience please! We will understand this better when momentum is introduced in Chapter 29 and a formalism to treat systems of particles is developed in Chapters 30 and beyond.

- Too Large

IF the applied force is larger than its critical value, THEN the inclined plane will slide forward under the block. The block will then appear to slide UP the incline.

> This is not unlike the frictionless table accelerating underneath
> the particle set in motion in the *Non-Inertial Express* examples in Chapter 9.

The acceleration of the block up the incline is reduced to zero as the magnitude of the applied force approaches its critical value.

The block moves down the incline for small applied force and up the incline for sufficiently large applied force. *Ergo* [by the INTERMEDIATE VALUE THEOREM], there exists at least one critical value of the applied force for which the block remains at rest on the plane.

Thus far, we have proven the existence of F_{A0}, but have not actually determined its value. This we shall do now via two separate analyses. The first is more intuitive involving a pseudoforce, while the second is more proper and considers only *bona fide* forces.

METHOD ONE: [*A Pseudoanalysis*]

Let's consider the motion of the block from the viewpoint of the non-inertial frame of reference which is at rest with respect to the inclined plane. From this perspective, the block feels a pseudoforce on account of its inertia.

ASIDE: The block has a tendency to stay in a state of uniform motion, but the inclined plane pushes it forward. Consequently, the block feels as though it is being pushed horizontally into the surface on which it rides.

Our [distinctly pseudo-] analysis adheres to the standard recipe.

1. Sketch—see Figure 14.3.

2. The block is acted upon by two real forces and a pseudoforce.

Block

Weight ○ $\vec{W} = M\,g$ [vertically down] $= M\,g$ [↓].

Normal ○ $\vec{N} = N$ [perpendicular to incline, up and right] $= N$ [↗].

Pseudo ○ The pseudoforce, $\vec{F_P'}$, is equal in magnitude and opposite in direction to the [horizontal] "$M\,\vec{a}$" of the block as measured in a true IRF.

$$\vec{F_P'} = M\,(-\vec{a}) = -M|\vec{a}|\ [\rightarrow] = M\,a\ [\leftarrow] = F_P'\ [\leftarrow].$$

Figure 14.4 contains a drawing of the (pseudo)FBD for the block.

3. Coordinates

$(x\,,\,y) = (\ \text{IN PLANE [positive downward]}\,,\,\text{OFF PLANE [positive outward]}\ )$.

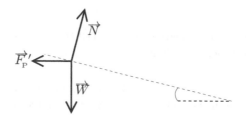

FIGURE 14.4   A PseudoFBD for the Block on an Accelerating Incline

4. (Pseudo) Newton's Second Law

   Block

OFF PLANE: $M\,a_y = F_{\text{NET}y} = M\,g\,\cos(\theta) + F_{\text{P}}'\,\sin(\theta) - N.$

   Since the block remains at rest with respect to the incline, a_y and the perpendicular component of the net force both vanish.

$$0 = M\,g\,\cos(\theta) + F_{\text{P}}'\,\sin(\theta) - N \qquad \Longrightarrow \qquad N = M\,g\,\cos(\theta) + F_{\text{P}}'\,\sin(\theta)\,.$$

   The normal force cancels the OFF PLANE components of the weight of the block and the pseudoforce.

IN PLANE: $M\,a_x = F_{\text{NET}x} = M\,g\,\sin(\theta) - F_{\text{P}}'\,\cos(\theta).$

   Since the block remains at rest, the IN PLANE component of the net force vanishes too. Therefore

$$0 = M\,g\,\sin(\theta) - M\,a\,\cos(\theta) \qquad \Longrightarrow \qquad a = g\,\tan(\theta)\,.$$

5. The magnitude of the horizontal acceleration that must be imparted to the block–incline system in order to keep the block relatively at rest on the surface of the incline is $g\,\tan(\theta)$.

6. A few moments' thought confirms the veracity of this result.

   SCALING   The acceleration due to gravity, g, sets the scale.

   ANGULAR VARIATION   At small angles, a should be small, since it is easy to imagine how a "too large" acceleration would cause the block to "float up the incline."

   [For small angles, $\tan(\theta)$ is small.]

   As θ grows, the needed acceleration must increases non-linearly, and indeed without bound. One can recognise the abject futility of attempting to keep the block at rest on a steeply sloped plane solely by pushing sideways.

   [As $\theta \to \pi/2$, $\tan(\theta)$ BLOWS UP ($\to \infty$).]

   MAGNITUDE   For all angles in the range $0 \le \theta \le \frac{\pi}{2}$, $\tan(\theta) > \sin(\theta)$. Thus, the magnitude of the horizontal acceleration that must be imparted to the incline-and-block system is greater than the magnitude of the acceleration that the block itself would experience were the incline at rest.

   NORMAL FORCE   The increase in the normal force, N, from that which would occur on a plane at rest, is consistent.

METHOD TWO: [*A Real and Proper Analysis*]

Let's re-analyse the situation, this time considering only real forces.

1. Sketch—see Figure 14.3.

2. The block is acted upon by gravity and the normal force.

Block

Weight ∘ $\vec{W} = M\,g\ [\downarrow]$.

Normal ∘ $\vec{N} = N\ [\,\text{up, perpendicular to incline}\,] = N\ [\nearrow]$.

With only two forces on the block, the FBD in Figure 14.5 is simple.

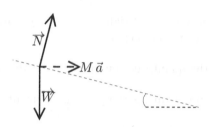

FIGURE 14.5 The FBD for the Block on an Accelerating Incline

3. Coordinates

   We'll employ block–incline adapted coordinates, $(x\,,y) = (\,\text{IN PLANE}\,,\text{OFF PLANE}\,)$.

4. $m\,\vec{a} = \vec{F}_{\text{NET}}$

   The block, when it is at rest relative to the incline, actually experiences constant acceleration in the horizontal forward direction.

   $$\vec{a} = a\ [\rightarrow] = (a_x\,,a_y) = (\,a\,\cos(\theta)\,,a\,\sin(\theta)\,)\,.$$

   Block

   OFF PLANE: $M\,a_y = M\,a\,\sin(\theta) = F_{\text{NET}y} = -M\,g\,\cos(\theta) + N$.
   Hence,
   $$N = M\big(g\,\cos(\theta) + a\,\sin(\theta)\big)\,.$$

   This normal force is larger than in the case in which the incline is not accelerated, as it must *cause* the net horizontal acceleration of the block in the forward direction (in addition to cancelling the perpendicular component of the weight of the block).

   [The weight force, acting vertically, cannot produce horizontal acceleration.]

   IN PLANE: $M\,a_x = M\,a\,\cos(\theta) = F_{\text{NET}x} = M\,g\,\sin(\theta)$.
   After cancelling and rearranging, the acceleration of the incline required to keep the block at rest on the surface is
   $$a = g\,\tan(\theta)\,.$$

5. The result of the pseudoanalysis, *i.e.*, $a = g \tan(\theta)$ is reaffirmed.

6. Let's first survey the range of possibilities in order to validate the arguments which preceded the dynamical analysis.

   ◇ **Underpowered**
   IF the x-component of the weight exceeds $M\,a_x$, THEN there exists a non-zero net force acting DOWN the incline which is not subsumed into the forward acceleration, AND the block will slide down the incline.

   ◇ **Overpowered**
   IF the x-component of the weight is less than $M\,a_x$, THEN there exists a non-zero residual acceleration directed UP the incline, AND the block will slide up the incline.

   ◇ **Goldilocks**
   IF the overall acceleration is "just right" (*i.e.*, $a = g \tan(\theta)$), THEN the block remains at rest relative to the incline.

Second, let's investigate the quantitative predictions of the model.

   ◇ IF $\theta \to 0$ (there is no incline as the plane is horizontal), THEN the needed acceleration is

$$a\big|_{\theta \to 0} = \lim_{\theta \to 0} g \tan(\theta) = 0\,,$$

precisely as one would expect.

   ◇ IF $\theta \to \frac{\pi}{4}$ (the incline is at 45°), THEN the system must be accelerated at g in order to keep the block relatively at rest on the incline.

   [This makes a lot of sense when you think about it.]

   ◇ IF $\theta \to \frac{\pi}{2}$ (a vertical cliff), THEN

$$a\big|_{\theta \to \frac{\pi}{2}} = \lim_{\theta \to \frac{\pi}{2}} g \tan(\theta) = \infty\,,$$

confirming one's suspicion that there is no possibility of holding the block fixed in place on a vertical incline by accelerating the system.

Chapter 15

Spring Fever

It may come as somewhat of a surprise to realise that all of the forces entertained thus far have been constant in magnitude and direction. We shall now enlarge our dynamics toolkit to include a particular type of force with position-dependent magnitude.

HOOKE'S LAW   Robert Hooke (*circa* 1635–1703) vied with Newton for the honour of being considered the pre-eminent natural philospher of his era, and with Christopher Wren as the premier architect in the rebuilding of London following the Great Fire of 1666. His eponymous force law concerns the behaviour of springs and elastic materials, and it reads as follows:

$$\vec{F_{\text{s}}} = -k\,\Delta\vec{x}\,.$$

There are three important elements in this force law.

— The negative sign indicates that the spring force is RESTORATIVE. The spring force acts so as to attempt to return the spring to its **equilibrium** [*a.k.a.* natural] **length**, *i.e.*, its length when no external forces act upon it. To summarise:

> IF one pulls on the spring [extending it], **THEN** the spring pulls back,
>
> **WHEREAS**
>
> IF one pushes on the spring [compressing it], **THEN** the spring pushes back.

k   The spring constant, k, is a measure of the stiffness of the spring. The SI units associated with k are

$$[\,k\,] = \frac{\text{N}}{\text{m}} = \frac{\text{kg}}{\text{s}^2}\,.$$

The spring constant is a phenomenological parameter which depends on the geometry of the spring [its size and shape], and on its material composition.

$\Delta\vec{x}$ This factor represents the amount by which the length of the spring has changed from its equilibrium length. This change may arise from extension or compression. Most often, one end of the spring is fixed. In such cases, the distinction between the equilibrium length of the spring and the equilibrium position of its free endpoint, $\vec{x}_0$, is elided. Hence,

$$\Delta\vec{x} = \vec{x} - \vec{x}_0\,.$$

Ideal Spring   An ideal spring is **sizeless, inertialess**, and produces a **Hooke's Law** force.

- Sizelessness eliminates any and all effects which might arise from the finite size and shape of the spring.

    Real spring coils collide when subjected to copious compression, and they deform on extreme extension.

- Inertialessness ensures the absence of any tendency to resist acceleration.

    [Therefore, no amount of force is consumed imparting acceleration to parts of the spring.]

- Conformity with Hooke's Law ensures that the magnitude of the spring force increases linearly with displacement from equilibrium, while the direction in which the force acts is always toward the equilibrium position.

    [The non-linearities exhibited by real springs make for complicated analyses.]

EXAMPLE [*A Mass–Spring System*]

One end of an ideal spring, with spring constant k, is attached to a fixed anchor point, while the other is affixed to a block of mass M riding on a frictionless horizontal plane.

Q: What sorts of dynamical behaviour does this system exhibit?
A: Let's consider several *Gedanken* experiments, described in **Parts A** through **D** below.

Part A: Suppose that the spring is at its equilibrium length and the block is at rest.

1. See the sketch in Figure 15.1.

2. The sole dynamical constituent, the block, experiences weight, normal, and spring forces.

    Block

    Weight ∘ $\vec{W} = M\,g\,[\,\text{vertically down}\,] = M\,g\,[\downarrow]$.

    Normal ∘ $\vec{N} = N\,[\,\text{vertically up}\,] = N\,[\uparrow]$.

    Spring ∘ $\vec{F}_{\text{s}} = -k\,(\vec{0}) = \vec{0}$.

    The forces acting on the block are shown in the FBD in Figure 15.1.

3. Coordinates

    The natural choice for coordinates is $(x\,,\,y) = (\,\text{horizontal}\,,\,\text{vertical}\,) = (\rightarrow\,,\,\uparrow)$, as indicated in the sketch. Simplicity dictates that the origin coincide with the location of the block when the spring assumes its equilibrium length, *i.e.*, $\vec{x}_0 = \vec{0}$.

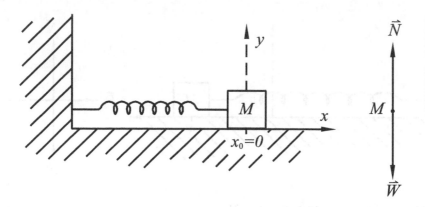

FIGURE 15.1   Sketch and FBD for Part **A**

4. $m\,\vec{a} = \vec{F}_{\text{NET}}$

    Block

    Vertical: $M\,a_y = F_{\text{NET}y} = N - M\,g.$
    > The absence of OFF PLANE acceleration constrains the normal force to precisely cancel the weight.

    Horizontal: $M\,a_x = F_{\text{NET}x} = 0.$
    > There are no forces acting in the horizontal direction.

5. The net force acting on the block vanishes and its motion is inertial, in accord with N1. [The block is originally at rest, and it remains at rest.]

Part B: Suppose that the block is displaced from its equilibrium position, $\vec{x}_0 = \vec{0}$, and brought to rest at $\vec{x} = x\,[\rightarrow]$. The spring remains affixed to the block. An applied force, $\vec{F}_{\text{A}}$, is needed to keep the block at rest.

1. Sketch—see Figure 15.2.

2. The block experiences weight, normal, spring, and applied forces.

    Block

    Weight $\circ$ $\vec{W} = M\,g\ [\downarrow].$

    Normal $\circ$ $\vec{N} = N\ [\uparrow].$

    Spring $\circ$ $\vec{F}_{\text{s}} = -k\,\Delta\vec{x} = -k\,(\vec{x} - \vec{0}) = -k\,x\ [\rightarrow] = k\,x\ [\leftarrow].$

    Applied $\circ$ $\vec{F}_{\text{A}} = F_{\text{A}}\ [\rightarrow].$

    The FBD for this situation appears in Figure 15.2.

3. Coordinates
    $(x\,,\,y) = (\rightarrow\,,\,\uparrow)$, of course!

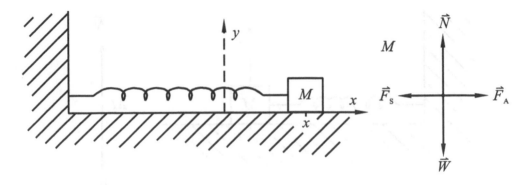

FIGURE 15.2 Sketch and FBD for Part **B**

4. $m\,\vec{a} = \vec{F}_{\text{NET}}$

Vertical: $0 = F_{\text{NET}y} = N - M\,g$, as usual.

Horizontal: $M\,a_x = F_{\text{NET}x} = F_{\text{B}} - k\,x$.

The block is stationary, and it remains so. This requires cancellation of the applied and spring forces,

$$0 = F_{\text{A}} - k\,x \qquad \Longrightarrow \qquad F_{\text{A}} = k\,x\,.$$

5. In order for the block to remain at rest, an applied force is needed to cancel the spring force. This applied force must then inherit proportionality to both the spring constant, k, and the extension[1] of the spring, x.

Part C: Suppose that the block described in Part **A** is displaced from its equilibrium position to $\vec{x} = x\ [\leftarrow] = -x\ [\rightarrow]$, and brought to rest.

ASIDE: In the present context, x is the magnitude of the displacement. [The direction is specified by "$-[\rightarrow]$."] One might elect to incorporate the direction algebraically by writing $\vec{x} = \tilde{x}\ [\rightarrow]$, where $\tilde{x}$ is the [horizontal component of the] displacement. Consistency is maintained by the identification $\tilde{x} = -x$.

An applied force, $\vec{F}_{\text{A}}$, is needed in order to keep the block at rest.

1. Sketch—see Figure 15.3.

2. The block is the sole dynamical constituent and experiences four forces.

Block

Weight ∘ $\vec{W} = M\,g\ [\downarrow]$.

Normal ∘ $\vec{N} = N\ [\uparrow]$.

Spring ∘ $\vec{F}_{\text{S}} = -k\,\Delta\vec{x} = -k\left(\vec{x} - \vec{0}\right) = -k\,x\ [\leftarrow] = k\,x\ [\rightarrow]$.

Applied ∘ $\vec{F}_{\text{A}} = F_{\text{A}}\ [\leftarrow]$.

The relevant FBD appears in Figure 15.3.

---

[1] One might say "x-tension," eh?

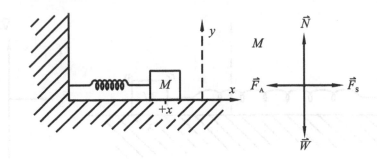

FIGURE 15.3 Sketch and FBD for Part **C**

3. Coordinates

   $(x, y) = (\rightarrow, \uparrow)$ suffice nicely.

4. $m\,\vec{a} = \vec{F}_{\text{NET}}$

   Block

   Vertical: *Ho hum.*

   Horizontal: $M\,a_x = F_{\text{NET}x} = +k\,x - F_{\text{A}}.$

   That the block is held at rest requires $F_{\text{A}} = k\,x$.

5. An applied force is needed to cancel the restorative force of the spring in order that the block remain at rest. The strength of the applied force depends on the spring constant and the distance by which the spring is compressed.

Part D: Suppose that the block–spring system is in the state depicted in Part **B**, and that the applied force is suddenly reduced to zero.

ASIDE: Effectively, the block was pulled away from equilibrium and then released.

1. Sketch—see Figure 15.4.

2. Weight, normal, and spring forces act on the block.

   Block

   Weight ∘ $\vec{W} = M\,g\ [\downarrow].$

   Normal ∘ $\vec{N} = N\ [\uparrow].$

   Spring ∘ $\vec{F}_{\text{s}} = -k\,\Delta\vec{x} = -k\,(\vec{x} - \vec{0}) = -k\,x\ [\rightarrow] = k\,x\ [\leftarrow].$

   The FBD in this case appears alongside the sketch in Figure 15.4.

3. Coordinates

   $(x, y) = (\rightarrow, \uparrow),$ *encore.*

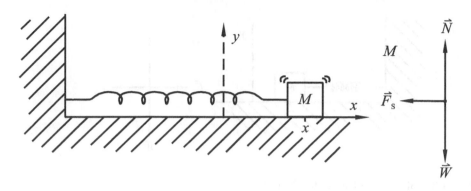

FIGURE 15.4   Sketch and FBD for Part **D**

4. $m\,\vec{a} = \vec{F}_{\text{NET}}$

   Block

   Vertical: *Whatever!*

   Horizontal: $M\,a_x = F_{\text{NET}x} = F_{\text{s}} = -k\,x$.

   The spring provides the net force acting on the block.

   [The minus sign arises from the restorative property of the spring force.]

5. The dynamical equation governing the motion of the block reads

$$M\frac{d^2x}{dt^2} = -k\,x\,.$$

This equation of motion is of precisely the same form as the defining equation for SIMPLE HARMONIC OSCILLATION [SHO], mentioned in Chapter 4 and discussed extensively in VOLUME II. Therefore, the trajectory of the block, in this instance, is necessarily sinusoidal.

   ◇   The block OSCILLATES about its equilibrium position.

   ◇   The AMPLITUDE of the block's motion is the difference between its equilibrium and release positions, *i.e.*, $x - x_0$. [In general, the amplitude is fixed by the operative initial conditions.]

   ◇   The FREQUENCY of the motion depends on invariant physical parameters of the system: the mass of the block, M, and the spring constant, k.

Chapter 16

Fact and Friction

Until now, all of the forces that have been considered were independent of the velocity of the particle upon which they act. Friction exhibits [a trivial form of] velocity dependence.

Friction   Friction is a contact interaction occurring between particles/systems at mutual rest, **static friction**, or in relative motion, **kinetic friction**.

> [Herein lies the velocity dependence: not moving *vs.* moving.]

- Friction is operative wherever and whenever surfaces are in contact with one another. It is directed **parallel** to the plane common to the contacting surfaces.

- Friction is a resistive force. It is oriented in such a way as to **oppose**

STATIC   unrealised motion [*i.e.,* motion which would occur if friction were absent], or

KINETIC   actual motion.

- Static friction is a REACTIVE force. Its magnitude, f_{s}, is just sufficient to ensure that there is zero net force acting on the particle.
  The magnitude of the static frictional force is bounded above $f_{\mathrm{s}} \leq f_{\mathrm{s,max}}$.

- Kinetic friction is modelled as a constant force, with magnitude f_{K}.

- The magnitudes of the frictional forces, whether static or kinetic, depend on intrinsic and extrinsic aspects of the contacting surfaces.

INTRINSIC   Effects due to the material compositions of the surfaces and their conditions [especially microscopic roughness] are parameterised by **the coefficients of static** $[\mu_{\mathrm{s}}]$ and   **kinetic** $[\mu_{\mathrm{k}}]$ **friction**.

EXTRINSIC   The normal force of contact [acting between the surfaces], N, has direct bearing on the magnitude of the frictional forces.

Incorporation of the intrinsic and extrinsic aspects into a model for friction leads to the following formulae for the magnitudes of the frictional forces:

$$f_{\mathrm{s}} \leq f_{\mathrm{s,max}} = \mu_{\mathrm{s}} N \qquad \text{and} \qquad f_{\mathrm{K}} = \mu_{\mathrm{k}} N.$$

EXAMPLE [*A Thought Experiment Involving Friction*]

This model for friction is investigated in the following series of *Gedanken* experiments: **Parts A** through **D**.

Part A: A block of mass M is at rest on a horizontal frictional plane surface. IF no external forces are applied to the block, THEN, *unsurprisingly* [in accord with N1], nothing happens. The block remains at rest on the surface.

 ASIDE: There are no horizontal forces acting. The weight of the block is cancelled by the normal force of contact produced by the surface.

Part B: A block of mass M is at rest on a horizontal frictional plane surface. IF a "gentle" horizontal force, $\vec{F_A} = F_A [\rightarrow]$, is applied to the block, THEN—*quelle surprise*—again nothing happens. Absence of motion in the presence of an applied force corroborates the existence of a static frictional force. Let's verify this by following the dynamical recipe.

1. Sketch—see Figure 16.1.

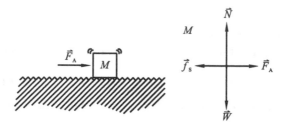

FIGURE 16.1 Sketch and FBD for the Gently Pushed Block on the Frictional Plane

2. The block is the only dynamical constituent. Four disparate forces act on it.

   Block

Weight ○ $\vec{W} = M g [\downarrow]$.

Normal ○ $\vec{N} = N [\uparrow]$.

Applied ○ $\vec{F_A} = F_A [\rightarrow]$.

Friction ○ $\vec{f_s} = f_s [\leftarrow]$. IF there were no frictional force operative, THEN the block would accelerate to the right under the influence of the applied force. The resistive aspect of friction dictates its opposition to this unrealised motion, and hence the force of static friction acts to the left.

3. Coordinates: $(x, y) = (\rightarrow, \uparrow)$.

4. $m \vec{a} = \vec{F}_{\text{NET}}$

   Block

Vertical: $M a_y = F_{\text{NET}y} = N - M g$.
            Since $a_y = 0$, $N = M g$.

Horizontal: $M a_x = F_{\text{NET}x} = F_A - f_s$.
            The absence of motion requires $a_x = 0$ and thus $f_s = F_A$, for consistency with N1 and N2.

5. The block does not accelerate, because the force of static friction adjusts its magnitude so as to precisely counter the applied force.

6. In the light of Part **B** of the *Gedanken* experiment, several properties of the model for static frictional force are seen to be eminently reasonable:

- It acts parallel to the surface.

- It is directed in opposition to the unrealised motion.

- Its magnitude is only as large as necessary to cancel what would otherwise be some non-zero net force.

Q: Did it matter that the applied force acted to the right?

A: Of course not! IF the gentle force had instead acted to the left, THEN the force of static friction would act to the right in order to counter the [unrealised] leftward motion of the block.

Part C: A block of mass M is at rest on a horizontal frictional plane surface. The magnitude of the gentle force applied to the block in Part **B** is increased until the block is **on the verge of sliding**.

ASIDE: By this turn of phrase it is meant that "If an additional jot or iota of force to the right were applied, then the block would suddenly start moving."

The applied force attains a critical value,[1] $\vec{F}_A = \vec{F}_{A0} = F_{A0} \ [\rightarrow]$ [signified by the subscripted naught], when the block is on the verge of sliding,

Since the block remains at rest [*i.e.*, does not accelerate], the dynamics are precisely the same as in Part **B** quoted above, with the static frictional force precisely cancelling the applied force. At the critical applied force, the static frictional force is equal to [*a.k.a.* saturates] its upper bound. Therefore,

$$F_{A0} = f_{S,\max} = \mu_S N \,.$$

Part D: The magnitude of the applied force is increased beyond its critical value, F_{A0}, determined in Part **C**. In accord with the definition of the critical force, the block must begin to slide and hence kinetic friction is operative.

1. Sketch—see Figure 16.2.

2. The block is the only dynamical constituent. Four distinct forces act on it.

Block

Weight ∘ $\vec{W} = M g \ [\downarrow]$.

Normal ∘ $\vec{N} = N \ [\uparrow]$.

Applied ∘ $\vec{F}_A = F_A \ [\rightarrow]$.

Friction ∘ $\vec{f}_K = f_K \ [\leftarrow] = \mu_K N \ [\leftarrow]$.

    The frictional force acts in opposition to the actual motion of the block.

---

[1] The behaviour of the block when the applied force is less than F_{A0} is markedly different from that when the applied force exceeds F_{A0}.

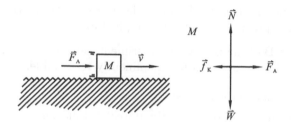

FIGURE 16.2   Sketch and FBD for the Moving Block on the Frictional Plane

3. Coordinates: $(x, y) = (\rightarrow, \uparrow)$.

4. $m\,\vec{a} = \vec{F}_{\text{NET}}$

   Block

   Vertical: $M\,a_y = F_{\text{NET}y} = N - M\,g.$ As before, $N = M\,g.$

   Horizontal: $M\,a_x = F_{\text{NET}x} = F_{\text{A}} - \mu_{\text{k}}\,N.$

5. The block accelerates in response to the net force which acts upon it,

$$M\,a_x = F_{\text{A}} - \mu_{\text{k}}\,N = F_{\text{A}} - \mu_{\text{k}}\,M\,g.$$

   In the last equality we invoked the expression obtained for the normal force via analysis of the block's vertical [y-component] equation of motion.

6. The dynamical response of the block, when the applied force is "ungentle" enough to cause it to move, is in accord with one's intuitions and expectations.

   ◇   The block moves.

   ◇   The block accelerates at a rate less than it would experience if the plane were frictionless.

   ◇   Were there no applied force acting, the moving block would slow and eventually halt.

   [It is not so surprising that Aristotle did not discern Newton's First Law, eh?]

This sequence of *Gedanken* experiments confirms the reasonableness of the friction model.

Q: What relation, if any, exists between μ_{s} and μ_{k} for a block–surface system?

A: Self-consistency and empirical observation demand that $\mu_{\text{s}} > \mu_{\text{k}}$.

The model for the force of friction is compactly expressed by:

$$f(\vec{v}) \quad \begin{cases} \leq \mu_{\text{s}}\,N \ [\text{opposed to unrealised motion}]\,, & \text{for } \vec{v} = \vec{0} \\ = \mu_{\text{k}}\,N \ [\text{opposed to actual motion}]\,, & \text{for } \vec{v} \neq \vec{0} \end{cases}.$$

Chapter 17

Fun with Friction

Our model for the frictional force exhibits a crude form of velocity dependence.

$$\text{IF} \quad \vec{v} = \vec{0}, \quad \text{THEN} \quad |\vec{f}| \leq f_{\text{S,max}} = \mu_{\text{S}} N,$$
$$\text{WHEREAS} \quad \text{IF} \quad \vec{v} \neq \vec{0}, \quad \text{THEN} \quad |\vec{f}| = f_{\text{K}} = \mu_{\text{K}} N.$$

Furthermore, accurate phenomenology[1] militates for the relation $\mu_{\text{S}} > \mu_{\text{K}}$. The leftmost sketch in Figure 17.1 shows the crude model, while a more realistic illustration of frictional behaviour appears on the right.

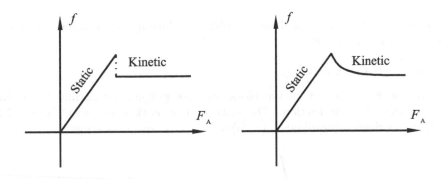

FIGURE 17.1 Frictional Force *vs.* Applied Force: Our Model and an Improvement

The coefficients of static/kinetic friction are phenomenological [must be measured]. All relevant information about the surfaces themselves is subsumed into the values of μ_{S} and μ_{K}.

ASIDE: Tribologists[2] have developed empirical rules for estimating some frictional coefficients.

EXAMPLE [*Measuring the Coefficient of Static Friction*]

To determine μ_{S} for a given pair of surfaces [*e.g.*, steel on steel, cardboard on concrete, wood on brick, *etc.*], one might adopt the following experimental procedure.

μ_{S}-i   Sheet the surface of a plane, with variable and calibrated inclination angle, with one of the materials, and fabricate a block from [or sheeted with] the other material.

---

[1]It requires less force to keep an object sliding against friction than it does to get the same object started from rest.

[2]These are members of the *group* or *clan* who study friction for a living.

μ_{s}-ii   Place the block on the incline while the incline is at a sufficiently small angle that the block does not slide.

μ_{s}-iii   Increase the angle until the block is on the verge of sliding. The MAXIMUM angle for which the block is able to remain at rest is the **angle of repose**,[3] θ_r.

[Repeat multiple times to obtain an accurate and precise value for θ_r.]

FIGURE 17.2   Determining the Coefficient of Static Friction

μ_{s}-iv   Perform the dynamical analysis in order to relate μ_{s} to the values of experimental parameters, M, θ_r.

(1) Sketch—see Figure 17.2.

(2-3) The forces which act on the block are: its weight, the normal force, and the force of static friction. The FBD for the block appears in Figure 17.3. Incline-oriented coordinates, $(x\,,\,y) = (\searrow,\,\nearrow)$, are to be employed.

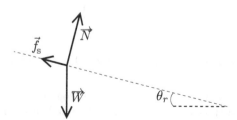

FIGURE 17.3   The FBD for the Block

(4-5) Application of $m\,\vec{a} = \vec{F}_{\text{NET}}$ and analysis

IF the angle of inclination is less than or equal to the angle of repose for the particular block–incline system, THEN the block remains at rest, and thus does not accelerate, *viz.*,

$$0 = M\,a_x = M\,g\,\sin(\theta) - f_{\text{s}}\,,\quad\text{and}\quad 0 = M\,a_y = N - M\,g\,\cos(\theta)\,.$$

At the angle of repose, the force of static friction attains its maximum value. Hence,

$$0 = M\,g\,\sin(\theta_r) - f_{\text{s,max}} = M\,g\,\sin(\theta_r) - \mu_{\text{s}}\,N\,,$$

---

[3]The angle of repose is a critical value of the inclination, since the behaviour of the block is markedly different for angles above θ_r than for those below.

AND
$$N = M\,g\,\cos(\theta_r)\,.$$

Combining these results yields a precise determination of $\mu_{\!s}$ in terms of the measured angle of repose:

$$0 = \sin(\theta_r) - \mu_{\!s}\,\cos(\theta_r) \qquad \Longrightarrow \qquad \mu_{\!s} = \frac{\sin(\theta_r)}{\cos(\theta_r)} = \tan(\theta_r)\,.$$

(6) **Q:** Is this a reasonable result?

A: Yes.

⬦ This expression for $\mu_{\!s}$ [parameterising the intrinsic aspect of the static frictional force] does not depend on the extrinsic factors $\{M, g\}$.

⬦ Surfaces which slide easily are expected to have small values of θ_r. In this limit,

$$\lim_{\theta_r \to 0} \mu_{\!s} = \lim_{\theta_r \to 0} \tan(\theta_r) = 0^+\,,$$

the coefficient, $\mu_{\!s}$, is correspondingly small.

⬦ IF the angle of repose is $\theta_r = \frac{\pi}{4} = 45°$, THEN

$$\mu_{\!s} = \lim_{\theta_r \to \frac{\pi}{4}} \tan(\theta_r) = 1\,.$$

In this instance, the frictional force acting between the surfaces has the same magnitude as the normal force of contact.

⬦ Steeper angles of repose, θ_r greater than $\frac{\pi}{4}$, are possible, provided that the frictional force is greater in magnitude than the normal force acting between the surfaces. Such "sticky" surfaces have $\mu_{\!s} > 1$.

ASIDE: In Chapter 14, we maintained the relative position of a block situated on a frictionless incline by accelerating the incline–block system. There we descried that the ratio of the needed acceleration to that of gravity was

$$\frac{a}{g} = \tan(\theta)\,.$$

It is amusing that this ratio corresponds exactly to the minimum value of $\mu_{\!s}$ needed to hold the block at rest on the frictional incline.

EXAMPLE [*Measuring the Coefficient of Kinetic Friction*]

To determine $\mu_{\!k}$ for a pair of surfaces [*e.g.*, steel on steel, cardboard on concrete, wood on brick, *etc.*], one might adopt the following experimental procedure.

$\mu_{\!k}$**-i** Sheet a horizontal plane surface with one of the materials, and prepare a block made from [or sheeted with] the other material.

$\mu_{\!k}$**-ii** Place the block on the plane. Apply an initial horizontal force, $\vec{F}_{A,\text{initial}}$, to set the block in motion.

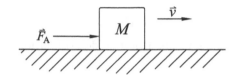

FIGURE 17.4 Determining the Coefficient of Kinetic Friction

μ_k-**iii**   Decrease the applied force from its initial value, $\vec{F}_{\mathrm{A,initial}}$, to its critical value, $\vec{F}_{\mathrm{A,critical}}$, such that the block is moving uniformly [at constant velocity].

[Repeat numerous times to obtain an accurate estimate of the critical applied force.]

μ_k-**iv**   Perform the dynamical analysis to relate the coefficient, μ_k, to experimental parameters.

(1) Sketch—See Figure 17.4

(2-3) The weight, normal, applied, and kinetic friction forces all act on the block, and its FBD is displayed in Figure 17.5. Let's adopt $(x\,,\,y) = (\rightarrow,\uparrow)$ coordinates.

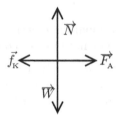

FIGURE 17.5 The FBD for the Block

(4-5) Application of $m\,\vec{a} = \vec{F}_{\mathrm{NET}}$ and analysis

While the block is moving on the surface, N2 provides the following generic set of dynamical equations:

$$M\,a_x = F_\mathrm{A} - f_\mathrm{K} = F_\mathrm{A} - \mu_\mathrm{k}\,N\,,\quad\text{and}\quad 0 = M\,a_y = N - M\,g\,.$$

For the block to move uniformly, its acceleration must vanish. This requires that the [critical value of the] applied force precisely cancel the force of kinetic friction. Thus, the horizontal component of the equation of motion becomes

$$0 = M\,a_x = F_{\mathrm{A,critical}} - \mu_\mathrm{k}\,N\,,$$

while the vertical component of the equation of motion is unaffected. Taken together, these imply that

$$\mu_\mathrm{k} = \frac{F_{\mathrm{A,critical}}}{M\,g}\,.$$

(6) **Q:** Is this a reasonable result?

A: Yes, sort of. It is consistent with the phenomenological underpinnings of the model for the kinetic friction force.

ASIDE: Truth be told, this isn't the usual way of determining μ_k. It is more straightforward to employ a known applied force, measure the net acceleration, and then deduce the magnitude of the [retarding] kinetic frictional force.

EXAMPLE [*A Sled Problem with Friction*]

Suppose that a sled, with a child onboard, is pulled by means of a rope as in Figure 17.6.

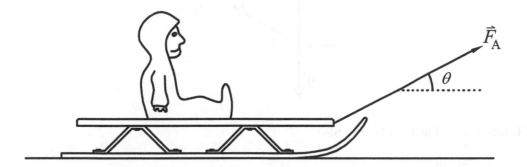

FIGURE 17.6   The Canonical Sled Problem

The physical model of this system is predicated upon the following assumptions:

• The mass, M, of the sled and rider system [hereafter "the sled"] remains constant.

• The rope is ideal.

• The rope forms a constant angle, θ, with the horizontal everywhere along its length.

• The sled is moving in the forward direction.

• The sled remains in contact with the horizontal ground.

• The coefficient of kinetic friction, μ_k, is constant.

• The acceleration due to gravity is constant.

• No drag forces act.

ASIDE: These conditions guard against instances in which the dynamics dramatically change, *e.g.*, going from snow cover to a salted roadway, or applying a force so great as to cause the front of the sled to rise off the ground.

Many questions might be asked. For this example, let's assume that the applied force and the angle are fixed and inquire as to the acceleration of the sled.

1. Sketch—see Figure 17.6.

2. There is but one dynamical constituent: the sled. It experiences four forces.

   Sled

   Weight ∘ $\vec{W} = M\,g\ [\downarrow]$.

   Normal ∘ $\vec{N} = N\ [\uparrow]$.

   Friction ∘ $\vec{f}_{\text{K}} = \mu_{\text{K}}\,N\ [\leftarrow]$.

   Applied ∘ $\vec{F}_{\text{A}} = F_{\text{A}}\,[\,\text{forward and up at angle }\theta\,] = F_{\text{A}}\,[\nearrow]$.

   This enumeration of the forces underlies the FBD in Figure 17.7.

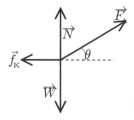

FIGURE 17.7   The FBD for the Sled

3. Coordinates

   $(x\,,\,y) = (\text{horizontal}\,,\,\text{vertical}) = (\rightarrow\,,\,\uparrow)$.

4. $m\,\vec{a} = \vec{F}_{\text{NET}}$

   Sled

   Vertical: $M\,a_y = F_{\text{NET}y} = N + F_{\text{A}y} - M\,g = N + F_{\text{A}}\,\sin(\theta) - M\,g$.

   Horizontal: $M\,a_x = F_{\text{NET}x} = F_{\text{A}x} - \mu_{\text{K}}\,N = F_{\text{A}}\,\cos(\theta) - \mu_{\text{K}}\,N$.

5. A trivial linkage has already been effected by setting the tension in the [ideal] rope equal to the applied force.

   The more significant constraint is that the sled remain on the horizontal surface, *i.e.*, $a_y = 0$. This in turn requires, via N2, that the y-component of the net force acting on the sled be zero, which determines the magnitude of the normal force of contact:

   $$N = M\,g - F_{\text{A}}\,\sin(\theta)\,.$$

   As expected, the normal force is diminished [from what it would have been in the absence of the applied force] by the upward component of the applied force.

   Substitution of this expression for the normal force into the horizontal equation of motion yields

   $$M\,a_x = F_{\text{A}}\,\cos(\theta) - \mu_{\text{K}}\big[M\,g - F_{\text{A}}\,\sin(\theta)\big] = F_{\text{A}}\big[\cos(\theta) + \mu_{\text{K}}\,\sin(\theta)\big] - \mu_{\text{K}}\,M\,g\,.$$

   Hence, the acceleration of the sled is

   $$a_x = \frac{F_{\text{A}}}{M}\Big[\cos(\theta) + \mu_{\text{K}}\,\sin(\theta)\Big] - \mu_{\text{K}}\,g\,,$$

expressed entirely in terms of sled–system parameters: mass, applied force, angle, and the coefficient of kinetic friction.

[All of these parameters are amenable to accurate measurement.]

An important dynamical aspect of this result is that [under the set of assumptions underpinning the model] the acceleration is constant in time.

6. Rather than argue in favour of this result for the acceleration, let's instead consider some of the interesting questions which might have been asked!

- The mathematical form of the expression for the acceleration in the forward direction, a_x, suggests that it may be POSITIVE, ZERO, or NEGATIVE.

   Q: Are all three possibilities physically allowed? What does each entail?

- The angle is often dictated by the comfort of the person pulling the sled.

   Q: For a particular fixed angle, what magnitude of the applied force is needed to maintain uniform motion?

- Suppose instead that the magnitude of the applied force is fixed.

   Q: At what angle should one pull so as to maximise the forward acceleration of the sled?[4]

   Q: At what angle should one pull to move the sled uniformly?

TEASER FOR CHAPTER 18

Oftentimes friction is, not unjustly, regarded as having an undesirable influence on the motion of objects. In many facets of everyday life one strives to reduce, mitigate, or obviate its more pernicious effects.

Q: Under what circumstances might friction be actually useful?

A1: Every time that one takes a step while walking, or rides in an automobile or on a bicycle, one is **impelled** by the force of static friction.

A2: Every time that one turns a corner while walking, or driving, or bicycling, one ought to acknowledge the **vital necessity** of the force of static friction. [In Chapter 18, we shall investigate the role of the force of static friction in the production of the centripetal acceleration experienced by a car which is rounding a curve of constant radius.]

INCORRECT REASONING

A person attempting to dismiss the above motivation for turning to the next chapter might attempt to reason *falsely* thus:

"It is impossible for static friction to play a significant role in either locomotion or turning, since both imply movement, while static friction acts only to *prevent* motion."

Q: Is this a valid rationale for *sliding past* the next chapter?

A: Nope. In the usual situation (*i.e.*, no *skidding*[5]) a rolling car or bicycle tire is instantaneously at rest with respect to that part of the roadway with which it is in contact. In walking, one rolls one's body forward above a stationary foot planted on the surface.

---

[4]One may be surprised to learn that the angle for maximum acceleration is not $\theta_0 = 0$. While it is true that at $\theta = 0$ all of the applied force acts in the direction of motion, pulling upward on the sled has the beneficial effect of diminishing the kinetic friction force.

[5]No *kidding*, either.

Chapter 18

Cornering: Flat and Banked

In Chapter 17 we made the audacious claim that it is static friction, acting between the car's tires and the road surface, which provides the force responsible for the centripetal acceleration of a car rounding a curve. In this chapter, we will explore this phenomenon in some detail.

EXAMPLE [*Rounding a Flat Corner*]

A sports car rounds a turn on a flat, horizontal road.

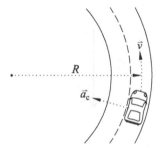

FIGURE 18.1 A Sports Car Rounding a Curve on a Flat, Horizontal Road

Q: With what maximum speed can the car *safely* navigate the curve?
A: Our condition for safety is that the car not skid.
In order to estimate the safe speed, we must construct a model [approximately] describing the situation. The assumptions which enter into the mathematical model are that:

○ The mass of the car, M, is constant.

○ The acceleration due to gravity, g, is constant.

○ The curve has constant radius of curvature, R.

○ Static friction is the dominant force acting between the road and tires.

○ The coefficient of static friction, μ_s, is constant.

○ Any and all drag forces are ignored.

○ The car remains at a constant height. [It doesn't rise up on two wheels, *etc.*]

ASIDE: The assumptions of constancy can be relaxed if one is willing to think instantaneously, as in the case of *The Long and Winding Road* mentioned in Chapter 8. We shall do this in the next chapter when we consider an instance of non-uniform circular motion.

Being as we are *driven* to assess the reasonableness of the model, we shall not be satisfied with only an algebraic [formal] solution in this case, but will obtain a numerical value for the speed [for representative data] also. For this computation, we shall set $M = 10^3 \, \text{kg} = 1$ tonne, $g = 10 \, \text{m/s}^2$, $R = 50 \, \text{m}$, and $\mu_{\text{s}} = 0.5$.

ASIDE: One metric tonne roughly corresponds to the mass of a modern compact car or an older-model sports car. The value for g is close enough given the crudeness of the model. The radius suggests the scale of a highway on-ramp. The coefficient of static friction is a convenient underestimate under normal conditions. Perhaps the road is wet, or has a dusting of sand or gravel on the surface, or the tires are somewhat worn.

The answer to the question is obtained by following the usual dynamical recipe.

1. Sketch—see Figure 18.1.

2. There is one dynamical constituent: the sports car. Three forces act upon it.

   Sports car

   Weight ∘ $\vec{W} = M \, g \; [\downarrow]$.

   Normal ∘ $\vec{N} = N \; [\uparrow]$.

   Friction ∘ $\vec{f_{\text{s}}} = f_{\text{s}} \; [\leftarrow]$, with $f_{\text{s}} \leq f_{\text{s,max}} = \mu_{\text{s}} \, N$.

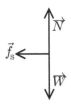

FIGURE 18.2   FBD for the Sports Car

3. Coordinates

   $(x\,,\,y) = (\rightarrow\,,\,\uparrow) = (\text{ radial }\,,\,\text{ vertical })$

   More accurately, one should regard the x-coordinate as instantaneously radial.

4. $m \, \vec{a} = \vec{F}_{\text{NET}}$

   Sports car

   Vertical: $M \, a_y = F_{\text{NET}y} = N - M \, g.$

   > That the car remains on the level road requires that $a_y = 0$, and therefore the normal force of contact between the tires and the road precisely cancels the weight of the sports car.

   Radial: $M \, a_x = F_{\text{NET}x} = -f_{\text{s}}.$

   > [The negative sign is consistent with the centripetal acceleration of the car.]

   For the sports car to navigate the curve without skidding, the magnitude of the force of static friction must not exceed its upper bound,[1]

   $$f_{\text{s}} \leq f_{\text{s,max}} = \mu_{\text{s}} \, N\,.$$

---

[1] CAVEAT: Taking explicit account of direction through the sign(s) of the component(s), one might obtain $-f_{\text{s}} \geq -f_{\text{s,max}} = -\mu_{\text{s}} \, N$. Provided that one performs the analysis self-consistently at all stages, the physical results shall be in agreement.

5. No [formal] linkage is needed as there is only one dynamical constituent. The maximum safe speed is determined by the constraint that the car not skid.

   i The normal force exactly cancels the weight, *viz.*, $N = M g$.

   ii The upper bound to the static frictional force is $f_{s,\text{max}} = \mu_s N = \mu_s M g$.

   iii Recognition that the acceleration of the sports car is centripetal leads to the following identification of its magnitude:

   $$M \, a_x = -M \, a_c = -M \, \frac{v^2}{R}.$$

   [The negative sign signifies that the centre of curvature of the road is to the left of the car at the instant shown in Figures 18.1 and 18.2.] Furthermore, static friction is the sole cause of this centripetal "$M \, a$", whence

   $$M \frac{v^2}{R} = f_s \le \mu_s M g \qquad \Longrightarrow \qquad v^2 \le \mu_s g R.$$

   iv WHEN the maximum safe speed is reached (*i.e.*, the tires are on the verge of slipping), THEN the last expression, above, attains equality. Thus:

   $$v_{\text{max}}^2 = \mu_s g R \qquad \Longrightarrow \qquad v_{\text{max}} = \sqrt{\mu_s g R}.$$

6. Is this result [obtained from such a crude model] reasonable?

   M The maximum safe speed is independent of the mass of the car.
   That is, this speed is the same for our sports car and a transport truck.

   [Not really reasonable, but this is a first approximation.]

   μ_s IF the tires are "stickier" (larger μ_s), THEN v_{max} increases, WHEREAS IF the road is "slick" (smaller μ_s), THEN v_{max} is decreased. [This stands to reason.]

   g On the flat and level curve, g is a measure of how tightly the tires are pressed against the road. Should this roadway be found on the surface of the Moon, one must round the corner more slowly since the lunar "g" is smaller than its terrestrial counterpart.

   R The dependence on R is eminently sensible. Tight corners must be navigated at lower speeds than gentle curves under the same road conditions.

Getting back to our particular example, when $R = 50\,\text{m}$, $\mu_s = 0.5$, and $g = 10\,\text{m/s}^2$, the maximum safe speed is computed to be

$$v_{\text{max}} = \sqrt{(0.5) \cdot 10 \cdot 50} = 5\sqrt{10} \simeq 15.8\,\text{m/s} \simeq 57\,\text{km/h} \simeq 36\,\text{miles per hour}.$$

This value seems consistent with the posted speed limits found on tight highway ramps. The model employed constitutes a zeroth-order analysis [*a.k.a.* "back of the envelope calculation"] which provides a check of the more sophisticated models for computing safe vehicle speeds favoured by civil/transportation engineers.

EXAMPLE [*Rounding a Banked Corner*]

To enable faster cornering, racetracks [and some roadways!] are **banked**. Let's see how this works, using the same assumptions and parameters as in the previous example, except that the road surface is no longer flat. Instead, it is banked at $30° = \pi/6\,\text{rad}$.

[This is a very steep banking angle chosen for ease of calculation and for dramatic effect.]

Retaining in this model the constraint that the car remain at a constant height requires that it track evenly through the curve.[2]

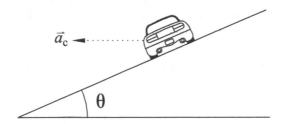

FIGURE 18.3   A Sports Car Rounding a Banked Curve

Q: With what maximum speed can the car safely navigate this banked curve?
A: Our condition for safety is that the car not skid, just as in the previous example. Let's redo the dynamic analysis.

1. Sketch—see Figure 18.3.

2. There is one dynamical constituent: the sports car. It experiences three forces.

   Sports car

   Weight ∘ $\vec{W} = M\,g\,[\downarrow]$.
   Normal ∘ $\vec{N} = N\,[\text{angled up and left, perpendicular to the roadway}] = N\,[\nwarrow]$.
   Friction ∘ $\vec{f}_{\text{s}} = f_{\text{s}}\,[\text{angled down and left, along the roadway}] = f_{\text{s}}\,[\swarrow], \quad f_{\text{s}} \leq \mu_{\text{s}}\,N$.

   Figure 18.4 shows the FBD for the sports car on the banked turn at a particular instant.

3. Coordinates

   $(x\,,\,y) = (\rightarrow,\,\uparrow) = (\text{ radial },\text{ vertical })$, just as in the previous example.

   ASIDE:

   Q: The normal force is perpendicular to, and the frictional force is parallel to, the surface of the road. Shouldn't we employ 'incline–oriented' coordinates?

   A: Nope. The simplicity of the forces is trumped by the fact that the overall $m\,\vec{a}$ is directed horizontally inward toward the centre of the turn.

---

[2] As race car drivers navigate banked turns, they typically move the cars inward (down) and outward (up) along the road surface, thus enlarging the effective radius of curvature of the turn and enabling them to safely maintain a higher speed.

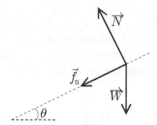

FIGURE 18.4 The FBD for the Sports Car on a Banked Track

4. $m\,\vec{a} = \vec{F}_{\text{NET}}$

   Sports car

Vertical: $M\,a_y = F_{\text{NET}y} = N_y - f_{S_y} - M\,g$.

With a bit of squinting at the FBD, the y-components of the normal and frictional forces can be expressed in terms of their magnitudes:

$$N_y = N\,\cos(\theta) \qquad \text{and} \qquad f_{S_y} = f_S\,\sin(\theta).$$

That the sports car tracks through the turn at fixed height constrains the acceleration, $a_y = 0$, and hence

$$0 = N\,\cos(\theta) - f_S\,\sin(\theta) - M\,g.$$

Radial: $M\,a_x = F_{\text{NET}x} = -N_x - f_{S_x}$.

The x-components of the normal and friction forces are both directed toward the left, and they have respective magnitudes $N\,\sin(\theta)$ and $f_S\,\cos(\theta)$, so

$$M\,a_x = -N\,\sin(\theta) - f_S\,\cos(\theta).$$

5. The condition that the car not skid fixes the maximum[3] safe cornering speed. Let's follow the same path as in the previous example.

   i   The vertical equation of motion determines: $N\,\cos(\theta) - f_S\,\sin(\theta) = M\,g$.

   ii   The radial equation of motion reads: $M\,a_x = -N\,\sin(\theta) - f_S\,\cos(\theta)$.

   iii   The acceleration of the sports car is centripetal, and hence [with minus signs signifying "inward, toward the centre"],

$$M\,a_x = -M\,\frac{v^2}{R} = -N\,\sin(\theta) - f_S\,\cos(\theta)$$

$$\implies \quad N\,\sin(\theta) + f_S\,\cos(\theta) = M\,\frac{v^2}{R}.$$

---

[3] For highly banked or slippery tracks one can also compute the minimum speed at which one must travel to avoid sliding down the inclined roadway.

iv The tires are on the verge of slipping when the force of static friction attains its maximum value $f_{\mathrm{s,max}} = \mu_{\mathrm{s}}\, \tilde{N}$. The particular value of the normal force in this expression must be distinguished [in this case by a tilde], because the normal force is presumed to vary with the speed of the car. When the car is on the verge of sliding, the equations of motion become

$$\left\{ \begin{array}{l} \tilde{N}\,\left[\sin(\theta) + \mu_{\mathrm{s}}\,\cos(\theta)\right] = M\,\dfrac{v_{\mathrm{max}}^2}{R} \\[2ex] \tilde{N}\,\left[\cos(\theta) - \mu_{\mathrm{s}}\,\sin(\theta)\right] = M\,g \end{array} \right\} \quad \text{two equations for unknowns } \{v_{\mathrm{max}}^2, \tilde{N}\}.$$

Solving this set of two equations is merely a technical problem. The simplest way to determine the speed is to divide one equation by the other.

[This has the salutary effect of completely cancelling the $\tilde{N}$ and M dependences.]

$$\frac{x\text{-eqn}}{y\text{-eqn}} = \frac{\sin(\theta) + \mu_{\mathrm{s}}\,\cos(\theta)}{\cos(\theta) - \mu_{\mathrm{s}}\,\sin(\theta)} = \frac{v_{\mathrm{max}}^2}{g\,R}$$

$$\implies \quad v_{\mathrm{max}} = \sqrt{\frac{\sin(\theta) + \mu_{\mathrm{s}}\,\cos(\theta)}{\cos(\theta) - \mu_{\mathrm{s}}\,\sin(\theta)}\,g\,R}\,.$$

6. Let's *kick the tires* on this expression for v_{max}.

M, g, R The dependence on these three factors is the same as on the flat road.

θ The dependence on the banking angle is complicated.[4]

    $\theta \to 0$ IF the curve is not banked, THEN

$$\lim_{\theta \to 0} v_{\mathrm{max}} = \lim_{\theta \to 0} \sqrt{\frac{\sin(\theta) + \mu_{\mathrm{s}}\,\cos(\theta)}{\cos(\theta) - \mu_{\mathrm{s}}\,\sin(\theta)}\,g\,R} = \sqrt{\frac{(0) + \mu_{\mathrm{s}}\,(1)}{(1) - \mu_{\mathrm{s}}\,(0)}\,g\,R} = \sqrt{\mu_{\mathrm{s}}\,g\,R}\,,$$

in exact accord with the flat [unbanked] road result.

    $0 \approx \theta \ll 1$ IF the banking angle is "small," THEN the maximum safe speed at which the curve may be rounded is increased from its flat-track value, as the term inside the square root containing all of the angular dependence is bounded below by μ_{s} [its limiting value at $\theta = 0$].

    $\theta \uparrow$ WHEN θ increases [from zero], THEN the factor expressing the angular dependence grows [monotonically] AND thus the maximum safe cornering speed increases too.

    $\theta_{\mathrm{critical}}$ Close examination reveals that there exists a critical angle,[5]

$$\theta_{\mathrm{c}} = \cot^{-1}\left(\mu_{\mathrm{s}}\right), \text{ for which } \cos(\theta_{\mathrm{c}}) - \mu_{\mathrm{s}}\,\sin(\theta_{\mathrm{c}}) = 0 \quad \implies \quad \lim_{\theta \to \theta_{\mathrm{c}}^-} v_{\mathrm{max}} = \infty\,.$$

Wow! This seems amazing, but it is pathological.

[The maximum speed is imaginary for angles greater than the critical angle.]

---

[4] To avoid some confusion we shall, for now, eschew reverse banking, in which case $\theta < 0$.
[5] Here the critical angle is approached from below, and thus it is necessarily positive.

Let's reconsider the specific data: $R = 50\,\mathrm{m}$, $\mu_{\mathrm{s}} = 0.5$, $g = 10\,\mathrm{m/s^2}$, and $30°$ banking pertinent to the particular stretch of roadway. Before proceeding apace with the analysis, it is prudent to ensure that the actual banking angle does not exceed the "critical banking angle." In this instance, $\theta_{\mathrm{c}} = \cot^{-1}(0.5) = \tan^{-1}(2) \simeq 1.107\,\mathrm{radians} \simeq 63.4°$, which is well in excess of the actual banking angle. Hence our analysis is not invalid. Substituting these parameters into the expression for the maximum safe speed yields

$$v_{\mathrm{max}} = \sqrt{\frac{\frac{1}{2} + (0.5)\,\frac{\sqrt{3}}{2}}{\frac{\sqrt{3}}{2} - (0.5)\,\frac{1}{2}} \cdot 10 \cdot 50} \simeq 27\,\mathrm{m/s} \simeq 98\,\mathrm{km/h} \simeq 60\,\text{miles per hour}\,.$$

This is MUCH faster than the maximum safe speed on the level road!

Q1: How exactly does banking work?

A1: Banking the track gives rise to two effects enabling the car to navigate the curve at a higher safe speed. Both involve the normal force. Recall that

$$N\cos(\theta) = M\,g + f_{\mathrm{s}}\sin(\theta) \qquad \text{and} \qquad M\,a_{\mathrm{centripetal}} = N\sin(\theta) + f_{\mathrm{s}}\cos(\theta)\,,$$

at speeds [somewhat] below the maximum safe speed. Two important observations are:

- The normal force is greater than the weight, $M\,g$, and hence the maximum force of static friction is increased beyond that attained on a level road. Banking improves the "road-holding performance" of the tires.

- The centripetal component of the normal force assists the car through the turn.

Q2: What does the driver of the car feel when rounding the curve [at the maximum speed]?

A2: The [as yet still unknown] value of the normal force at maximum speed is $\tilde{N}$. At this speed, the force of static friction reaches its maximum value, $f_{\mathrm{s}} = \mu_{\mathrm{s}}\,\tilde{N}$, and the [constrained] y-equation of motion becomes

$$N\cos(\theta) = M\,g + f_{\mathrm{s}}\sin(\theta) \quad \Longrightarrow \quad \tilde{N}\cos(\theta) = M\,g + \mu_{\mathrm{s}}\,\tilde{N}\sin(\theta)$$

$$\Longrightarrow \quad \Big(\cos(\theta) - \mu_{\mathrm{s}}\sin(\theta)\Big)\,\tilde{N} = M\,g\,.$$

Therefore, the magnitude of the normal force [while the car maintains constant height] is:

$$\tilde{N} = \frac{M\,g}{\cos(\theta) - \mu_{\mathrm{s}}\sin(\theta)}\,.$$

Don't be misled! It might appear from the above expression that the normal force does NOT depend on the speed of the car. This is not correct, as the constraint introduced implicit dependence on the speed.

IF the car's speed is less than v_{max}, THEN the normal force differs from $\tilde{N}$.

Plugging in $\theta = 30°$ and $\mu_s = 0.5$, the normal force at maximum speed in this instance is

$$\widetilde{N} = \frac{M\,g}{\frac{\sqrt{3}}{2} - 0.5 \cdot \frac{1}{2}} \simeq 1.57\,M\,g\,.$$

Thus, the force exerted by the road on the car exceeds its weight by approximately $57\,\%$. Anyone and anything riding along in the car [passengers and contents] endures a proportionate increase in the normal force that they experience from those parts of the car [seats and cupholders and the like] with which they are in contact.

Let's *kick the tires on* our expression for $\widetilde{N}$.

M, g   The dependence on these factors is as expected: the weight sets the scale.

θ   The dependence on the banking angle is complicated.

   $\theta \to 0$ IF the curve is not banked, THEN

$$\lim_{\theta \to 0} \widetilde{N} = \lim_{\theta \to 0} \frac{M\,g}{\cos(\theta) - \mu_s \sin(\theta)} = \frac{M\,g}{(1) - \mu_s\,(0)} = M\,g\,.$$

   in exact accord with the level road result.

   $0 \approx \theta \ll 1$ IF the banking angle is "small," THEN the normal force is slightly greater than the weight of the car. The factor appearing in the denominator is bounded above by 1 [its limiting value at $\theta = 0$]. Small-angle banking increases the normal force of contact between the car and the road.

   $\theta \uparrow$ WHEN θ increases [from zero], THEN the denominator decreases monotonically, AND thus the normal force increases monotonically.

   θ_{critical}   The normal force diverges at θ_c, the same critical angle as appeared in the analysis of the maximum safe speed.

$$\lim_{\theta \to \theta_c^-} \widetilde{N} = \lim_{\theta \to \theta_c^-} \frac{M\,g}{\cos(\theta) - \mu_s \sin(\theta)} \to \frac{M\,g}{\cos(\theta_c) - \mu_s \sin(\theta_c)} \to \frac{M\,g}{0+} \to \infty\,.$$

   Wow! This IS really pathological!
   [The magnitude of the normal force is negative for angles greater than θ_c.]

   ASIDE: It's quite hard to imagine how the roadway surface could produce an infinite normal force and harder still to imagine how the car could survive[6] it!

Q: What gives rise to the [pathologies at the] critical angle?
A: Implicit in the maximum speed analysis was the constraint that the frictional force attains its upper bound. With the increase in the normal force due to banking, it is possible that the frictional force never reaches its limiting value, and this constraint breakdown is signalled by the aberrant appearance of θ_{critical}.

While thirty-degree curve banking is unrealistic for a highway, one commonly finds such turns on a rollercoaster like that featured in Chapter 19!

---

[6] A snarky person might affect an ersatz Scots accent so as to say,

   Aye, Capt'n Kirk, sir. The Enterprise canna' withstand another of those corners.

Chapter 19

Non-Uniform Circular Motion

Non-uniform circular motion occurs when the trajectory of a particle lies along a circle, or a segment of circular arc, but the speed of the particle is not constant, *i.e.,* both centripetal and tangential components of acceleration are present. The standard illustration of non-uniform circular motion is a loop-the-loop found in modern rollercoasters.[1]

EXAMPLE [*A Rollercoaster Loop-the-Loop: Equations of Motion*]

Five of the *Steinein* siblings, *Albert, Bertal, Clande, Declan,* and *Ecland,* henceforth "A," "B,", "C," "D," and "E," venture to an amusement park to ride on rollercoasters. After a particularly thrilling experience, they breathlessly discuss their favourite parts of the loop-the-loop portion of the ride.

A   waxes ecstatic about the strong feeling of acceleration and compression into his seat upon entering the loop.

B   is most excited when the car is pointed straight up and slowing fastest.

C   *loves* that feeling of lightness when upside-down, traversing the top of the loop.

D   enjoys the sensation of rapid acceleration when the car is pointed straight down.

E   breathes a sigh of relief when rocketing out of the loop, back onto the level track.

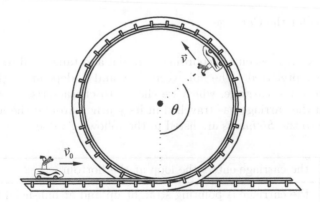

FIGURE 19.1   A Rollercoaster Track with a Loop-the-Loop

Q: How does the carriage navigate the loop-the-loop? Can the various sensations experienced by the riders be explained?
A: Let's model the system dynamically in order to appreciate the *Steineins'* experiences.

---

[1]Although, in fact, the situation depicted in the *Long and Winding [and flat] Road* Addendum to Chapter 8 is a far more commonplace, if much less dramatic, realisation of non-uniform circular motion.

Assumptions entering into the making of the mathematical model are that:

o   The rollercoaster carriage and passengers together have total mass M.

o   The track is smooth, horizontal, and level entering and leaving the loop.

o   The loop is circular[2] with [constant] radius R.

o   The track is frictionless, and no drag forces act.

o   The acceleration due to gravity, g, is constant.

o   The speed of the carriage entering the loop-the-loop is v_0.

1. Sketch—see Figure 19.1.

2. The carriage experiences two forces: its weight and the normal force.

   Carriage

Weight o $\vec{W} = M\,g\,[\downarrow]$.

Normal o $\vec{N} = N\,[\,\text{radially inward, toward the centre of the loop}\,]$.

FIGURE 19.2   FBD for the Carriage

3. Coordinates   Carriage-centred coordinates radial and tangential to the track at each instant are preferred. The unit vectors, $\hat{r}$ and $\hat{t}$, depend on the instantaneous position of the carriage, which we choose to characterise by θ, the angle through which the carriage has travelled in its journey around the loop. Angles of significance to the *Steineins* are listed in the following table.

$\theta = 0$	the carriage enters the loop at the bottom	A
$\theta = \frac{\pi}{2}$	the carriage is pointing straight up and at half-height	B
$\theta = \pi$	the carriage is upside-down and at maximum height	C
$\theta = \frac{3\pi}{2}$	the carriage is pointed straight down again at half-height	D
$\theta = 2\pi$	the carriage returns to the bottom and exits the loop	E

---

[2]The track must twist out of the plane in order to avoid self-intersection, but we will pretend that the loop is circular.

4. $m\,\vec{a} = \vec{F}_{\text{NET}}$

Carriage

Radial: $M\,a_r = F_{\text{NET}r} = -N + M\,g\,\cos(\theta)$.

The radial component of the acceleration is centripetal, and therefore

$$M\,a_r = -\frac{M\,v^2}{R} = -N + M\,g\,\cos(\theta)\,.$$

Tangential: $M\,a_t = F_{\text{NET}t} = -M\,g\,\sin(\theta)$.

The minus sign arises from the choices inherent in the angular parameterisation, and conforms with our expectations that the carriage slow down on its way up $[0 < \theta < \pi]$ and speed up on its way down $[\pi < \theta < 2\pi]$.

5. The equations of motion governing the movement of the carriage are:

$$\frac{M\,v^2}{R} = N - M\,g\,\cos(\theta) \qquad \text{and} \qquad a_t = -g\,\sin(\theta)\,.$$

These dynamical equations, along with boundary/initial conditions, provide all of the information needed to determine the dynamical behaviour of the carriage and the sensations of the *Steineins*.

EXAMPLE [*Loop-the-Loop Carriage Speed: Qualitative*]

Swayed by kinematical modes of thinking, one might propose to determine the speed of the carriage by first finding its trajectory, computing the velocity by differentiation with respect to time, and forming the speed from the velocity components. While this method is fruitful, it is hardly optimal since it requires two integrations with respect to t and follows them with a t derivative. The more direct approach begins with the tangential acceleration, regarded as the time rate of change of [tangential] speed,

$$a_t = -g\,\sin(\theta)\,.$$

The time dependence of the tangential acceleration is buried in the angle denoting the instantaneous position of the carriage, $\theta = \theta(t)$. Figure 19.3 illustrates the manner in which the tangential acceleration of the carriage varies with its position in the loop.

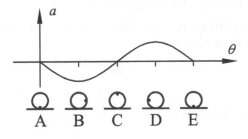

FIGURE 19.3 The Tangential Acceleration of the Rollercoaster Carriage *vs.* Position

Inspection of the formula and figure reveals complete concordance with the sensations experienced by the riders.

A   Upon entering the loop, the carriage is at its highest speed. The carriage then begins to decelerate at an increasing rate.

B   The tangential deceleration is maximised when the car is pointing straight up, *i.e.*, at $\theta = \pi/2$. Beyond this point, the carriage continues to decelerate, but at a reduced rate.

C   The carriage moves with least speed while at the top of the loop, $\theta = \pi$.

[Clande must keep her wits about her to perceive this!]

D   All the way back down the loop, the carriage's forward speed increases. The maximum acceleration occurs when the car is pointing straight down, *i.e.*, at $\theta = 3\pi/2$. Beyond this point the acceleration is still forward, but reduced in magnitude.

E   The tangential acceleration returns to zero as the carriage exits the loop.

These inferences concerning the tangential acceleration enable qualitative deduction of the approximate form of a graph of speed *vs.* position.

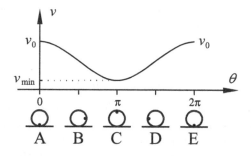

FIGURE 19.4   The Tangential Speed of the Rollercoaster Carriage *vs.* Position

Inspection of the figure reveals that it is in accord with the riders' perceptions. We shall forego the explicit demonstration of qualitative agreement and proceed directly to a quantitative analysis.

EXAMPLE [*Loop-the-Loop Carriage Speed: Quantitative*]

Q: What is the speed of the carriage as a function of its angular position?
A: The tangential equation of motion reads

$$a_t = -g\,\sin(\theta)\,.$$

The LHS is the time rate of change of the tangential speed, $a_t = \frac{dv_t}{dt}$. When the motion is circular with fixed radius, R, the tangential and angular speeds are related by

$$v = \omega\,R \qquad \text{with} \qquad \omega = \frac{d\theta}{dt}\,.$$

This realisation enables the following expansion of the tangential acceleration:

$$a_t = \frac{dv}{dt} = \frac{dv}{d\theta}\frac{d\theta}{dt} = \frac{d\,(\omega R)}{d\theta}\,(\omega) = R\,\omega\,\frac{d\omega}{d\theta} = \frac{1}{2}\,R\,\frac{d\omega^2}{d\theta}\,.$$

Substituting this last expression into the tangential equation of motion yields

$$\frac{1}{2} R \frac{d\omega^2}{d\theta} = -g \sin(\theta).$$

There are several equivalent ways in which one may proceed to integrate this form of the equation of motion to relate the instantaneous angular speed of the carriage with its angular position.

METHOD ONE: [*The fancy approach*]
Choosing to express the `sine` function as the derivative of minus `cosine`, leads to

$$\frac{1}{2} R \frac{d\omega^2}{d\theta} = +g \frac{d\cos(\theta)}{d\theta}.$$

Simplifying ["cancelling" the $d\theta$ differentials, dividing both sides by R, and multiplying by 2] results in

$$d\left(\omega^2\right) = \frac{2g}{R} d\cos(\theta).$$

Each side of this differential equation can be self-consistently integrated: the LHS, from some initial angular speed–squared, ω_i^2, to an unspecified final angular speed–squared, ω^2; the RHS, over the corresponding range of angles, $\theta_i \to \theta$. The once-integrated tangential equation of motion reads

$$\omega^2 - \omega_i^2 = \frac{2g}{R} \Big[\cos(\theta) - \cos(\theta_i) \Big].$$

The boundary/initial conditions [appropriate for this example] remain to be incorporated into the derived formula. In the present case, the carriage enters the loop at angle $\theta_i = 0$, with initial speed $v_i = v_0 = \omega_0 R$. Thus,

$$\omega^2 - \omega_0^2 = \frac{v^2}{R^2} - \frac{v_0^2}{R^2} = \frac{2g}{R} \Big[\cos(\theta) - 1 \Big].$$

Multiplying by R^2 and rearranging leads to our final result,

$$v^2 = v_0^2 - 2gR\big[1 - \cos(\theta)\big],$$

in which the speed–squared of the carriage is determined as a function of its initial [entering] speed and the angle defining the position of the carriage.

METHOD TWO: [*Plain and simple integration*]
Multiplying the adapted dynamical equation by $2\,d\theta$, and dividing by R, yields

$$d\left(\omega^2\right) = -\frac{2g}{R} \sin(\theta)\,d\theta.$$

The LHS is integrated from some initial angular speed–squared, ω_i^2, to a final angular speed–squared, ω^2, while the RHS is integrated over the corresponding range of angles, $\theta_i \to \theta$. The result of these integrations is

$$\omega^2 - \omega_i^2 = \frac{2g}{R} \Big[\cos(\theta) - \cos(\theta_i) \Big],$$

which is exactly the same result as was obtained above.

METHOD THREE: [*A variation on the theme*]
Performing an indefinite integration of the tangential equation of motion yields

$$\omega^2 = \frac{2\,g}{R}\cos(\theta) + C\,,$$

where C is a constant of integration. That the angular speed is ω_0 when $\theta = 0$ fixes the value of the constant, $C = \omega_0^2 - \frac{2g}{R}$, and hence

$$\omega^2 - \omega_0^2 = -\frac{2\,g}{R}\left[1 - \cos(\theta)\right]\,,$$

encore.

[In Chapter 27, this result shall be rederived using energy methods.]

....

EXAMPLE   [*Loop-the-Loop Normal Force*]

Q: How does the normal force contribute to the sensations that the *Steinein* siblings experience while riding in the carriage?

A: We can formally express the magnitude of the normal force as a function of angular position by mere rearrangement of the radial component of N2,

$$N(\theta) = M\,g\,\cos(\theta) + \frac{M v^2}{R}\,.$$

This formula is deceptively simple, however, because an implicit angular dependence enters via the v^2 term. Fortunately, we have just computed $v^2(\theta)$. Including this information yields a more complete expression for the magnitude of the normal force,

$$N(\theta) = M\left[\frac{v_0^2}{R} - 2\,g + 3\,g\,\cos(\theta)\right]\,,$$

which conforms exactly to our intuitive expectations.

MAX   The normal force is maximised at $\theta = 0,\, 2\,\pi$.

   ○ $\cos(0) = 1 = \cos(2\,\pi)$ is the maximum value of the **cosine** function.

   ○ $v\big|_{\theta=0,2\,\pi} = v_0$ is the highest speed, maximising the centripetal $M a$ at these instants.

These facts, taken together, ensure that the maximal normal force occurs at **A** and **E**, just as the carriage enters and exits the loop.

MIN   The normal force is minimised at $\theta = \pi$.

   ○ $\cos(\pi) = -1$ is the minimum value of the **cosine** function.

   ○ $v\big|_{\theta=\pi}$ is the lowest speed, minimising the concurrent centripetal $M a$.

   ○   The entire weight of the carriage acts centripetally.

These facts convince us that the magnitude of the normal force reaches its minimum when the carriage is at the top of the loop.

Chapter 20

Drag Forces

Drag   Drag is a resistive force arising whenever an object moves through a fluid medium (either gaseous or liquid). [An object does not follow the parabolic path predicted by our simple model of projectile motion, developed in Chapter 7, primarily because of drag forces.]

Drag forces

- Oppose the motion of the particle on which they act.

- Depend on the size, shape, and surface properties[1] of the moving object.

- Depend on certain fluid properties of the medium, *e.g.,* viscosity.

- Depend on the speed of the moving particle. There are two important special cases which are commonly used to model drag.

    L   In the case of LINEAR DRAG, the magnitude of the drag force is proportional to the speed of the moving object with respect to the medium. Linear drag is dominant in situations like that of a small ball bearing falling through still air. The *gory details*—the size, shape, and composition of the particle, and the relevant properties of the fluid—are all subsumed into a single drag coefficient, b_1.

    Q   For QUADRATIC DRAG, the magnitude of the drag force is proportional to the relative speed–squared of the object. Quadratic drag is appreciable in the motions of an automobile, skydiver, or larger ball through air. Again, all of the details are encapsulated in a coefficient, b_2.

    C   CONSTANT DRAG [the third(!) of the two significant cases] is indistinguishable from kinetic friction!

EXAMPLE [*Linear Drag Force*]

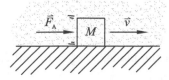

FIGURE 20.1   A Particle Moving through a Draggy Medium

---

[1]A great deal of research and development effort is directed toward the design of low-drag materials and the devising of efficient methods for their manufacture.

Let's encounter linear drag through a *Gedanken* experiment involving a particle moving through a fluid medium under an applied force. The underlying assumptions are that:

- The particle has [constant] mass M.

- The particle rides along a frictionless horizontal plane surface.

- A constant applied force, $\vec{F_A}$, acts on the particle in the direction of its motion.

- The velocity of the particle at time t is $\vec{v}(t)$.

- The motion of the particle through the fluid gives rise to a linear drag force,

$$\vec{F}_{D1} = -b_1\,\vec{v}\,,$$

for some [well-defined] drag coefficient b_1. [That the drag force acts in opposition to the motion of the particle is made explicit in the formula above.]

Q: What sort of dynamics will the particle experience?
A: Let's follow the usual recipe to explore the various possibilities for the motion of the particle.

1. Sketch—see Figure 20.1.

2. Four forces act on one particle.

   Particle

   Weight ○ $\vec{W} = M\,g$ [↓].
   Normal ○ $\vec{N} = N$ [↑].
   Applied ○ $\vec{F_A} = F_A$ [→].
   L-Drag ○ $\vec{F}_{D1} = -b_1\,\vec{v} = b_1\,v$ [←].

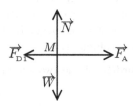

FIGURE 20.2   FBD for the Particle Moving in a Draggy Medium

3. Coordinates $(x,\,y) = (\rightarrow,\,\uparrow)$.

4. $m\,\vec{a} = \vec{F}_{NET}$

   Particle

   Vertical: $M\,a_y = F_{NETy} = N - M\,g$.
       The particle remains on the plane, so $a_y = 0$, and $N = M\,g$.
   Horizontal: $M\,a_x = F_{NETx} = F_A - b_1\,v$.
       Thus, the horizontal component of the acceleration of the block is

$$a_x = \frac{1}{M}\left(F_A - b_1\,v\right) = -\frac{b_1}{M}\left(v - \frac{F_A}{b_1}\right).$$

5. The linear drag coefficient, b_1, must be positive [and finite], and thus, the sign of the x-component of the acceleration of the particle depends crucially on the relative magnitudes of the applied and drag forces. Under the assumptions made at the outset [notably: the particle remains on the plane, and the motion is parallel to the applied force], the particle's motion is effectively 1-d. Hence,

$$a_x = \frac{dv}{dt}.$$

For convenience,[2] we introduce the positive quantity, $v_{\mathrm{T}} = F_{\mathrm{A}}/b_1$ [with the dimensions of velocity], and proceed with the analysis.

$$\frac{dv}{dt} = -\frac{b_1}{M}\left(v - v_{\mathrm{T}}\right) \qquad \Longrightarrow \qquad \frac{dv}{v - v_{\mathrm{T}}} = -\frac{b_1}{M}\,dt.$$

This latter equation may be integrated from an initial forward speed v_0, at an initial time t_0, to speed v, at a later time t.

$$\ln\left[v' - v_{\mathrm{T}}\right]\Big|_{v'=v_0}^{v'=v} = -\frac{b_1}{M}\,t'\Big|_{t'=t_0}^{t'=t} \qquad \Longrightarrow \qquad \ln\left[\frac{v - v_{\mathrm{T}}}{v_0 - v_{\mathrm{T}}}\right] = -\frac{b_1}{M}\left(t - t_0\right).$$

Exponentiating both sides of this relation will eliminate the logarithm, yielding

$$\frac{v - v_{\mathrm{T}}}{v_0 - v_{\mathrm{T}}} = e^{-\frac{b_1}{M}(t-t_0)} \qquad \Longrightarrow \qquad v(t) = v_{\mathrm{T}} + \left(v_0 - v_{\mathrm{T}}\right)e^{-\frac{b_1}{M}(t-t_0)}.$$

We may choose [to set our watches so that] $t_0 = 0$, and consider the behaviour of our model system for times $t > 0$. There are two distinct ranges of possibilities separated by a special case.

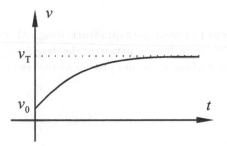 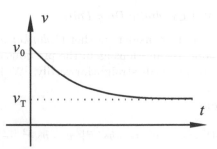

FIGURE 20.3 Speed *vs.* Time when $v_0 < v_{\mathrm{T}}$ and $v_0 > v_{\mathrm{T}}$

< IF the initial [forward] speed, v_0, is less than v_{T}, THEN the speed of the particle asymptotically approaches v_{T} from below.

= IF v_0 is precisely equal to v_{T}, THEN the speed remains constant at v_{T}.

> IF the initial speed, v_0, is greater than v_{T}, THEN the speed of the particle asymptotically approaches v_{T} from above.

The constant v_{T} is called the **terminal velocity**, as it is the speed to which the system tends asymptotically as $t \to \infty$, irrespective of its initial speed.

---

[2]In a moment, we will expose the true motivation.

6. Here are a few comments pertinent to the results of this *Gedanken* experiment.

- The terminal velocity may be discerned from the equation of motion [N2]. IF $v = v_T = F_A/b_1$, THEN the drag force exactly cancels the applied force, AND the block moves inertially [with constant velocity].

- IF $F_A = 0$, *i.e.*, the applied force vanishes, THEN the terminal velocity is zero, AND the particle moves forward *eternally* and yet traverses only a finite distance. This is the ZENOish motion studied in Chapter 4.

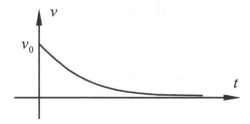

FIGURE 20.4   A Particle Coasting in a Resistive Medium Experiences ZENOish Motion

- The [forward] speed, as an explicit function of time, can be integrated to obtain the trajectory of the particle,

$$x(t) - x(t_0) = v_T \left(t - t_0 \right) - \frac{M}{b_1} \left(v_0 - v_T \right) \left\{ e^{-\frac{b_1}{M}(t - t_0)} - 1 \right\}.$$

EXAMPLE [*Quadratic Drag Force*]

Let's now briefly consider another *Gedanken* experiment to investigate quadratic drag. Making the same assumptions as in the previous example [except for the nature of the drag force], the analysis proceeds straightforwardly. [We'll skip most of the steps and just quote the result.]

Particle

Q-Drag ∘ $\vec{F}_{D2} = -b_2 |\vec{v}|^2 \hat{v} = b_2 v^2 [\leftarrow]$.

The FBD is the same as in Figure 20.2, and applying N2 yields the equation of motion

$$a_x = \frac{1}{M} \left(F_A - b_2 v^2 \right) = -\frac{b_2}{M} \left(v^2 - \frac{F_A}{b_2} \right).$$

This is a **second-order *non*-linear ordinary differential equation.**

ASIDE: To solve the equation of motion for the trajectory would require the introduction of mathematical techniques best left to a differential equations class.

From N2 the terminal velocity is determined to be

$$v_T = \sqrt{\frac{F_A}{b_2}},$$

as this particular value of the speed leads to cancellation of the applied and quadratic-drag forces acting on the particle.

Chapter 21

Work and Energy

ENERGY is perhaps[1] the most important concept in physics! [For the remainder of this VOLUME (and VOLUMES II and III), we shall rather rigorously—and very vigorously—apply energy methods in our analyses of particular phenomena.]

Three foundational/axiomatic properties are that:

○ **ENERGY may exist in a variety of forms.** For example,

Mechanical: Kinetic	Thermal	Chemical
Mechanical: Potential	Solar	Nuclear
Electro-Magnetic	*Toddler*	*Etc.*

[We are being slightly and intentionally glib here.]

○ **ENERGY in one form may be converted [*transformed*] into another.**

It may not be easy, but, in principle, it can always be done.

ASIDE: Perhaps you are reading these notes by the light of an electric lamp. It may be that your favourite purveyor of *energy services* obtains its electricity from a nuclear or natural gas power plant. At these facilities, nuclear energy or chemical energy is converted to heat, which is used to boil water. The steam expands and powers a turbine connected to an electrical generator. The electrical energy so obtained is then sent through the power grid and eventually directed to the lamp, which produces the light illuminating the page.

[Believe it or not, we've seriously glossed over the steps that are involved!]

● **ENERGY is conserved.**

That is, regardless of its form, **the total amount of energy associated with a system and its environment is fixed.** In yet other words: IF an isolated system loses energy in one form, THEN it must gain an equal amount in other forms.

Energy Conservation is the paramount feature of energy.

---

[1]The CAVEAT "perhaps" appears because, in relativistic mechanics, energy and momentum are inextricably entwined and thereby share the honour of primacy.

For now we will confine ourselves to consideration of mechanical energy, starting with a precise technical definition of MECHANICAL WORK.

MECHANICAL WORK   The mechanical work is the (line-)integral of a force, $\vec{F}$, as it acts along the trajectory associated with a displacement.

$$W_{if}\left[\,\vec{F}\,\right]_{(\text{path})} = \int_{\text{initial pos'n}}^{\text{final pos'n}} \vec{F}\cdot d\vec{s}\,.$$

Each part of this defining formula will be explained below.

The symbol, $W_{if}\left[\,\vec{F}\,\right]_{(\text{path})}$, on the LHS denotes the work performed by the force $\vec{F}$ acting along the particular path **from** the initial position, represented by "i," **to** the final position, denoted by "f." The amount of work done depends crucially on the force, $\vec{F}$; on the initial and final (end-)points; and on the path taken. Our generic symbol for work makes explicit reference to all of these expected dependencies. However, we shall elide these *accoutrements* whenever possible.

$\vec{F}$ The work done depends linearly on the force, which may exhibit any manner of dependence on time, position, velocity, and particle properties.

$d\vec{s}$ The integration occurs over each of the differential vector displacements comprising the entirety of the path. An intuitive way to visualise $d\vec{s}$ is as a single step along the particular path leading from the initial to final positions.

> MAGNITUDE $\left|d\vec{s}\right| = ds$ is the length of the particular[2] step. By adding together the lengths of all of the individual steps, one gets the total distance travelled in the course of the net displacement from the initial to the final position [*a.k.a.* the arc length of the path].

> DIRECTION $d\vec{s}$ points along the forward tangent to the path.

•   The DOT PRODUCT [of two vectors, yielding a scalar] was introduced in Chapter 5. In the present context, the input vectors, $\vec{F}$ and $d\vec{s}$, are those found locally at points along the path.

$\int \ldots d\vec{s}$ The line-integral is along the path [through space] from the initial position to the final position. Invoking the familiar notion of the Riemann Sum, we can envision $\Delta\vec{s}$ as a small [finite] step, and the integral as marching along the path while accumulating[3] the integrand.

---

[2] The steps can be of uniform or variable length.

[3] Imagine possessing an **accumulator** [a machine that takes successive inputs and adds them to produce a running sum stored in its memory] which is set to zero at the initial position. Take the first step $\Delta\vec{s}$, and at some point (beginning, middle, or end) in the step, form the dot product of the step vector with the force vector at that point. Enter the result of the computation in the accumulator. Take the second step, form the dot product of the second step vector with the local force vector, and toss this result into the accumulator. Keep doing this until the entire path is traversed and the final point is reached. The value in the accumulator is the Riemann Sum approximation to the line integral expressing the mechanical work done by the force $\vec{F}$ acting through the displacement from the initial to the final position along the specified path. [Sorry for the apparent tedium, but there are many instances in which the line integral cannot be done in closed form and recourse to numerical approximation is the only way to accurately determine the amount of mechanical work done.]

The SI unit of work is the [N · m]. Mechanical work is a form of energy, and the SI unit of energy is the joule [J], so-named in honour of James Prescott Joule (1818–1889). Consistency then demands that

1 joule ≡ 1 newton (force) acting through 1 metre (displacement).

There are two important things to note which follow directly from the definition of work.

W   Work is a scalar quantity. It has magnitude only, and no directional aspect.

W   Work respects the PRINCIPLE OF LINEAR SUPERPOSITION.

The net work done by a collection of forces acting through a common displacement is precisely equal to the work done by the net force.

To render all of this (slightly) less opaque, let's consider a bunch of examples.

EXAMPLE [*An Energetic Block on a Horizontal Frictionless Plane*]

A block of mass M is pushed along a frictionless horizontal plane with constant force $\vec{F}_A$ through a displacement $\vec{s}$. Suppose that the block follows a straight-line path directly from its initial position, x_i, to its final position, x_f.

Part A: Suppose that the force is horizontal and aligned with the particle's displacement, $\vec{F}_A \parallel \vec{s}$. [This situation is effectively one-dimensional.]

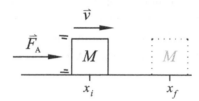

FIGURE 21.1   Pushing a Block through a Displacement

The work done by the applied force is

$$W_{if}\left[\vec{F}_A\right] = \int_i^f \vec{F}_A \cdot d\vec{s} = \int_{x_i}^{x_f} F_A \, ds = F_A \int_{x_i}^{x_f} dx = F_A\,(x_f - x_i) = F_A\, s \,,$$

where $s = x_f - x_i$ is the distance through which the block moved. The integrand simplified because the applied force was parallel to each and every differential displacement comprising the trajectory of the block. The constancy of the applied force then rendered the integration trivial.

Q: Are there circumstances in which the work done is equal to zero?
A: There are two distinct possibilities:

$F_A = 0$ IF the particle coasts from x_i to x_f with no applied force acting on it, THEN the work done [by the non-existent force!] on the particle is zero. This stands to reason: No work is done unless a force acts.

$s = 0$ IF the particle does not move, THEN no work is done.
In other words, IF you push on something and it doesn't move, THEN, despite your efforts, you have not done work on [transferred energy to] it.

In order to lift the block quasi-statically one must maintain an upward force which precisely cancels the weight of the block: $\vec{F}_A = M\,g$ [upward]. The trajectory of the block is along a vertical straight-line path from an initial height, y_i, to a final height, $y_f = y_i + h$, as shown in Figure 21.4.

To determine the work done by the applied force, one forms the integrand,

$$\vec{F}_A \cdot d\vec{s} = M\,g\,[\uparrow] \cdot ds\,[\uparrow] = M\,g\,dy\,,$$

and computes the work to be

$$W_{if}\left[\,\vec{F}_A\,\right] = \int_{y_i}^{y_f} \vec{F}_A \cdot d\vec{s} = M\,g \int_{y_i}^{y_f} dy = M\,g\left[(y_i + h) - y_i\right] = M\,g\,h\,.$$

The work done by the applied force is positive, consistent with the fact that the block moves in the direction of the force.

The integrand for the work done by the weight force is

$$\vec{W} \cdot d\vec{s} = M\,g\,[\downarrow] \cdot ds[\uparrow] = M\,g\,ds\,[\downarrow] \cdot [\uparrow] = -M\,g\,dy\,.$$

Integrating this yields the work done by the weight of the block as it is lifted:

$$W_{if}\left[\,\vec{W}\,\right] = \int_{y_i}^{y_f} \vec{W} \cdot d\vec{s} = -M\,g \int_{y_i}^{y_f} dy = -M\,g\,h\,.$$

This work is negative, since the block moves upward while the weight force is directed downward.

It is no accident that the magnitude of the work done by the weight of the block is the same as that done by the applied force. Recall that the adverb *quasi-statically* implies that $M\vec{a} \simeq \vec{0}$ throughout the motion. According to N2, this requires that the net force is [nearly] zero. Consequently, the net work must vanish. Direct computation of the net work, by both means,

$$\text{Net Work} = W_{if}\left[\,\vec{F}_{\text{NET}}\,\right] = W_{if}\left[\,\vec{F}_A + \vec{W}\,\right] = W_{if}\left[\,\vec{0}\,\right] = 0\,,$$
$$\text{Net Work} = W_{if}\left[\,\vec{F}_A + \vec{W}\,\right] = W_{if}\left[\,\vec{F}_A\,\right] + W_{if}\left[\,\vec{W}\,\right] = M\,g\,h - M\,g\,h = 0\,,$$

assures us that it does vanish in this case.

....

EXAMPLE [*Energetic Sledding*]

A sled[5] with total mass M is pulled through a straight-line displacement, $s = x_f - x_i$, along a horizontal frictional surface, by means of a rope held at fixed angle θ.

The situation is illustrated in Figure 21.5, along with the FBD depicting all of the forces acting on the sled. Making the natural choice of coordinates: $(x,\,y) = (\rightarrow,\,\uparrow)$, and applying N2, we obtain the following dynamical equations:

$$M\,a_x = F_A \cos(\theta) - \mu_k\,N\,,$$
$$M\,a_y = N + F_A \sin(\theta) - M\,g\,.$$

---

[5] Recall that in Chapter 17 we defined "sled" to mean the sled plus its rider(s).

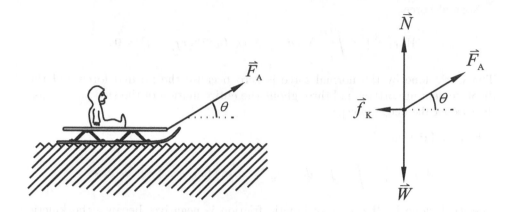

FIGURE 21.5 Pulling a Sled across a Horizontal Frictional Surface

The constraint that the sled remain on the surface, $a_y = 0$, requires that the normal force cancel that part of the weight which is in excess of the upward component of the applied force. *I.e.,*

$$N = M g - F_A \sin(\theta).$$

With the normal force thus ascertained, the horizontal equation of motion can be re-expressed entirely in terms of known [or measurable] quantities:

$$M a_x = F_A \left(\cos(\theta) + \mu_k \sin(\theta)\right) - \mu_k M g.$$

The net work can be computed directly from the net force:

$$\begin{aligned}
\text{Net Work} &= W_{if}\left[\vec{F}_{\text{NET}}\right] \\
&= W_{if}\left[M a_x \,\hat{\imath}\right] \\
&= W_{if}\left[\left\{F_A \left(\cos(\theta) + \mu_k \sin(\theta)\right) - \mu_k M g\right\}\hat{\imath}\right] \\
&= \left\{F_A \left(\cos(\theta) + \mu_k \sin(\theta)\right) - \mu_k M g\right\} s.
\end{aligned}$$

The net work may also be obtained as the sum of the amounts of work done by each force.

$\vec{F}_A$ Applied Force

$$W_{if}\left[\vec{F}_A\right] = \int_{x_i}^{x_f} \vec{F}_A \cdot d\vec{s} = F_A \cos(\theta) \left(x_f - x_i\right) = F_A \cos(\theta)\, s.$$

The work done by the applied force is positive, which is consistent with the fact that the sled moves in the forward direction.

$\vec{W}$ Weight

$$W_{if}\left[\vec{W}\right] = \int_{x_i}^{x_f} \vec{W} \cdot d\vec{s} = M g \cos(-\pi/2) \left(x_f - x_i\right) = 0.$$

The work done by the weight is zero, since the sled moves in the forward direction while the weight acts straight down.

$\vec{N}$ Normal Force

$$W_{if}\left[\vec{N}\right] = \int_{x_i}^{x_f} \vec{N} \cdot d\vec{s} = N\,\cos(\pi/2)\,(x_f - x_i) = 0\,.$$

The work done by the normal force is zero because the normal force and the displacement are orthogonal throughout the entire motion of the sled [as was also the case for the weight force].

$\vec{f}_{\text{K}}$ Kinetic Friction

$$W_{if}\left[\vec{f}_{\text{K}}\right] = \int_{x_i}^{x_f} \vec{f}_{\text{K}} \cdot d\vec{s} = \mu_{\text{k}}\,N\,\cos(\pi)\,(x_f - x_i) = -\mu_{\text{k}}\,N\,s\,.$$

The work done by the force of kinetic friction is negative, because the kinetic friction acts resistively, *i.e.*, in opposition to the forward motion.

Thus, superposing the various and sundry works,

$$\begin{aligned}
\text{Net Work} &= W_{if}\left[\vec{F_{\text{A}}}\right] + W_{if}\left[\vec{W}\right] + W_{if}\left[\vec{N}\right] + W_{if}\left[\vec{f}_{\text{K}}\right] \\
&= F_{\text{A}}\,\cos(\theta)\,s + 0 + 0 - \mu_{\text{k}}\,N\,s \\
&= \left\{F_{\text{A}}\,\cos(\theta) - \mu_{\text{k}}\,N\right\}s \\
&= \left\{F_{\text{A}}\,\cos(\theta) - \mu_{\text{k}}\,(M\,g - F_A\,\sin(\theta))\right\}s \\
&= W_{if}\left[\vec{F_{\text{A}}} + \vec{W} + \vec{N} + \vec{f}_{\text{K}}\right].
\end{aligned}$$

In this sequence of equalities, the fact that the sled does not experience net y-acceleration determined the unique value of the magnitude of the normal force in terms of known [externally prescribed] quantities. The important bit is that the efficacy of the PRINCIPLE OF LINEAR SUPERPOSITION applied to the computation of net work has been demonstrated.

Stay tuned for more fun and exciting examples of work and energy.

Chapter 22

All Work and Some Play

Recall that the MECHANICAL WORK performed by a force, $\vec{F}$, is the line integral of the force as it acts through a displacement:

$$W_{if}\left[\vec{F}\right]_{(\text{path})} = \int_{\text{initial pos'n}}^{\text{final pos'n}} \vec{F} \cdot d\vec{s}.$$

In this chapter, we shall compute the mechanical work performed by various forces as the particles upon which they act undergo displacements.

EXAMPLE [*Energetically Pushing a Block on an Inclined Plane*]

A block of mass M riding on an inclined plane is pushed with a constant force, $\vec{F}_A$, acting parallel to the incline and in the downward sense. The block experiences a net displacement $\vec{s}$ down the incline. The coefficient of kinetic friction between the surface of the incline and the block is μ_k. Let's assume that the block is in motion down the incline during the entire time interval under consideration.

ASIDE: Should this assumption prove false,[1] then the analysis would have to be performed in two regimes. The first regime commences at the initial time and ends when the block stops [at the instant that it attains its maximum height up the incline]. The second regime starts at the end of the first and continues to the time when the block reaches its final position.

In these circumstances, the applied force is parallel to the net displacement, $\vec{F}_A \parallel \vec{s}$, while the kinetic friction force acts anti-parallel to the same displacement, $\vec{f}_K \parallel -\vec{s}$ throughout the motion of the block.

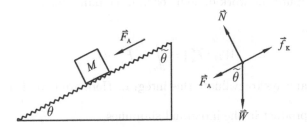

FIGURE 22.1 Pushing a Block down a Frictional Incline

Q: What is the total amount of mechanical work performed by the collection of forces which act upon the sliding block?

A: Let's figure it out, eh?
Incline-oriented coordinates, in which the x-axis is directed down the incline while the y-axis is perpendicular to the plane surface, prove to be the most advantageous choice.

---

[1]That is, if the block is moving uphill at the start [then rising, stopping, falling, and passing by the initial position, *en route* to its final position].

ASIDE: To characterise the frictional force in terms of the parameters provided in the statement of the problem,[2] we must first uncover the constraints inherent in the dynamics.

We proceed expeditiously through the steps of the dynamical recipe to obtain the following equations of motion:

$$M\,a_x = F_{\text{NET}x} = F_{\text{A}} + M\,g\,\sin(\theta) - f_{\text{K}}$$
$$M\,a_y = F_{\text{NET}y} = N - M\,g\,\cos(\theta)$$

That the block slides along the surface, $a_y = 0$, constrains the magnitude of the normal force and thereby determines the magnitude of the frictional force:

$$N = M\,g\,\cos(\theta), \quad \text{and hence} \quad f_{\text{K}} = \mu_{\text{K}}\,N = \mu_{\text{K}}\,M\,g\,\cos(\theta).$$

This result can be substituted into the x-equation of motion,

$$F_{\text{NET}x} = F_{\text{A}} + M\,g\,\sin(\theta) - \mu_{\text{K}}\,M\,g\,\cos(\theta) = F_{\text{A}} + M\,g\left(\sin(\theta) - \mu_{\text{K}}\,\cos(\theta)\right),$$

to yield an expression for the net force acting on the block which is given entirely in terms of presumably known quantities.

Returning to the question at hand, here are two approaches to computing the net mechanical work performed on the block.

METHOD ONE: [*Computing the Work Done by the Net Force*]
Given the net force, determined above, the net work is

$$\text{Net Work} = W_{if}\left[\vec{F}_{\text{NET}}\right]$$
$$= W_{if}\left[F_{\text{A}} + M\,g\left(\sin(\theta) - \mu_{\text{K}}\,\cos(\theta)\right)\right]$$
$$= \left\{F_{\text{A}} + M\,g\left(\sin(\theta) - \mu_{\text{K}}\,\cos(\theta)\right)\right\}\,s\,,$$

where $s = x_f - x_i$ is the distance that the block moves down the incline.

METHOD TWO: [*Computing the Net Work Done*]
One may choose to compute the work done by each force individually.

$\vec{F}_{\text{A}}$ Applied Force

$$W_{if}\left[\vec{F}_{\text{A}}\right] = \int_{x_i}^{x_f} \vec{F}_{\text{A}} \cdot d\vec{s}\,.$$

There are several ways to evaluate this integral. Here are two of them.

    o   The dot product in the integrand simplifies,

$$\vec{F}_{\text{A}} \cdot d\vec{s} = \left|\vec{F}_{\text{A}}\right| ds \cos(0) = F_{\text{A}}\,ds\,,$$

because the applied force and displacement vectors are parallel everywhere. The constancy of F_{A} and the straight-line trajectory of the block together enable the integration to be performed:

$$W_{if}\left[\vec{F}_{\text{A}}\right] = \int_{x_i}^{x_f} F_{\text{A}}\,ds = F_{\text{A}} \int_{x_i}^{x_f} dx = F_{\text{A}}\left(x_f - x_i\right) = F_{\text{A}}\,s\,.$$

---

[2]These are the block mass, M; the angle of inclination, θ; and the acceleration due to gravity, g.

o   Linearity of both integration and the dot product ensures that

$$W_{if}\left[\vec{F_A}\right] = \vec{F_A} \cdot \int_{x_i}^{x_f} d\vec{s} = \vec{F_A} \cdot \vec{s} = F_A\,[\nearrow] \cdot s\,[\nearrow] = F_A\,s\,[\nearrow] \cdot [\nearrow] = F_A\,s\,,$$

when the applied force is constant throughout the displacement.

The work done by the applied force is positive, as expected.

$\vec{W}$ Weight

$$W_{if}\left[\vec{W}\right] = \int_{x_i}^{x_f} \vec{W} \cdot d\vec{s} = \int_{x_i}^{x_f} M\,g\,[\downarrow] \cdot d\vec{s} = M\,g \int_{x_i}^{x_f} [\downarrow] \cdot d\vec{s}\,.$$

This integral may be evaluated by recourse to any of the following three means.

o   The geometric definition of the dot product allows expansion and simplification of the integrand,

$$[\downarrow] \cdot d\vec{s} = (1)\,ds\,\cos\left(\widetilde{\theta}\right) = \cos\left(\widetilde{\theta}\right)\,ds\,.$$

The facts underlying this simplification are: the magnitude of any unit vector is equal to 1; ds is the magnitude of the differential displacement down the incline; and $\widetilde{\theta}$ is the angle between $d\vec{s}$ and $[\downarrow]$, as shown in Figure 22.1. The angles $\widetilde{\theta}$ and θ are related by geometric theorems[3] which guarantee that $\widetilde{\theta} = \pi/2 - \theta$. Furthermore,[4] $\cos\left(\pi/2 - \theta\right) \equiv \sin(\theta)$. Thus,

$$W_{if}\left[\vec{W}\right] = M\,g\,\sin(\theta)\,s\,,$$

in perfect conformity with our expectations.

o   As the weight of the block is a constant force, and the displacement is along a straight-line path, the work simplifies to

$$W_{if}\left[\vec{W}\right] = M\,g\,[\downarrow] \cdot \int_{x_i}^{x_f} d\vec{s} = M\,g\,[\downarrow] \cdot \vec{s} = M\,g\,[\downarrow] \cdot s\,[\nearrow]\,.$$

The dot product of [vertically down] and [down along the incline] is

$$[\downarrow] \cdot [\nearrow] = (1)\,(1)\,\cos\left(\widetilde{\theta}\right) = (1)\,(1)\,\cos\left(\frac{\pi}{2} - \theta\right) = +\sin(\theta)\,,$$

since $\widetilde{\theta}$ is the interior angle between the two unit vectors. Hence,

$$W_{if}\left[\vec{W}\right] = M\,g\,\sin(\theta)\,s\,.$$

o   The horizontal run of the block lies orthogonal to the weight force, and therefore cannot contribute to the work done by the weight. The vertical drop is parallel to the weight force and has magnitude $h = s\,\sin(\theta)$. Thus, the entire amount of work done by the weight is

$$W_{if}\left[\vec{W}\right] = M\,g\,h = M\,g\,s\,\sin(\theta)\,.$$

This work is positive and independent of the means by which it is computed.

---

[3] COMPLEMENTARITY and SUPPLEMENTARITY theorems, to be precise.

[4] This identity is corroborated by the study of right triangles, inspection of the graphs of **sine** and **cosine**, and examination of the expansion formulae for trig functions of sums and differences of angles.

$\vec{N}$ Normal

$$W_{if}[\vec{N}] = \int_{x_i}^{x_f} \vec{N} \cdot d\vec{s}.$$

This integral, too, can be evaluated in a variety of ways.

- The definition of the dot product enables simplification of the integrand,

$$\vec{N} \cdot d\vec{s} = |\vec{N}|\, ds \, \cos(\pi/2) = 0 \,,$$

since the displacement is along the inclined surface, while the normal force is perpendicular to the same surface.

- The normal force is constant, while the block's trajectory is straight. Hence,

$$W_{if}[\vec{N}] = N\,[\diagdown] \cdot \int_{x_i}^{x_f} d\vec{s} = N\,[\diagdown] \cdot \vec{s} = N\,[\diagdown] \cdot s\,[\diagup] = N\,s\,[\diagdown]\cdot[\diagup] = 0\,,$$

because the dot product vanishes when the vectors are orthogonal.

One cannot imagine circumstances in which the normal force does non-zero work.

$\vec{f}_{\text{K}}$ Kinetic Friction

$$W_{if}[\vec{f}_{\text{K}}] = \int_{x_i}^{x_f} \vec{f}_{\text{K}} \cdot d\vec{s}.$$

This integral is amenable to different approaches as well.

- The integrand simplifies,

$$\vec{f}_{\text{K}} \cdot d\vec{s} = |\vec{f}_{\text{K}}|\, ds \, \cos(\pi) = -\mu_{\text{K}} N\, ds = -\mu_{\text{K}} M\, g\, \cos(\theta)\, ds\,,$$

because the resistive force of kinetic friction is anti-parallel to each infinitesimal displacement along the inclined surface. Owing to the constancy of the frictional force and the straight trajectory of the block, the work becomes

$$W_{if}[\vec{f}_{\text{K}}] = \int_{x_i}^{x_f} -\mu_{\text{K}} N\, ds = -\mu_{\text{K}} N \int_{x_i}^{x_f} ds = -\mu_{\text{K}} N\,(x_f - x_i) = -\mu_{\text{K}} M\, g\, \cos(\theta)\, s\,.$$

- The work done by the constant kinetic friction force is equal to the dot product of the force and the net displacement of the block, *viz.*,

$$W_{if}[\vec{f}_{\text{K}}] = \vec{f}_{\text{K}} \cdot \int_{x_i}^{x_f} d\vec{s} = \vec{f}_{\text{K}} \cdot \vec{s} = \mu_{\text{K}} M\, g\, \cos(\theta)\,[\diagup]\cdot s\,[\diagup]$$

$$= \mu_{\text{K}} M\, g\, s\, \cos(\theta)\,[\diagup]\cdot[\diagup] = -\mu_{\text{K}} M\, g\, s\, \cos(\theta)\,.$$

This result agrees precisely with that obtained by expansion of the integrand.

The work done by kinetic friction is negative, consistent with its dissipative nature.

Thus, the total amount of work done by this collection of forces is,

$$\text{Net Work} = W_{if}[\vec{F}_{\text{A}}] + W_{if}[\vec{W}] + W_{if}[\vec{N}] + W_{if}[\vec{f}_{\text{K}}]$$

$$= F_{\text{A}}\, s + M\, g\, \sin(\theta)\, s + 0 - \mu_{\text{K}} M\, g\, \cos(\theta)\, s$$

$$= \left\{ F_{\text{A}} + M\, g\left(\sin(\theta) - \mu_{\text{K}} \cos(\theta)\right) \right\}\, s\,.$$

We note, *encore*, that the work done by the net force is precisely equal to the net work done by all of the forces considered separately.

The net work may turn out to be POSITIVE, ZERO, or NEGATIVE. We'll consider each possibility in turn.

+   *The net work may be positive.*

In this instance, the net force acts down the incline, and the block accelerates [increases its speed] throughout the time interval in which it passes from its initial position to its final position.

0   *The net work may be zero.*

Here, the net force vanishes, and the block coasts [moves with constant velocity] along the incline *en route* from its initial to its final position.

−   *The net work may be negative.*

In this case, the net force is directed upward along the incline, and the block decelerates [its speed decreases] as it proceeds down the slope.

These three classes of behaviour are significant, and our distinguishing them here is in anticipation of the WORK–ENERGY THEOREM, to be presented in Chapter 23.

CAVEAT: One may have been lulled into a false sense of complacency by the examples considered thus far, because each integration became trivial owing to the constancy of the forces acting on the block. Let's now investigate the work done by a position-dependent force acting through various displacements.

EXAMPLE [*Mechanical Work Performed by a Spring Force*]

An ideal spring, anchored at one end, has its free end attached to a block of mass M which rides on a frictionless horizontal surface, as pictured in Figure 22.2. Hooke's Law expresses the magnitude and direction of the spring force,

$$\vec{F}_{\text{s}} = -k\,\vec{x}\,,$$

where k, the spring constant, is associated with the geometry [size and shape] and material composition of the spring.

Let's suppose that the motion of the block is confined to one dimension, which we choose to identify as the x-direction [with increasing x away from the wall anchoring the spring]. Furthermore, let's set the origin of coordinates so that the block is at $x_0 = 0$ when the spring is at its equilibrium length.

[The analysis is simplified by virtue of these choices.]

The block starts at an initial position, x_i, and arrives at a final position, x_f. In the course of this displacement, the amount of work done on the block by the spring force may be readily and accurately computed:

$$W_{if}\left[\vec{F}_{\text{s}}\right] = \int_{x_i}^{x_f} \vec{F}_{\text{s}} \cdot d\vec{s} \qquad \left[\text{where } \vec{F}_{\text{s}} \cdot d\vec{s} = F_{\text{s}x}\,dx = -k\,x\,dx\right]$$

$$= -k \int_{x_i}^{x_f} x\,dx = -\frac{1}{2}\,k\,x^2\bigg|_{x_i}^{x_f} = -\frac{1}{2}\,k\left(x_f^2 - x_i^2\right).$$

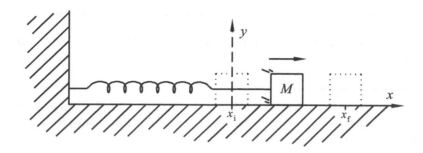

FIGURE 22.2   A Block–Spring System in Which the Block Moves from x_i to x_f

One *must*[5] realise that the very fact that we were able to perform the integration and thereby determine an explicit formula for the work done by the spring as the block moved from x_i to x_f is exceptional!

Let's pause a moment to examine the implications of this latest result.

> ASIDE: For this next bit we shall rely heavily on the interpretation of the integral as the area under a section of a curve.

Most crucially, the work done by the spring depends only on the initial and final points defining the displacement. In Figure 22.3 we have plotted the spring force along with a set of points, A, B, ..., F.

> ASIDE: The linearity of the spring force is concomitant with Hooke's Law. That the force is NEGATIVE for positive x and POSITIVE for negative x is a manifestation of its restorative nature. Implicit in Figure 22.3 is the identification of the origin, O, as that position at which the spring force vanishes.

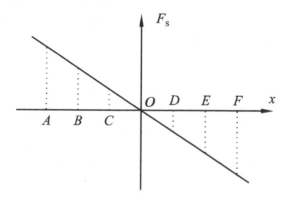

FIGURE 22.3   Areas under the Force Function

AB Upon going from A to B, the spring force acts in the same direction as the displacement. Consequently, the work done on the block is positive:

$$W_{AB}\left[\overrightarrow{F_{\mathrm{s}}}\right] > 0\,.$$

---

[5]We are being overly dramatic to foreshadow a sequence of important results beginning in Chapter 24.

BC When the block moves from B to C, the same consequences ensue as in AB, but the amount of work done is less than in the previous case:

$$W_{AB}\left[\vec{F_s}\right] > W_{BC}\left[\vec{F_s}\right] > 0.$$

CO Should the block move from C to the origin O, the situation is essentially the same as in AB and BC, except that the amount of work done is less than in both previous cases:

$$W_{AB}\left[\vec{F_s}\right] > W_{BC}\left[\vec{F_s}\right] > W_{CO}\left[\vec{F_s}\right] > 0.$$

OD Under displacement of the block from O to D, the spring force acts in the opposite direction, and thus the work done on the block is negative:

$$W_{OD}\left[\vec{F_s}\right] < 0.$$

CD The work done by the spring on the block as it proceeds from C to D is the sum of the work done through the two path segments $C \to O$ and $O \to D$. Consideration of Figure 22.3 leads us to expect cancellation of the positive and negative contributions to the total work associated with the two segments of the path. This is corroborated by the explicit integration result

$$W_{CD}\left[\vec{F_s}\right] = W_{CO}\left[\vec{F_s}\right] + W_{OD}\left[\vec{F_s}\right] \approx 0.$$

> ASIDE: A way to appreciate this result is that we partitioned the overall path **from C to D** into segments, $\{\ C \to O\ ,\ O \to D\ \}$, and expressed the total work as the sum of the work performed along each of the elements of the partition. This is a well-defined mathematical strategy provided that the integrand does not exhibit singular behaviour within the intervals.

DE The work done by the spring on the block is negative, and its magnitude is increasingly negative compared to that computed from $O \to D$. Even though the magnitudes of the displacements are comparable,

$$W_{DE}\left[\vec{F_s}\right] < W_{OD}\left[\vec{F_s}\right] < 0,$$

on account of the increasing magnitude of the spring force.

The function which describes the work done by the spring force acting upon the block possesses several rather interesting properties.

- The work done by the spring is UNCHANGED IF $x_i \to -x_i$ OR $x_f \to -x_f$.

- The work done by the spring is ZERO when $x_f = \pm x_i$.

  – Contributions of equal magnitude and opposite sign from the two constituent paths [**from** x_i **to** the equilibrium point $x_0 = 0$, and **from** $x_0 = 0$ **to** $-x_i$] cancel.

  + **Q:** Eh? Can we really start and end at the same place?

    A: You betcha!

    Mass–spring systems are realisations of the SIMPLE HARMONIC OSCILLATOR. [The kinematics of SHO were encountered in Chapter 4. The dynamics were exposed in Chapter 15.] Since the system is oscillatory, it is inevitable that the block return to its initial position.

    ASIDE: If the point is at the extreme range of the oscillatory motion, then it is visited once per cycle. Otherwise, it is visited twice per cycle.

We'll finish with a few generic computations of the net work done by the spring from particular chosen starting points $x_i \in \{A, E, O\}$ to an arbitrary end point x.

$$W_{Ax} = -\frac{1}{2}\,k\left(x^2 - A^2\right) = \frac{1}{2}\,k\left(A^2 - x^2\right)$$

$$W_{Ex} = -\frac{1}{2}\,k\left(x^2 - E^2\right) = \frac{1}{2}\,k\left(E^2 - x^2\right)$$

$$W_{Ox} = -\frac{1}{2}\,k\left(x^2 - 0^2\right) = -\frac{1}{2}\,k\,x^2$$

We see from the formulae, and their curves sketched below, that they are all merely shifted versions of the same concave–downward parabola.

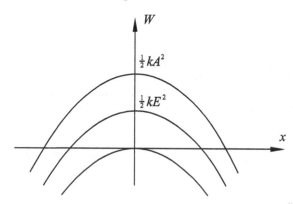

FIGURE 22.4 Functions Expressing the Work Done by the Spring on the Block for Fixed Initial Positions, $\{A, E, O\}$, and Variable Final Positions x

Chapter 23

The Work–Energy Theorem

In all of the examples in the previous two chapters, careful observation reveals that:

$$\text{WHEN the net work is } \begin{Bmatrix} \text{positive} \\ \text{zero} \\ \text{negative} \end{Bmatrix}, \text{ THEN the particle } \begin{Bmatrix} \text{speeds up} \\ \text{coasts} \\ \text{slows down} \end{Bmatrix}.$$

The Work–Energy Theorem relates the mechanical work done on a particle to changes in its state of motion.

WORK–ENERGY THEOREM   The net work done on a particle manifests itself as a net change in the particle's KINETIC ENERGY.

KINETIC ENERGY   Kinetic energy is a form of mechanical energy ascribed to an object by virtue of its motion.

PROOF of the WORK–ENERGY THEOREM

We shall prove the theorem by quantifying kinetic energy. The net work is equal to the work done by the net force,

$$W_{if}\left[\text{NET}\right]_{(\text{path})} = W_{if}\left[\vec{F}_{\text{NET}}\right]_{(\text{path})} = \int_{\text{initial pos'n}}^{\text{final pos'n}} \vec{F}_{\text{NET}} \cdot d\vec{s}.$$

In general, the path can meander, and the force can be variable [dependent on time, position, velocity, *etc.*], as indicated in Figure 23.1. This computation looks rather hopeless.[1]

Rather than despair of ever calculating the net work in the face of these hurdles, let's think more generally. Appealing to N2 gives us our first, most important step:

$$m\,\vec{a} = \vec{F}_{\text{NET}} \quad \text{AND} \quad \vec{a} = \frac{d\vec{v}}{dt} \quad \Longrightarrow \quad W_{if}\left[\vec{F}_{\text{NET}}\right]_{(\text{path})} = \int_{\text{initial pos'n}}^{\text{final pos'n}} m\,\frac{d\vec{v}}{dt} \cdot d\vec{s}.$$

The second step is to regard the velocity vector in its geometric sense as a directed magnitude:

$$\vec{v} = v\,\hat{v} = (\text{speed})\left[\text{direction in which it is moving}\right].$$

---

[1] "Oh no, it's not!" one might cheerfully exclaim. "*All* that we have to do is *parameterise* the path, *construct* the pull-back of the one-form $dW = \vec{F}_{\text{NET}} \cdot d\vec{s}$, and *solve* the resulting integral." Right! However, difficulties may arise at the *parameterising*, *constructing*, and *solving* stages which render impossible direct analytic computation of the work.

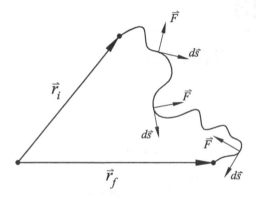

FIGURE 23.1 Computing Work

Hence, the acceleration of the particle receives separate contributions from changes in its speed and direction:

$$\vec{a} = \frac{d\vec{v}}{dt} = \frac{d(v\,\hat{v})}{dt} = \frac{dv}{dt}\,\hat{v} + v\,\frac{d\hat{v}}{dt}\,.$$

The time rate of change of the speed, $\frac{dv}{dt}$, is directed parallel to the velocity of the particle, while the rate at which the direction is changing, $\frac{d\hat{v}}{dt}$, is necessarily perpendicular to the velocity. Furthermore, the motion of the particle is always parallel to its instantaneous velocity. Thus, the integrand in the expression for net work becomes

$$\vec{F}_{\text{NET}} \cdot d\vec{s} = m\left(\frac{dv}{dt}\,\hat{v} + v\,\frac{d\hat{v}}{dt}\right) \cdot d\vec{s}.$$

These observations about the directions of the respective terms in the acceleration make it incontrovertible that $\frac{d\hat{v}}{dt} \cdot d\vec{s} \equiv 0$ and $\hat{v} \cdot d\vec{s} \equiv ds$ [where ds is the differential arc length]. Thus, the integrand simplifies to

$$\vec{F}_{\text{NET}} \cdot d\vec{s} = m\,\frac{dv}{dt}\,ds\,.$$

The third step is to employ a familiar calculus trick—the CHAIN RULE.

$$m\,\frac{dv}{dt}\,ds = m\left(\frac{dv}{ds}\,\frac{ds}{dt}\right)\,ds = m\,v\,\frac{dv}{ds}\,ds = m\,v\,dv\,,$$

where $\frac{ds}{dt}$, the time rate of change of arc length, has been identified as the speed of the particle. The last equality in the sequence above did not arise from cancellation of the differentials "ds" in the numerator and the denominator [as is often and misleadingly portrayed]. Instead, we have effected a **change of variables** from arc length to speed. The original integral was along the trajectory of the particle and had as its limits the initial and final [vector] positions. This integral was pulled back to an integral over the arc length of the particular path. After the application of the chain rule, the integral is with respect to speed, and its limits are the initial and final [scalar] speeds.

Finally then, the net work done on the particle during its displacement from an initial position, x_i, to a final position, x_f, under the influence of a net external force is

$$W_{if}\left[\text{NET}\right]_{\text{(path)}} = \int_{\text{initial speed}}^{\text{final speed}} m\,v\,dv = \int_{v_i}^{v_f} m\,v\,dv = \frac{1}{2}\,m\,v^2\bigg|_{v_i}^{v_f} = \frac{1}{2}\,m\left(v_f^2 - v_i^2\right)\,.$$

We now pause to redefine [more accurately, **quantify**] kinetic energy.

KINETIC ENERGY   A particle of mass m experiencing translational motion with speed v has BULK[2] kinetic energy

$$K = \frac{1}{2}\,m\,v^2 = \frac{1}{2}\,m\,|\vec{v}|^2 = \frac{1}{2}\,m\,\vec{v}\cdot\vec{v}\,.$$

The units of kinetic energy are indeed joules, since

$$1\,\text{J} = 1\,\text{N}\cdot\text{m} = 1\,\frac{\text{kg}\cdot\text{m}^2}{\text{s}^2} = 1\,\text{kg}\left(\frac{\text{m}}{\text{s}}\right)^2\,.$$

Returning to the formal expression for the net work done on the particle, one sees that it is equal to the change in the kinetic energy of the particle:

$$W_{if}\left[\text{NET}\right]_{(\text{path})} = \frac{1}{2}\,m\,\left(v_f^2 - v_i^2\right) = K_f - K_i = \Delta K\,.$$

The steps taken to arrive at this expression constitute an irrefutable proof of the Work–Energy Theorem.

A few comments are in order. [*As usual, eh?*]

- The derivation seemed tautological, since kinetic energy was defined in precisely the manner necessary to ensure the veracity of the theorem.

Consider the alternatives.

- **Q:** What if the powers of m and v were different, *i.e.*, $\widetilde{K} = \frac{1}{2}\,m^a\,v^b$?
  A: The dimensions of $\widetilde{K}$ do not work out to joules unless $a = 1$ and $b = 2$.

- **Q:** Could one avoid the inevitability alluded to above by having a dependence on some other dimensional parameter?
  A: Perhaps. However, then the quantity could not accurately be described as "energy of motion" because it would depend on more than the state of motion of the particle.

- **Q:** How about the factor of one-half? Could that have been different?
  A: Well, yeah, ... But then the Work–Energy Theorem would have to have a compensating factor built into it. Suppose that the "kinetic energy" were defined to be $\widetilde{K} = \frac{1}{3}\,m\,v^2$. Then one would be forced to write:

$$W_{if}\left[\text{NET}\right]_{(\text{path})} = \frac{3}{2}\,\Delta\widetilde{K}\,.$$

---

[2]Here is another *weaselly word* needed to satisfy the lawyers, whose purpose will be explained anon. [Oooops! Please pardon the semantic ambiguity. To clarify: we shall explain the qualifier "BULK," whereas the *purpose of lawyers* is still open to question.]

- The Work–Energy Theorem may be used to rederive many of the dynamical results that we've obtained for blocks on planes and inclines. In many instances the energy methods are more easily employed than the dynamical analysis based directly on N2.

- Recall that in Chapter 3 a third [and unexpected] kinematical formula holding for one-dimensional motion under constant acceleration was revealed. There, we had to meekly accept that the relation

$$v_f^2 = v_i^2 + 2\,a\,\Delta x$$

exists, but were at a loss as to its physical interpretation. Also, upon introducing vectors we realised that there was no natural extension of this relation to motions under constant acceleration occurring in more than one dimension.

In the light of the present chapter, we are now able to recognise that the mysterious kinematical formula is equivalent to the expression of the Work–Energy Theorem in the particular case of a constant net external force. First, observe that constant net external force gives rise to constant acceleration, which greatly simplifies computation of the net work:

$$W_{if}[\vec{F}_{\text{constant}}] = W_{if}[m\,\vec{a}] = \int_{\vec{r}_i}^{\vec{r}_f} m\,\vec{a}\cdot d\vec{s} = m\,\vec{a}\cdot \int_{\vec{r}_i}^{\vec{r}_f} d\vec{s} = m\,\vec{a}\cdot\Delta\vec{s},$$

where $\Delta\vec{s} = \vec{r}_f - \vec{r}_i$ is the net displacement of the particle. The second step is to express the net change in the kinetic energy in the time interval in which the particle progresses from $\vec{r}_i$ to $\vec{r}_f$ and obtain

$$\Delta K = K_f - K_i = \frac{1}{2}\,m\,\left(v_f^2 - v_i^2\right).$$

The third step is to assemble the results of the first two steps, via the Work–Energy Theorem, to yield the following relation:

$$v_f^2 - v_i^2 = 2\,\vec{a}\cdot\Delta\vec{s},$$

in any number of dimensions, whenever the acceleration, $\vec{a}$, is strictly constant.

> ASIDE: This explication of the third 1-d kinematical relation is sufficient to convince us that any attempts we might have made to generalise it to accommodate vectors [except in the precise manner quoted above] would have been futile!

Chapter 24

Conservative Forces

In all[1] of our explicit analyses of work,

$$W_{if}\left[\vec{F}\right]_{(\text{path})} = \int_{\vec{r}_i}^{\vec{r}_f} \vec{F} \cdot d\vec{s},$$

paths were chosen which led directly from the initial to the final position, $\vec{r}_i \rightarrow \vec{r}_f$.

[This certainly made it easier to evaluate the integrals, eh?]

And yet, the expression for work can accommodate any path, however complicated, subject to a few technical conditions. These conditions, cursorily summarised, amount to:

- Continuity[2]
- Differentiability with respect to the path's parameter[3]

One expects that the amount of work done by a force acting through a given displacement depends crucially on the path taken from the initial to the final position. However, there exists a class of forces, called CONSERVATIVE FORCES, whose members belie this expectation.

CONSERVATIVE FORCE   A force is conservative IFF the work done by the force acting through any particular displacement is independent of the path.

Path independence is a very restrictive condition, requiring that the work done by a conservative force acting through a displacement from a particular inital point to a certain final point be the same, irrespective of the actual sequence of intermediate points encountered *en route* from start to finish.

However, IF either the initial point OR the final point changes,
THEN the amount of work is expected to differ.

Path independence of the work done by a conservative force, $\vec{F_c}$, is expressed by writing:

$$W_{if}\left[\vec{F_c}\right]_{(\text{path 1})} = W_{if}\left[\vec{F_c}\right]_{(\text{path 2})},$$

with reference to Figure 24.1 for clarity.

An equivalent formulation is that **a given force is conservative IFF the net work done by that force is zero when the initial and final points are the same,** *viz.*, the work vanishes for a closed path,

$$W_{ii}\left[\vec{F_c}\right]_{(\text{closed path})} \equiv 0.$$

---

[1]With the exception of our general discussion of the Work–Energy Theorem in Chapter 23.

[2]There must be no sudden hops, no disappearances and reemergences.

[3]This condition is necessary in all open intervals, but may be relaxed at a finite number of discrete points.

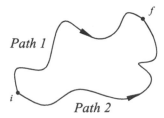

FIGURE 24.1   Two Distinct Paths from i to f

The [powerful] idea of path equivalence of work may also be formulated in terms of **compound** paths which go to, and then from, a particular intermediate point. A compound path which begins and ends at i, while passing through the point labelled f, is illustrated in Figure 24.2.

$$W_{if}\big[\overrightarrow{F_{\text{C}}}\big]_{(\text{path 1})} + W_{fi}\big[\overrightarrow{F_{\text{C}}}\big]_{(\text{path 1}')} = W_{ii}\big[\overrightarrow{F_{\text{C}}}\big]_{(\text{path 1}+\text{path 1}')} \equiv 0\,.$$

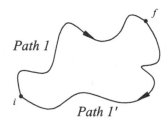

FIGURE 24.2   A Compound Path Starting and Ending at i and Including f

When the path is closed [*a.k.a.* a null displacement[4]], the work done by $\overrightarrow{F}$ is

$$W_{\mathcal{PP}}\big[\overrightarrow{F}\big]_{\gamma} = \oint_{\gamma} \overrightarrow{F} \cdot d\vec{s}\,,$$

where the trajectory of the particle, which starts and ends at a particular point $\mathcal{P}$, is denoted by γ. The modified integral sign "$\oint$" is a reminder that we are considering the very special case of closed paths.

> ASIDE: Closed trajectories generally must be **oriented** to distinguish between the two *a priori* different ways in which they may be traversed. Fortunately, it turns out that orientation considerations are moot for conservative forces!

CONSERVATIVE FORCE   [*the technical definition*]

LET γ denote a closed curve which includes the point $\mathcal{P}$. IF $W_{\mathcal{PP}}\big[\overrightarrow{F}\big]_{\gamma} = 0$ for all possible γ, THEN the force, $\overrightarrow{F}$, is conservative.

---

[4]The displacement is null in that the particle starts and stops at the same place.

The following pictorial analysis is to verify that the technical definition coincides with the prior definition of conservative forces based on path independence.

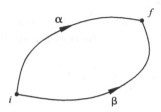

FIGURE 24.3 Points i and f Joined by Two Distinct Paths

An initial point, i, and a distinct final point, f, $i \neq f$, are joined by paths labelled α and β, as illustrated in Figure 24.3. Given α and β, one may construct a three-stage compound path γ from i to f, expressed[5] as follows and illustrated in Figure 24.4.

$$\gamma = \gamma_3 + \gamma_2 + \gamma_1 = \beta + (-\beta) + \alpha, \quad \text{where}$$

γ_1	*forward* from i to f along path α
γ_2	*backward* from f to i along path β, *i.e.,* $-\beta$
γ_3	*forward* from i to f along path β

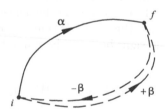

FIGURE 24.4 A Compound Path, $\gamma = \beta - \beta + \alpha$, from i to f

The plan is to argue separately that

$$W_{if}\left[\vec{F_c}\right]_\gamma = W_{if}\left[\vec{F_c}\right]_\alpha \quad \text{AND} \quad W_{if}\left[\vec{F_c}\right]_\gamma = W_{if}\left[\vec{F_c}\right]_\beta,$$

and infer, by transitivity,[6] that

$$W_{if}\left[\vec{F_c}\right]_\alpha = W_{if}\left[\vec{F_c}\right]_\beta.$$

Here is that plan put into effect. Expand the work done along the path γ into the sum of the amounts of work done along the three constituent stages ($\gamma = \beta - \beta + \alpha$):

$$W_{if}\left[\vec{F_c}\right]_\gamma = W_{if}\left[\vec{F_c}\right]_\alpha + W_{fi}\left[\vec{F_c}\right]_{-\beta} + W_{if}\left[\vec{F_c}\right]_\beta.$$

---

[5]The constituent sub-paths are "operator ordered," and read from right to left.
[6]Transitivity is the arithmetic property: If $A = B$ and $A = C$, then $B = C$.

The terms on the RHS may be summed in two distinct ways, according to associativity.[7]

$\beta + (-\beta + \alpha)$

The work done through the first two stages of the compound path, $-\beta + \alpha$, is

$$W_{if}\left[\vec{F_c}\right]_\alpha + W_{fi}\left[\vec{F_c}\right]_{-\beta} \equiv W_{ii}\left[\vec{F_c}\right]_{-\beta+\alpha} \equiv 0 \,,$$

since the force is [assumed to be] conservative and these two stages together constitute a closed loop starting and ending at i. Hence, the entire contribution to $W_{if}\left[\vec{F_c}\right]_\gamma$ comes from the last stage:

$$W_{if}\left[\vec{F_c}\right]_\gamma = W_{if}\left[\vec{F_c}\right]_\beta \,.$$

$(\beta - \beta) + \alpha$

The work done through the two-stage compound path $\beta - \beta$ is

$$W_{fi}\left[\vec{F_c}\right]_{-\beta} + W_{if}\left[\vec{F_c}\right]_\beta \equiv W_{ff}\left[\vec{F_c}\right]_{\beta-\beta} \equiv 0 \,,$$

for the same reasons as those quoted in the analysis made just previously. While this result may seem *too cute*, let's consider two compelling arguments for its veracity.

- ○ *Ploughing through the details*

  Let's compute the work along the paths β [going forward], and $-\beta$ [going backward] by employing the PARTITION [into single steps], COMPUTE [the work done in each step], and SUM [the stepwise contributions] strategy.

  [Work as a Riemann Sum was discussed (in a footnote) in Chapter 21.]

  We may choose exactly the same partition of the forward and reverse paths. [An inessential difference is that the steps occur in the opposite order in the respective cases.] The crucial consequence of this choice is that reversing the direction of the steps produces a change in the sign, but not the magnitude,[8] of each and every *accumulated* term. Therefore

  $$W_{fi}\left[\vec{F_c}\right]_{-\beta} = -W_{if}\left[\vec{F_c}\right]_\beta \qquad \text{and hence} \qquad W_{fi}\left[\vec{F_c}\right]_{-\beta} + W_{if}\left[\vec{F_c}\right]_\beta = 0 \,.$$

- ○ *Using one's little grey cells*[9]

  Treading back and forth along β amounts to traversing a degenerate[10] loop. From the assumption that the force is conservative, it follows directly that the net work must be ZERO.

However one chooses to think about this combination of paths, the inescapable conclusion is that

$$W_{if}\left[\vec{F_c}\right]_\gamma = W_{if}\left[\vec{F_c}\right]_\alpha \,.$$

---

[7] Associativity of addition is that $(A + B) + C = A + (B + C)$.

[8] CAVEAT: We are making the [reasonable] assumption that the force acting is not time-dependent, so the inbound and outbound work at the same arc length element in the partition must have equal magnitude and opposite sign.

[9] This is the *modus operandi* of M. Hercule Poirot, Agatha Christie's fictional detective.

[10] The loop is degenerate in that it is closed and yet does not bound any 2-d surface, not that it is of dubious character!

The two results obtained directly above,

$$W_{if}\left[\vec{F_c}\right]_\gamma = W_{if}\left[\vec{F_c}\right]_\alpha \quad \text{AND} \quad W_{if}\left[\vec{F_c}\right]_\gamma = W_{if}\left[\vec{F_c}\right]_\beta,$$

together imply that

$$W_{if}\left[\vec{F_c}\right]_\alpha \equiv W_{if}\left[\vec{F_c}\right]_\beta.$$

This equivalence affirms the mutual consistency of the colloquial and the technical formulations of CONSERVATIVE.

To ensure that the defining conditions are indeed met, conservative forces must possess[11] the following features.

1. The force must be independent of time.

   IF the force varies in time, THEN it becomes impossible to guarantee that the work will vanish over every closed path.

   > ASIDE: An extreme counter-example is provided by an out-and-back path in which the force is reversed during one leg of the trip.

2. The force must be independent of velocity.

   IF the force depends on the velocity [of the particle upon which the force acts], THEN it becomes impossible to guarantee that the work will vanish over every closed path.

   > ASIDE: An extreme counter-example is provided by an out-and-back path in which the speed is doubled during one leg of the trip.

3. The force must be **irrotational.**

   To understand this condition, let's look carefully at a counter-example. Consider a force field[12] in 2-d:

   $$\vec{F}(\vec{r}) = \vec{F}(x,y) = -y\,\hat{\imath} + x\,\hat{\jmath} = \left(-y\,,\,x\right).$$

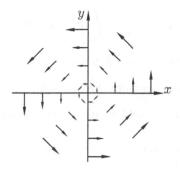

FIGURE 24.5   An Illustration of a Force Field Which is Not Irrotational

In Figure 24.5, the force is represented by a collection of arrows.[13] The length of

---

[11]Recall that it was stated that conservative forces are exceptional rather than commonplace.

[12]One may think of a force field as a vector-valued function of position determining the magnitude and direction of the force at each and every point. We will have much more to say about fields later.

[13]The figure provides an incomplete description of the field in that only a few such arrows are drawn.

each arrow is proportional to the magnitude of $\vec{F}$, and its direction is parallel to $\vec{F}$ at the point in space where the tail begins.

Let's compute the work done by this force along a circular closed path centred upon the origin, γ_0, shown in Figure 24.6. Suppose that the path is traversed in the anti-clockwise sense.

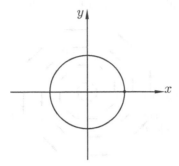

FIGURE 24.6   A Particular Closed Path, γ_0, in the Same xy-plane as the Force Field

The total amount of work performed by the force must be independent of the specific starting point. However, we will now argue that the work,

$$W\left[\vec{F}\right]_{\gamma_0} = \oint_{\gamma_0} \vec{F} \cdot d\vec{s}\,,$$

cannot possibly be equal to zero for some paths. This, in turn, shows that the amount of work done depends crucially on the path [the set of all of the points through which the trajectory passes]. From Figures 24.5 and 24.6, it is evident that the force is tangent to the path every step of the way around the circle. Thus, $\vec{F} \cdot d\vec{s}$, the integrand, is positive definite everywhere.[14] IF the integrand is positive definite throughout a domain of integration, THEN the integral must be positive,

$$W\left[\vec{F}\right]_{\gamma_0} = \oint_{\gamma_0} \vec{F} \cdot d\vec{s} > 0\,,$$

and therefore it is impossible for this field to be conservative.

Q: Which aspect of this field presents the obstacle?

A: The impediment is that the field wraps about a point.[15]

ASIDE: A force field that does not wrap completely around any point is said to be **irrotational**. This is necessary for a force to be conservative.

---

[14]It is necessary [to appease the lawyers] that we add "except perhaps at a set of measure zero (*i.e.,* isolated points) for which $\frac{d\vec{s}}{dt} = \vec{0}$, where t is the parameter marking our progress around the circular path."
[15]In this case, the origin. Any point will suffice.

Chapter 25

Potential Energy

Q: What's so special about conservative forces?
A: The work done by such a force depends solely on the initial and final points of the putative trajectory of a particle on which the force acts. This feature enables construction of a **potential energy function**, $U(\vec{r})$, associated with the conservative force.

POTENTIAL ENERGY   Potential energy is energy possessed by a particle by virtue of its position in space. Only changes in potential energy are meaningfully determined. Such changes are associated with the work done by a conservative force, *viz.*,

$$\Delta U = U_{\text{final}} - U_{\text{initial}} = U(\vec{r}_f) - U(\vec{r}_i) = -W_{if}\left[\vec{F_{\text{c}}}\right] = -\int_{\vec{r}_i}^{\vec{r}_f} \vec{F_{\text{c}}} \cdot d\vec{s}.$$

Four features pertinent to potential energy are listed below.

First, the potential energy, $U(\vec{r})$, is a scalar [a magnitude with associated units], which often makes it easier to work with than [vectorial] forces.

Second, IF the physical system is one-dimensional, AND a force field, $\vec{F}(x)$, depends only on position, THEN it is always possible to construct a potential energy function, $U_F(x)$, associated with the force.

> ASIDE: This result has broader application than one might at first suspect, because highly symmetric systems in two and three dimensions are often governed by effectively one-dimensional dynamics.

Third, there are two [interrelated] reasons why only differences in potential energy matter.

    ○   Potential energy functions are only defined up to an additive constant.

        Suppose that for a particular conservative force, $\vec{F_{\text{c}}}$, we posit $U(\vec{r})$ as an associated potential energy function. Meanwhile, our friend *Mathilde*[1] proposes an alternative potential energy function, $\widetilde{U}(\vec{r})$,

$$\widetilde{U}(\vec{r}) = U(\vec{r}) + C.$$

        where C is a constant with units of energy.

        Q: How can this possibly be consistent?

---

[1]Mathilde puts a tilde on her potential energy function to distinguish it from ours.

A: Only differences in potential energy are meaningfully defined, and therefore:

$$\Delta U = U(\vec{r}_f) - U(\vec{r}_i) \qquad\qquad \Delta \widetilde{U} = \widetilde{U}(\vec{r}_f) - \widetilde{U}(\vec{r}_i)$$

$$= \left[U(\vec{r}_f) + C \right] - \left[U(\vec{r}_i) + C \right]$$

$$= U(\vec{r}_f) - U(\vec{r}_i).$$

Hence, $\Delta \widetilde{U} \equiv \Delta U$, irrespective of the value of C. Thus, we can *agree to disagree* with Mathilde about the value(s) of our respective potential energy functions, in complete confidence that everyone will always agree on items of true dynamical import.

○   Position is measured with respect to an arbitrarily chosen origin.

Different choices of the origin [of coordinates] are related by rigid translations which affect the values of the coordinates of all points in the same manner. Under rigid translations, any legitimate potential energy function will be simply shifted by a constant, but its overall shape will not be changed.[2]

> ASIDE: Another form of coordinate arbitrariness follows from the choice of the set of unit vectors which generate the axes. Under transformation among these sets of coordinates, the potential energy function changes in a regular, well-defined manner because it is a scalar function of position.

The fourth comment further clarifies the restriction that a conservative force field must be irrotational. [Recall that, in Chapter 24, it was demonstrated that a force field which is NOT irrotational cannot be conservative.]

Q: Okay, but why should this be regarded as a bad thing?

A: We'll demonstrate the pitfall by means of *reductio ad absurdum*.[3]

○   Suppose that $\vec{F}_{\text{rot}}(x, y)$, wrapping around a point in space (we'll make this point the origin), has an associated potential energy function, $U_{\text{rot}}(x, y)$. The potential energy function can be generated by picking a reference point and computing the work done by the force along paths leading from this point to all other points in space. [By definition, the shift in the potential energy away from the reference point is (-1) times the work done by the force.]

○   On the journey from the reference point to any other point, the path may go directly, or it may loop around the origin a total of $\mathcal{N}$ times. The work which is done by the force is different in each case, as can be seen most easily by consideration of a particular path, $\gamma_0^{\mathcal{N}}$, consisting of $\mathcal{N}$ loops of the same closed path, γ_0:

$$W\left[\vec{F}_{\text{rot}}\right]_{\gamma_0^{\mathcal{N}}} \equiv \mathcal{N} \times W\left[\vec{F}_{\text{rot}}\right]_{\gamma_0}.$$

In order that the potential energy function be single-valued,[4] the work done cannot depend on how many times the path loops around.

---

[2]We encountered this in the context of the Hookian spring–block system, discussed in Chapter 22. There, we were swept away with enthusiasm stemming from our discovery that we could calculate [in closed form] the mechanical work done by the variable spring force. When this was explicitly done for arbitrary displacements from various initial points, it was realised [recall Figure 22.4] that each of the resulting functions was a shifted version of the same parabola.

[3]That is, we shall first assume that a reasonable potential energy function can be constructed from such a force. Simple investigation of this putative potential energy function leads to an *absurd* and contradictory consequence. The logical inevitability of this erroneous conclusion nullifies the original hypothesis.

[4]That there be a uniquely specified value of potential energy associated with each point in space.

The putative potential energy function obtained from the non-irrotational force is not a function at all! Therefore, the assumption of its existence must be invalid.

True conservative forces evade these difficulties, because the work that they do in traversing any closed loop is identically equal to zero. Hence,

$$W\left[\vec{F_c}\right]_{\gamma_0^{\mathcal{N}}} \equiv \mathcal{N}\, W\left[\vec{F_c}\right]_{\gamma_0} \equiv \mathcal{N} \times 0 = 0\,,$$

forestalling contradiction.

> ASIDE: The local criterion that distinguishes an irrotational force is that it is **curl** free. In the language of vector calculus: $\vec{\nabla} \times \vec{F_{\mathrm{C}}} \equiv 0$. For the time being, we shall remain content with the (non-local) pictorial understanding obtained above.

EXAMPLE [*Potential Energy Associated with a Constant Gravitational Field*]

The weight force, introduced in Chapter 11, may be more generally regarded as a manifestation of local constant field gravity. In the case of a particle of mass m, this force is: $m\,\vec{g}$. In what follows, the subscript "g" is used to specify constant field gravity forces and potential energies.

$$\vec{F_{\mathrm{g}}} = \vec{W} = m\,g\,[\downarrow] \qquad \Longrightarrow \qquad W_{if}\left[\vec{F_{\mathrm{g}}}\right] = \int_{\vec{r}_i}^{\vec{r}_f} m\,g\,[\downarrow] \cdot d\vec{s}$$

$$\vdots \quad \ldots \text{ only the change in height matters } \ldots$$

$$= \int_{y_i}^{y_f} -m\,g\,dy = -m\,g\,(y_f - y_i)\,.$$

The y-coordinate specifies the vertical direction and is oriented so as to increase upward. The change in the potential energy associated with the gravitational force is

$$\Delta U_{\mathrm{g}} = U_{\mathrm{g},f} - U_{\mathrm{g},i} = U_{\mathrm{g}}(\vec{r}_f) - U_{\mathrm{g}}(\vec{r}_i) \equiv -W_{if}\left[\vec{F_{\mathrm{g}}}\right] = m\,g\,(y_f - y_i)\,.$$

Inspection of this chain of equalities leads to the realisation that the natural [MINIMAL] choice for the potential energy function associated with the weight force is

$$U_{\mathrm{g}}(\vec{r}) = U_{\mathrm{g}}(x, y, z) = m\,g\,y\,.$$

The potential energy vanishes, $U_{\mathrm{g}} = 0$, at reference height $y = 0$. Mathilde, whom we encountered earlier in the chapter, chose another value of y, say y_0, as the reference height by adding the constant term "$-m\,g\,y_0$" to the above minimal potential energy function, thus obtaining

$$\widetilde{U}_{\mathrm{g}}(\vec{r}) = \widetilde{U}_{\mathrm{g}}(x, y, z) = m\,g\,(y - y_0)\,.$$

EXAMPLE [*Potential Energy Associated with a Spring Force*]

A Hookian spring exerts a restoring force with magnitude proportional to the displacement from the equilibrium position of the free end of the spring:

$$\vec{F_s} = -k\,\vec{x}\,.$$

[Recall that k is the spring constant and that the equilibrium position of the spring corresponds[5] to $x = 0$.] The work done by the spring force acting through a displacement from $x_i \to x_f$ was computed in Chapter 22 to be

$$W_{if}\left[\vec{F_s}\right] = \int_{\vec{r_i}}^{\vec{r_f}} \vec{F_s} \cdot d\vec{s} = \ldots = -\frac{1}{2}\,k\left(x_f^2 - x_i^2\right)\,.$$

Thus, the change in the potential energy of the mass–spring system is

$$\Delta U_S = U_{S,f} - U_{S,i} = U_S(\vec{r_f}) - U_S(\vec{r_i}) \equiv -W_{if}\left[\vec{F_s}\right] = \frac{1}{2}\,k\left(x_f^2 - x_i^2\right)\,.$$

Hence, the natural choice for the potential energy function for a spring is:

$$U_S(\vec{r}) = U_S(x) = \frac{1}{2}\,k\,x^2\,.$$

In this final expression for the spring's potential energy we have

- Discarded all pretense that the system is anything but one-dimensional.

- Explicitly set the potential energy to vanish at the origin, *i.e.*,

$$U_S(x_{\text{reference}}) = 0 \quad \text{at} \quad x_{\text{reference}} = x_{\text{equilibrium}} = 0\,.$$

Q: Okay, we've seen how one goes from a conservative force to its associated potential energy function. Can one go the other way, *i.e.*, from a particular potential energy function to an expression for the associated conservative force?

A: You betcha! ... We'll do it in the next chapter.

---

[5]This correspondence is obtained and enforced by the choice of the origin of coordinates.

Chapter 26

Dynamics from Potential Energy

Suppose that a potential energy function, $U(\vec{r})$, is known, while its associated conservative force, $\vec{F_c}$, remains as yet unknown. By virtue of the fact that $U(\vec{r})$ is a potential energy function, its differences are uniquely specified by the work done by $\vec{F_c}$:

$$U_f - U_i = U(\vec{r}_f) - U(\vec{r}_i) \equiv -\int_{\vec{r}_i}^{\vec{r}_f} \vec{F_c} \cdot d\vec{s}.$$

IF the initial and final points are but an infinitesimal [single] step apart, $\vec{r}_f = \vec{r}_i + d\vec{r}$, THEN the LHS becomes[1]

$$U(\vec{r}_f) - U(\vec{r}_i) = U(\vec{r}_i + d\vec{r}) - U(\vec{r}_i) = dU(\vec{r}_i),$$

and the RHS reduces to

$$-\int_{\vec{r}_i}^{\vec{r}_i + d\vec{r}} \vec{F_c}(\vec{r}) \cdot d\vec{s} = -\vec{F_c}(\vec{r}_i) \cdot d\vec{r}.$$

Enforcing the equality LHS = RHS, without prescribing the initial point, yields

$$dU(\vec{r}) = -\vec{F_c}(\vec{r}) \cdot d\vec{r}.$$

ASIDE: One may be sorely tempted to write

$$\vec{F_c} = -\frac{dU(\vec{r})}{d\vec{r}}.$$

However, one must not succumb to this temptation because the dot product cannot be so cavalierly inverted, and differentiating with respect to a vector is as *dodgy* as dividing by one, *cf.* Chapter 6.

The expression for the work done by the conservative force acting through the infinitesimal displacement may be expanded in Cartesian coordinates using

$$\vec{F_c}(\vec{r}) = \left(F_{cx}(\vec{r}), F_{cy}(\vec{r}), F_{cz}(\vec{r}) \right), \quad \text{and} \quad d\vec{r} = \left(dx, dy, dz \right).$$

Thus,

$$dU(\vec{r}) = -\vec{F_c} \cdot d\vec{r} = - \left[F_{cx}\, dx + F_{cy}\, dy + F_{cz}\, dz \right].$$

IF the infinitesimal path extends solely in the x-direction, $d\vec{r} = dx\,\hat{\imath}$, THEN

$$dU(\vec{r}) \Big|_{d\vec{r} = dx\,\hat{\imath}} = - \left[F_{cx}\, dx + 0 + 0 \right] = -F_{cx}\, dx.$$

This equation may be solved for the x-component of the conservative force field. Explicitly, this reads [in non-standard notation],

$$F_{cx} = -\frac{dU(\vec{r})\Big|_{d\vec{r} = dx\,\hat{\imath}}}{dx}.$$

---

[1] Admittedly, the last equality is somewhat formal.

The conventional way to remind [or inform] ourselves that the path consists of a single-step path in the x-direction, is by writing

$$F_{cx} = -\frac{\partial U(x,y,z)}{\partial x}.$$

The symbol ∂ is read as *dye*, as in "*dye* U, *dye* x" for $\frac{\partial U}{\partial x}$.

Extension of the above arguments to single-step paths in the y- and z-directions yields

$$F_{cy} = -\frac{\partial U(x,y,z)}{\partial y}, \qquad \text{and} \qquad F_{cz} = -\frac{\partial U(x,y,z)}{\partial z},$$

for the y- and z-components of the conservative vector force field. The notion of single-step paths in the various coordinate directions is fully developed in the mathematical formalism of partial derivatives.

Partial Derivatives

The partial derivative of the function $f(x,y,z)$, with respect to x, is equal to the ordinary derivative with respect to the argument x, with the proviso that, for the purpose of taking this derivative, the y and z variables are held constant. Formally,

$$\frac{\partial f(x,y,z)}{\partial x} = \lim_{\epsilon \to 0} \frac{f(x+\epsilon, y, z) - f(x,y,z)}{\epsilon}.$$

The clause "*for the purpose of taking this derivative*" is to allow for successive partial differentiations with respect to different variables. A mathematical theorem asserts that [under certain restrictive conditions] the order of successive partial derivatives may be interchanged without affecting the result, *viz.*,

IF the function $f(x,y,z)$ is C^2, meaning that its second partial derivatives are continuous functions in their own right, THEN

$$\frac{\partial^2 f}{\partial x \, \partial y} = \frac{\partial^2 f}{\partial y \, \partial x}, \quad \frac{\partial^2 f}{\partial x \, \partial z} = \frac{\partial^2 f}{\partial z \, \partial x}, \quad \frac{\partial^2 f}{\partial y \, \partial z} = \frac{\partial^2 f}{\partial z \, \partial y}, \quad etc.$$

EXAMPLE [*Partial Derivatives of a Scalar Function of Three Variables*]

Consider a simple, but non-trivial, scalar function of three variables, $f(x,y,z) = x^2 \, y^3 \, z^4$. The partial first derivatives of f are

$$\frac{\partial f(x,y,z)}{\partial x} = 2\,x\,y^3\,z^4, \quad \frac{\partial f(x,y,z)}{\partial y} = 3\,x^2\,y^2\,z^4, \quad \text{and} \quad \frac{\partial f(x,y,z)}{\partial z} = 4\,x^2\,y^3\,z^3,$$

respectively. There are three [distinct, non-trivial] first partial derivatives of the function because it depends on three variables.

There are nine $[9 = 3 \times 3]$ second partial derivatives of f. Not all of these are distinct, since $f(x, y, z)$ is C^2. Four of these second partial derivatives are quoted below.

$$\frac{\partial^2 f(x, y, z)}{\partial x^2} = \frac{\partial}{\partial x} \left[\frac{\partial f(x, y, z)}{\partial x} \right] = \frac{\partial}{\partial x} \left[2 x y^3 z^4 \right] = 2 y^3 z^4 \,,$$

$$\frac{\partial^2 f(x, y, z)}{\partial x \, \partial y} = \frac{\partial}{\partial x} \left[\frac{\partial f(x, y, z)}{\partial y} \right] = \frac{\partial}{\partial x} \left[3 x^2 y^2 z^4 \right] = 6 x y^2 z^4 \,,$$

$$\frac{\partial^2 f(x, y, z)}{\partial z \, \partial y} = \frac{\partial}{\partial z} \left[\frac{\partial f(x, y, z)}{\partial y} \right] = \frac{\partial}{\partial z} \left[3 x^2 y^2 z^4 \right] = 12 x^2 y^2 z^3 \,,$$

$$\frac{\partial^2 f(x, y, z)}{\partial y \, \partial z} = \frac{\partial}{\partial y} \left[\frac{\partial f(x, y, z)}{\partial z} \right] = \frac{\partial}{\partial y} \left[4 x^2 y^3 z^3 \right] = 12 x^2 y^2 z^3 \,.$$

This subset of second-order partials illustrates the necessity of keeping careful track of the variables held constant in successive partial differentiations, and instantiates the interchangeability property.

GRADIENT OPERATOR   The gradient [operator] is a vector combination of partial derivatives. Acting on functions expressed in terms of Cartesian coordinates in three-dimensional space, the gradient assumes the form

$$\vec{\nabla} \equiv \hat{\imath} \, \frac{\partial}{\partial x} + \hat{\jmath} \, \frac{\partial}{\partial y} + \hat{k} \, \frac{\partial}{\partial z} \,.$$

The gradient acting on a particular scalar function f reads:

$$\vec{\nabla} f(x, y, z) = \left[\hat{\imath} \, \frac{\partial}{\partial x} + \hat{\jmath} \, \frac{\partial}{\partial y} + \hat{k} \, \frac{\partial}{\partial z} \right] f(x, y, z) = \left(\frac{\partial f}{\partial x} \,, \frac{\partial f}{\partial y} \,, \frac{\partial f}{\partial z} \right) \,.$$

Evaluation of the gradient of a scalar function at a point, $\mathcal{P}$, yields the magnitude and direction of the maximum rate of change of the function at $\mathcal{P}$.

ASIDE: A skiing analogy reveals the significance of the gradient, as (almost) every point on a ski hill lies on a unique "fall line."[2]

Local sections of the fall line in the neighbourhood of a point lie parallel to the direction in which a ball would roll were it released from rest at the point. It is amazing that these local "path beginnings" [GRADIENT VECTORS] stitch themselves together[3] into fall lines.

- The direction of the fall line at a point on the snowy surface is parallel to the gradient of the local elevation function (*i.e.*, the height above sea level or some other reference point) of the ski hill at that point.

- The magnitude of the gradient of the local elevation function at a particular point on the hill is proportional to the steepness of the fall line there.

Actual skiers' trajectories, as they *schuss* down the slope, do not quite coincide with the fall lines on account of inertial effects.

---

[2] Even when the hill has many twists and turns, and chutes and moguls, our legs "know"—and sometimes they tell us!—about the fall line.

[3] This is a distinctly non-trivial fact. Other examples of the integration of local vector fields into line-like structures are: the formation of streamlines in moving fluid, discussed in VOLUME II, and electric and magnetic field lines, VOLUME III.

Q: Must Cartesian coordinates be used to construct/formulate the gradient operator?

A: Nope. The gradient has a geometric meaning which transcends the coordinate scheme employed for its description. For example, in terms of spherical polar coordinates, (r, θ, ϕ), the gradient operator reads

$$\vec{\nabla} \equiv \hat{r}\,\frac{\partial}{\partial r} + \hat{\theta}\,\frac{1}{r}\,\frac{\partial}{\partial \theta} + \hat{\phi}\,\frac{1}{r\,\sin(\theta)}\,\frac{\partial}{\partial \phi}\,.$$

The gradient operator acting upon a scalar function f [written in terms of spherical polar coordinates (r, θ, ϕ)], yields the vector field

$$\vec{\nabla} f(r, \theta, \phi) = \hat{r}\,\frac{\partial f}{\partial r} + \hat{\theta}\,\frac{1}{r}\,\frac{\partial f}{\partial \theta} + \hat{\phi}\,\frac{1}{r\,\sin(\theta)}\,\frac{\partial f}{\partial \phi}\,.$$

The gradient can be expressed, albeit in more complicated form, in other sets of coordinates.

The foregoing analysis has irrefutably demonstrated that the gradient operator is the means of ascertaining the conservative force underlying a given potential energy function.

$$\vec{F}_{\text{c}}(\vec{r}) \equiv -\vec{\nabla} U(\vec{r})\,.$$

This equation holds true in all well-defined coordinate systems.

Dynamical Inferences from the Potential Energy Function

$\vec{\nabla}$ The gradient of the potential energy function [the steepest local rate of change] at a particular point in space determines the magnitude and the direction of the conservative force acting there.

$\vec{\nabla}$ The points at which the gradient vanishes, $\vec{\nabla} U(\vec{r}) = \vec{0}$, are significant for two related reasons. The conservative force vanishes [tautologically] AND the potential energy function is [instantaneously] flat, at such points. IF the net external force consists of the conservative force responsible for the potential energy function, THEN a particle placed at rest at such a point shall remain at rest at that point.

> ASIDE: With recourse once more to the skiing analogy: One may stand without planting one's poles or digging in one's edges at the very peak [where it is flat] and at the chalet [where it is also flat]. In addition, there are often flat spots [on the tops of moguls perhaps] where one may pause (momentarily) while changing direction, or (longer) to catch one's breath.

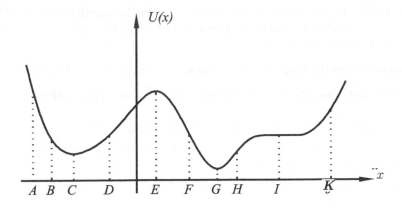

FIGURE 26.1 A One-Dimensional Potential Energy Ski Hill and Various *Steineins*

EXAMPLE [*A Potential Energy Function in One Dimension*]

The *Steineins* had an extended family gathering at a [1-d] ski hill. Along with *Albert, Bertal, Clande, Declan, Ecland,* viz., A, B, C, D, E, respectively, whom we have already met, were cousins *Fergus* (F), *Gusfer* (G), *Habodic* (H), *Ichabod* (I), and[4] *Kjerstin* (K). A cross-sectional view of the hill appears in Figure 26.1, with the positions of the various Steineins at a certain instant indicated.

Q1: Why the restriction to 1-d?
A1.1: It is difficult to draw 2-d or 3-d potential energy functions. Often one relies on projections of realistic higher-dimensional potential energy functions onto smaller subspaces.
A1.2: In 1-d, the CRITICAL POINTS of the potential energy function, $U(x)$, are simply classified into the categories MAXIMUM, MINIMUM, INFLECTION, and DISCONTINUOUS. In higher dimensions, the possibility of saddle points arises, complicating the analysis.

Q2: What can we infer about the dynamics of particles out on the ski hill?

A2: Let's consider carefully the potential energy function in neighbourhoods inhabited by the individual Steineins. By *fiat* we declare that all other forces cancel, and thus the conservative force giving rise to the potential energy is equal to the net force.

A   A particle adjacent to *A*lbert experiences a force acting to the right.

B   A particle beside *B*ertal experiences a force which acts to the right. The strength of this force is less than that experienced at *A*.

C   The force on a particle close to *C*lande vanishes under our assumptions.

   ○ IF the particle is at rest at *C*, THEN it will remain at *C* indefinitely.

   ○ IF the particle is displaced slightly to the left of *C*, THEN it will experience a force to the right [toward *C*].

   ○ IF the particle is displaced slightly to the right of *C*, THEN the force on it acts toward the left [toward *C*, once again].

---

[4]Cousin *Jerstink* did not ski that day as she had a cold. Remaining at her home, she wrapped herself in her warmest sweater, made herself a pot of tea, and indulged herself by reading over her favourite physics text.

The potential energy function has a point of **stable equilibrium** at C. The qualifier *stable* indicates that small displacements from C are met with a restoring force directed back toward C.

D   A particle directly alongside $\boldsymbol{D}$*eclan* feels a weakish force acting to the left.

E   At the exact spot where $\boldsymbol{E}$*cland* stands, the force is equal to ZERO.

- ○ IF a particle is at rest at $\boldsymbol{E}$, THEN it will remain at $\boldsymbol{E}$.
- ○ IF a particle is displaced slightly to the left of $\boldsymbol{E}$, THEN it will experience a force to the left, *i.e.*, away from $\boldsymbol{E}$.
- ○ IF it is displaced slightly to the right, THEN it will be pushed to the right.

The potential energy function has a point of **unstable equilibrium** at $\boldsymbol{E}$. *Unstable* indicates that particles undergoing small displacements from $\boldsymbol{E}$ experience forces which drive them further away.

F   A fellow standing with $\boldsymbol{F}$*ergus* feels a rightward force.

G $\boldsymbol{G}$*usfer* occupies another point of static equilibrium. The asymptotic trends in Figure 26.1 suggest that G resides at the $\boldsymbol{G}$LOBAL MINIMUM of the potential energy function.   [The chalet is where the skiers $\boldsymbol{G}$ather at the end of the day.]

H   A particle hanging out with $\boldsymbol{H}$*abodic* endures a force acting to the left.

I   A particle in $\boldsymbol{I}$*chabod*'s immediate vicinity experiences vanishing force.

- ○ IF the particle is at rest at $\boldsymbol{I}$, THEN it will remain at $\boldsymbol{I}$.
- ○ IF the particle is displaced slightly from $\boldsymbol{I}$ to EITHER the right OR the left, THEN it will still experience ZERO NET force.

Thus $\boldsymbol{I}$ lies within a region of **neutral equilibrium**. *Neutral* indicates that small displacements from $\boldsymbol{I}$ are met with dynamical indifference.

J $\boldsymbol{J}$*erstink* was just not able to come on this outing.

K   Finally, particles keeping close to $\boldsymbol{K}$*jerstin* experience forces to the left.

Chapter 27

Total Mechanical Energy

In the preceding chapters, two forms of mechanical energy, KINETIC and POTENTIAL, were identified and discussed.

Kinetic	$K = \frac{1}{2}mv^2$	Energy possessed by a particle by virtue of its motion through space.
Potential	$U = U(\vec{r})$	Energy possessed by a particle by virtue of its position in space.

Further consideration inspires us to subsume these into the TOTAL MECHANICAL ENERGY.

TOTAL MECHANICAL ENERGY   The total mechanical energy, E, of a particle[1] is the sum of its kinetic and potential energies, *viz.*,

$$E = K + U = \frac{1}{2}mv^2 + U(\vec{r}).$$

Provided that all of the forces which act are conservative,[2] the total mechanical energy is a CONSTANT OF THE MOTION. That is, the value of the total mechanical energy remains unchanged as the system undergoes dynamical evolution.

Q: Are you kidding? How does this work?

A: In the manner shown in the rigorous proof below.

PROOF that the Mechanical Energy is Constant When all Forces are Conservative

Suppose that a number of conservative forces act on a single particle of mass m. Each force possesses an associated potential energy function, and these are added together to yield an aggregate [TOTAL] potential energy function $U(\vec{r})$. At initial time t_i, the particle is located at position $\vec{r}_i$ and is moving with velocity $\vec{v}_i$. At final time t_f, the particle is found at $\vec{r}_f$ with velocity $\vec{v}_f$. The total mechanical energy has as its initial and final values:

$$E_i = K_i + U_i, \qquad \text{and} \qquad E_f = K_f + U_f,$$

respectively. Therefore, the change in the total mechanical energy of the particle during the time interval from $t_i \to t_f$ is

$$\Delta E = E_f - E_i = \left[K_f + U_f\right] - \left[K_i + U_i\right] = \left\{K_f - K_i\right\} + \left\{U_f - U_i\right\}.$$

---

[1]The dynamics of systems will be discussed anon beginning in Chapter 30.

[2]In fact, the condition is less restrictive than that stated. Better phrasing would be: "provided that zero net work is performed by any non-conservative forces present." At this stage, however, we'll insist that only conservative forces act in order to render more transparent the proof of the result.

where θ is measured anti-clockwise from the bottom of the loop, as shown in Figure 27.1. The initial value of the total mechanical energy [under these assumptions] is

$$E_i = K_i + U_{g,i} = \frac{1}{2} M \left| \vec{v}_0 \right|^2 + M g \times 0 = \frac{1}{2} M v_0^2 .$$

When the carriage is at angle θ, its total mechanical energy is

$$E_\theta = K_\theta + U_{g,\theta} = \frac{1}{2} M v_\theta^2 + M g R \left(1 - \cos(\theta) \right) .$$

Energy conservation ensures that the value of the total mechanical energy is the same at all points along the trajectory of the carriage. So,

$$E_i = E_\theta \qquad \Longrightarrow \qquad \frac{1}{2} M v_0^2 = \frac{1}{2} M v_\theta^2 + M g R \left(1 - \cos(\theta) \right) ,$$

and thus

$$v_\theta^2 = v_0^2 - 2 g R \left(1 - \cos(\theta) \right) .$$

This is the same relation for the speed–squared as was [laboriously] derived in Chapter 19!

[A hint of] **The power and generality of the energy approach to dynamics**

A particle of mass M, the constituent of a 1-d physical system, is subject to only conservative forces and possesses an aggregate potential energy function $U(x)$. The constant value of the total mechanical energy, E, is known [by experimental design or by measurement].

The conserved total mechanical energy is split between kinetic and potential forms in an *a priori* unknown manner,

$$E = \frac{1}{2} M v^2 + U(x) .$$

Algebraically solving this equation for the speed [velocity[3]] yields

$$v = \sqrt{\frac{2}{M} \left(E - U(x) \right)} .$$

The velocity is the time rate of change of the position function,

$$v = \frac{dx}{dt} \qquad \Longrightarrow \qquad \frac{dx}{\sqrt{E - U(x)}} = \sqrt{\frac{2}{M}} \, dt .$$

In the latter equality, the spatial and temporal dependencies have been separated. In principle,[4] these terms may be self-consistently integrated to yield an implicit expression, $t(x)$, for the trajectory of the particle. Furthermore, it may be possible to invert this to yield $x(t)$ and thus obtain (via energetic analysis) an explicit solution for the trajectory of the particle.

This IS amazing, eh?

---

[3] One must be careful, because the overall sign (in front of the square root) is ambiguous.
[4] CAVEAT: That which is possible in principle may well be impossible to effect in practice.

Chapter 28

Non-Conservative Forces and Power

In the last chapter, we demonstrated that IF only conservative forces act, THEN the total mechanical energy is conserved. A trivial extension of this result allows for non-conservative forces, provided that their net contribution to the total mechanical work is ZERO. Thus, a restatement of the last chapter's result is: IF only conservative forces act, OR all non-conservative forces combine to do ZERO net mechanical work, THEN the total mechanical energy is conserved.

Q: What about non-conservative[1] forces (*i.e.*, kinetic friction) in general?

A: Consider what happens when a block of mass m slides along a flat horizontal frictional surface from an initial point $\mathcal{P}$ to a final point $\mathcal{Q}$.

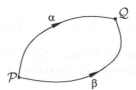

FIGURE 28.1 Sliding a Block from $\mathcal{P}$ to $\mathcal{Q}$ along Paths α and β

The normal force of contact between the surfaces of the block and plane is $N = m g$, and hence the magnitude of the force of kinetic friction acting on the block is

$$f_{\text{K}} = \mu_{\text{k}} N = \mu_{\text{k}} m g .$$

The work done by the force of kinetic friction is

$$W_{\mathcal{P}\mathcal{Q}}\big[\vec{f}_{\text{K}}\big]_{\text{path}} = \int_{\mathcal{P}}^{\mathcal{Q}} \vec{f}_{\text{K}} \cdot d\vec{s} .$$

The integrand, and hence the integral, is negative definite,

$$\vec{f}_{\text{K}} \cdot d\vec{s} = f_{\text{K}}\, ds\, \cos(\pi) = -\mu_{\text{k}}\, m\, g\, ds ,$$

as the force of kinetic friction is always directed opposite to the instantaneous motion of the block. The factors μ_{k}, m, and g are all constant, and thus

$$W_{\mathcal{P}\mathcal{Q}}\big[\vec{f}_{\text{K}}\big]_{\text{path}} = -\mu_{\text{k}}\, m\, g \int_{\mathcal{P}}^{\mathcal{Q}} ds = -\mu_{\text{k}}\, m\, g\, S_{\mathcal{P}\mathcal{Q},\text{path}} .$$

The symbol $S_{\mathcal{P}\mathcal{Q},\text{path}}$ denotes the arc length of the particular path from $\mathcal{P}$ to $\mathcal{Q}$. Kinetic friction is a non-conservative force.

---

[1] Dare we term these "liberal" forces?

ASIDE: Consider two different paths, $\{\alpha, \beta\}$, leading from $\mathcal{P}$ to $\mathcal{Q}$. Separate computations of the work done by kinetic friction along the paths yield

$$W_{\mathcal{PQ}}\big[\vec{f}_{K}\big]_{\alpha} = -\mu_{k}\, m\, g\, S_{\mathcal{PQ},\alpha} \qquad \text{and} \qquad W_{\mathcal{PQ}}\big[\vec{f}_{K}\big]_{\beta} = -\mu_{k}\, m\, g\, S_{\mathcal{PQ},\beta}\,,$$

respectively. As there is no compelling reason to expect the arc lengths of the two paths to be equal, it must be generally true that

$$W_{\mathcal{PQ}}\big[\vec{f}_{K}\big]_{\alpha} \neq W_{\mathcal{PQ}}\big[\vec{f}_{K}\big]_{\beta}\,.$$

The amount of work done by kinetic friction depends on the path taken by the block in its transit from $\mathcal{P}$ to $\mathcal{Q}$. Therefore, **kinetic friction cannot possibly possess an associated potential energy.**

In situations where both conservative and non-conservative forces are active, the total mechanical energy of the system is, generally, NOT conserved.

TOTAL ENERGY–WORK THEOREM   The amount by which the total mechanical energy of a system changes is equal to the net work done on the system by non-conservative forces,

$$\Delta E = E_f - E_i = W_{if}\big[\vec{F}_{\text{NET,non-C}}\big]\,.$$

PROOF of the Total Energy–Work Theorem

The change in the total mechanical energy during a time interval $\Delta t = t_f - t_i$, is [tautologically] the difference between the values of this quantity at the end and beginning of the interval:

$$\Delta E = E_f - E_i = \big[K_f - K_i\big] + \big[U_f - U_i\big]\,.$$

The total kinetic energy, K, is the sum of the kinetic energies of all of the particles in the system, while the total potential energy, U, is the aggregate of the various potential energy functions, each of which is associated with a conservative force.

It follows as a direct consequence of the WORK–ENERGY THEOREM, introduced in Chapter 23, that $K_f - K_i = W_{if}\big[\vec{F}_{\text{NET}}\big]$. WLOG, the net force, $\vec{F}_{\text{NET}}$, can be expressed as the sum of the net conservative and net non-conservative forces. Thus,

$$K_f - K_i = W_{if}\big[\vec{F}_{\text{NET}}\big] = W_{if}\big[\vec{F}_{\text{NET,C}} + \vec{F}_{\text{NET,non-C}}\big]$$
$$= W_{if}\big[\vec{F}_{\text{NET,C}}\big] + W_{if}\big[\vec{F}_{\text{NET,non-C}}\big]\,.$$

Meanwhile, the difference in potential energies in passing from the initial position to the final position[2] is determined by the work done by the associated conservative forces:

$$U_f - U_i \equiv -W_{if}\big[\text{NET,Cons}\big] = -W_{if}\big[\vec{F}_{\text{NET,C}}\big]\,.$$

Plugging these two results into the expression for ΔE yields

$$\Delta E = \big[K_f - K_i\big] + \big[U_f - U_i\big]$$
$$= W_{if}\big[\vec{F}_{\text{NET,C}}\big] + W_{if}\big[\vec{F}_{\text{NET,non-C}}\big] - W_{if}\big[\vec{F}_{\text{NET,C}}\big]$$
$$= W_{if}\big[\vec{F}_{\text{NET,non-C}}\big]\,.$$

Hence, the change in the total mechanical energy of a physical system is exactly equal to the net amount of work performed on the system by non-conservative forces!

---

[2]On occasion, we shall slip and say initial and final "states." This nomenclature admits more general situations in which a single position descriptor is insufficient to properly characterise the physical attributes of the system.

ASIDE: In many circumstances, these non-conservative forces are resistive [*e.g.*, kinetic friction, drag forces, *etc.*]. Such forces as these act to diminish the total mechanical energy of the system and are commonly termed **dissipative**.

The applied forces exerted by external agents on various and sundry blocks and sleds on planes and inclines have also been of the non-conservative type.[3]

Our studies of work and energy have involved the initial and final states of various systems, with little or no attention given to time. We have determined changes of energy, but have not yet converted these finite changes into average and instantaneous rates of change.

POWER Power is the rate at which work is done by a force. In more general terms: **power is the rate at which energy is transferred**.

The dimensions of power are energy per unit time. The SI unit is the watt, so-named in honour of James Watt (1736–1819). Watt was the fellow who developed/re-engineered the steam engine and thus *powered* the Industrial Revolution.

$$\text{One watt} = 1 \text{ W} = 1 \text{ J/s} = 1 \text{ joule per second.}$$

ASIDE: Mao Zedong's aphorism, "*Power flows from the barrel of a gun*," does not refer to energy transfer *per se*.

AVERAGE POWER The average power is the amount of work done during the time interval Δt, divided by Δt, *viz.*,

$$P_{\text{av}} = \frac{W_\Delta[\vec{F}]}{\Delta t} = \frac{\int_{\vec{r}_i}^{\vec{r}_f} \vec{F} \cdot d\vec{s}}{t_f - t_i}.$$

INSTANTANEOUS POWER The instantaneous power is the limiting value of the average power through an infinitesimally short time interval, *i.e.*, $\Delta t \to 0$,

$$P = \lim_{\Delta t \to 0} P_{\text{av}} = \frac{dW}{dt}.$$

The expression for instantaneous power can be further simplified by recalling the differential amount of work done by a force acting through a differential displacement [necessarily occurring through some vanishing time interval],

$$dW = d\left[\int_{\vec{r}_i}^{\vec{r}_f} \vec{F} \cdot d\vec{s}\right] = \vec{F} \cdot d\vec{r}.$$

In this last expression, $d\vec{r}$ is the difference $\vec{r}_f - \vec{r}_i$ in the limit as $\Delta t \to 0$. With this simplification for dW, the instantaneous power becomes

$$P = \frac{dW}{dt} = \frac{\vec{F} \cdot d\vec{r}}{dt} = \vec{F} \cdot \frac{d\vec{r}}{dt}.$$

The velocity is the instantaneous time rate of change of position, and hence

$$P = \frac{dW}{dt} = \vec{F} \cdot \vec{v}.$$

---

[3]Perhaps forces such as these might someday be called *augmentative*.

EXAMPLE [*Drag Force, Dissipation, and Power*]

A small [smooth and spherical] pebble is dropped through still air on a calm day. The pebble experiences two forces of significance: its weight, and linear drag air resistance.

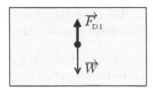

FIGURE 28.2   A Dropped Pebble Falling through Draggy Air

The system's dynamics are governed by N2,

$$F_{\text{NET}} = M g - b v ,$$

where $M g$ is the weight of the pebble, while b is a constant drag coefficient. We've dispensed with the vector designations, as the motion is strictly one-dimensional, and ensured that the drag force is acting resistively on the falling particle through our choice of signs.

The total mechanical energy of the system at a particular instant is given by the formula

$$E = K + U(x) = \frac{1}{2} M v^2 + U(x) ,$$

in which the values of E, x, and v are all expected to be time-dependent. The time rate of change of E may be consistently determined via the following sequence of steps. [This argument relies on properties of derivatives, definitions of kinematical quantities, the realisation that a conservative force field is determined by the gradient of its associated potential energy function, and N2.]

$$\frac{dE}{dt} = M v \frac{dv}{dt} + \frac{dU}{dx} \frac{dx}{dt} , \quad \text{with} \quad \frac{dx}{dt} = v , \quad \text{and} \quad \frac{dv}{dt} = a , \quad \text{while} \quad \frac{dU}{dx} = -F_{\text{NET,C}} ,$$

$$= M v a - F_{\text{NET,C}} v$$

$$= \left[M a - F_{\text{NET,C}} \right] v , \qquad \left[\text{however, by N2:} \quad M a = F_{\text{NET,C}} - b v \right]$$

$$= -b v^2 .$$

Two salient features of this result are noted below.

< 0 The time rate of change of the total energy is negative, as mechanical energy is converted to heat through the agency of the dissipative drag force.

v^2 The rate of energy loss is proportional to the speed–squared. [This effect drives the public policy arguments for a general lowering of highway speed limits to reduce the rate of consumption of fossil fuels.]

Chapter 29

Momentum and Impulse

Momentum is [nearly] as important as energy. Through consideration of momentum we shall finally be able to *throw off the shackles* restricting us to the analysis of single particles [and isolated particles within simple systems]. Furthermore, the "momentum point of view" will prove to be invaluable as we model the effective dynamics of more complicated systems of particles and rigid bodies.

LINEAR MOMENTUM   The linear momentum of a particle of mass M, moving with velocity $\vec{v}$ (measured with respect to some IRF), is

$$\vec{p} = M\,\vec{v}.$$

It is common practice to elide the adjective "linear" and simply speak of "the momentum" of a particle.

Momentum is a vector quantity, and its SI units are

$$[\vec{p}] = [M\,\vec{v}] = \text{kg} \cdot \text{m/s} = \text{N} \cdot \text{s}.$$

The $\text{kg} \cdot \text{m/s}$ form is obtained directly from the definition, while its "newton $\cdot$ seconds" re-expression is intended to foreshadow the introduction of **impulse**.

Consider a single particle of mass M, subjected to a net [external] force, $\vec{F}_{\text{NET}}$.

ASIDE: An implicit assumption which has heretofore commonly been made [and occasionally mentioned] is that the particle's mass is constant. The particle is neither gaining mass [*e.g.*, in the manner of a raindrop accreting water vapour] nor is it losing mass [*e.g.*, a rocket consuming and ejecting its fuel]. We draw attention to this assumption here while laying the groundwork for its relaxation!

It follows from N2 [written "backwards"] that

$$\vec{F}_{\text{NET}} \equiv M\,\vec{a} = M\,\frac{d\vec{v}}{dt} = \frac{d(M\,\vec{v})}{dt} = \frac{d\vec{p}}{dt},$$

provided that the mass is indeed constant, as was assumed. Thus, in this restricted case:

The time rate of change of momentum is equal to the net force!

Here are two items of note.

○   While far from obvious, it is nevertheless true that the above statement holds even when the mass of the particle is changing!

○   This re-statement in terms of momentum is a more powerful and general expression of N2 than that employed from Chapter 10 onward to this point. In fact, Newton's original formulation of the Second Law of Motion [for particles] was

$$\frac{d\vec{p}}{dt} = \vec{F}_{\text{NET}}, \qquad \text{rather than} \qquad M\,\vec{a} = \vec{F}_{\text{NET}}.$$

With momentum in hand, let's define IMPULSE.

IMPULSE   The impulse produced by a force, $\vec{F}$, is the time integral of the force as it acts through a time interval $t_i \rightarrow t_f$:

$$\vec{I}_{if}\left[\,\vec{F}\,\right] = \int_{t_i}^{t_f} \vec{F}\, dt\,.$$

Calculating the impulse for a 3-d force which depends on time, position, velocity, and particle properties may be daunting.

Two important properties of impulse are that

   ○   Impulse is a vector quantity.

   ○   The PRINCIPLE OF LINEAR SUPERPOSITION is operative. *I.e.*, the impulse of a sum of forces is equal to the sum of the impulses of the forces acting individually.

───────────────────

IMPULSE–MOMENTUM THEOREM   The impulse imparted by the net force acting on a particle throughout a time interval, $t_i \rightarrow t_f$, is equal to the change in the momentum of the particle during the same time interval, *viz.*,

$$\vec{I}_{if}\left[\,\vec{F}_{\text{NET}}\,\right] = \Delta\vec{p}\,.$$

PROOF of the Impulse–Momentum Theorem

Consider

$$\vec{I}_{if}\left[\,\vec{F}_{\text{NET}}\,\right] = \int_{t_i}^{t_f} \vec{F}_{\text{NET}}\, dt\,.$$

However, by N2, $\vec{F}_{\text{NET}} = \frac{d\vec{p}}{dt}$, and thus

$$\vec{I}_{if}\left[\,\vec{F}_{\text{NET}}\,\right] = \int_{t_i}^{t_f} \frac{d\vec{p}}{dt}\, dt\,.$$

Applying the inverse chain rule, with careful attention paid to the limits of integration,

$$\vec{I}_{if}\left[\,\vec{F}_{\text{NET}}\,\right] = \int_{\vec{p}_i}^{\vec{p}_f} d\vec{p} = \vec{p}\,\Big|_{\vec{p}_i}^{\vec{p}_f} = \vec{p}_f - \vec{p}_i = \Delta\vec{p}$$

completes the proof.

───────────────────

Impulse is to Momentum   as   Work is to Kinetic Energy.

A force acting in 1-d and exhibiting complicated time dependence is illustrated in Figure 29.1. This curve is intended to be suggestive of the force exerted by a bat upon a baseball when a "hit" occurs. There are three salient properties possessed by the force:

♭   **Slow Onset**   Photographs obtained using very high speed cameras show that struck baseballs deform considerably in the early stages of collision. This has the effect of lowering the force of contact [while increasing the area of contact] between the bat and the ball.

♭   **Large Peak Value**   A tremendous force is exerted by the bat. The peak value occurs around the time that the ball is maximally compressed/deformed.

> ASIDE: The force of the bat on the ball is [obviously] zero both before the onset of the collision and after it ends. If the force is assumed to be [mathematically speaking] CONTINUOUS, then [according to the MEAN VALUE THEOREM] a maximum value of the force must occur during the collision.

♭   **Abrupt Fall Off**   The ball springs back as it separates from the bat. The separation may occur before the ball has recovered its shape.

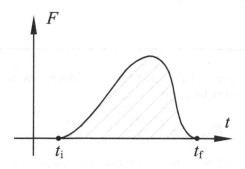

FIGURE 29.1   A Time-Varying Force in 1-d and Its Impulse

Perfect knowledge of a time-varying force requires an infinite amount of information. However, a single [meaningful] quantity that can characterise the force is its average value throughout a specified time interval.

AVERAGE FORCE   The average force is that constant force which produces the same impulse as the true force acting through the same time interval. These impulses, calculated separately, are:

$$\vec{I}_{if}\big[\vec{F}\,\big] = \int_{t_i}^{t_f} \vec{F}\,dt\,,$$

$$\vec{I}_{if}\big[\vec{F}_{av}\big] = \int_{t_i}^{t_f} \vec{F}_{av}\,dt = \vec{F}_{av}\int_{t_i}^{t_f} dt = \vec{F}_{av}\,\big[t_f - t_i\big] = \vec{F}_{av}\,\Delta t\,.$$

Equating these determines the value of the average force:

$$\vec{F}_{av} = \frac{\vec{I}_{if}\big[\vec{F}\,\big]}{\Delta t} = \frac{1}{\Delta t}\int_{t_i}^{t_f} \vec{F}\,dt\,.$$

A picture [see Figure 29.2] best conveys the idea that the average force is defined in terms of the impulse that it provides.

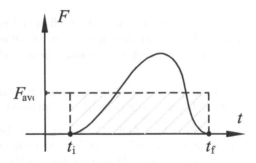

FIGURE 29.2   A Time-Varying Force in 1-d and Its Average Value

> ASIDE: The expression for the average force conforms exactly with what one might expect for the time average of a quantity. We will have more comments on this mode of averaging in the next chapter.

To obtain an estimate of the value of the **average net force**, one may avail oneself of the IMPULSE–MOMENTUM THEOREM to write

$$\vec{F}_{\text{NET,av}} = \frac{\vec{I}_{if}\left[\vec{F}_{\text{NET}}\right]}{\Delta t} = \frac{\Delta \vec{p}}{\Delta t}\,.$$

Of course this is correct and self-consistent, because it is the secant approximation to

$$\vec{F}_{\text{NET}} = \frac{d\vec{p}}{dt}\,.$$

Q: But how might all of this be useful?

A: Via the IMPULSE APPROXIMATION.

IMPULSE APPROXIMATION   The impulse approximation provides a rationale[1] for consistently discarding forces contributing relatively small amounts to the total impulse through some specified time interval. It reads:

> IF a particular force acts over a time interval which is much shorter than typical time scales [set by other forces], AND this force has a magnitude which is much larger than that of any other force acting [during this short time interval], THEN one may neglect the impulses produced by the other forces.

---

[1] Or perhaps a rationalisation.

EXAMPLE [*Baseball and the Impulse Approximation*]

Let's revisit the dynamics of a baseball as it is struck by a bat. Let's distinguish the regimes of: pre-, mid-, and post-contact, considering the time and force conditions necessary for application of the impulse approximation.

PRE-   The ball is pitched from the mound toward home plate. While the ball is in flight [approximately one-half second, or so], it is acted upon by gravity (its weight), and air resistance.

> **Q:** How did we arrive at this estimate of the time scale?
>
> **A:** A major-league pitcher can throw a fastball at about 90 miles per hour. This corresponds to $90 \times 5280/3600 = 132$ feet per second. The regulation distance from the mound to home plate is 66 feet. *Ergo*, it takes about one-half of a second[2] for the ball to travel from the pitcher's mound to home plate.

MID-   The time scale for the bat–ball collision is probably a few hundredths of a second or less. During the time that the bat is in contact with the ball, gravity and air resistance continue to act. The force of contact between the bat and the ball is much much greater than either the weight of the ball or the drag force which acts on the ball.

POST-   After the ball is struck it remains in flight for several seconds, passing over the playing field and sailing into the bleachers. Gravity and air resistance are the dominant forces on the ball in this regime.

This baseball scenario meets the conditions for application of the impulse approximation.

> ASIDE: A sports commentator may attempt to bolster a claim that a particular baseball player hits baseballs with prodigious force by quoting an impressive-sounding number. Whether consciously or not, the commentator is applying both the Impulse Approximation and the Impulse–Momentum Theorem.

MOMENTUM AND SYSTEMS

Up to this point we have exclusively focussed our attention on the kinematics and dynamics of single isolated point-like particles[3] or very simple systems of particles [*e.g.*, blocks joined by ropes or springs, or in contact with one another]. This is not enough! Nature teems with *systems of particles* whose dynamics we would like to study. To undertake these investigations requires the development of some high-powered machinery—with the salutary effect of leading us to a more complete understanding of the dynamics of single particles!

As our first step in this direction we note that earlier in this chapter, we defined the momentum of a single particle. In considering a system comprised of more than one particle, it is meaningful to construct the TOTAL MOMENTUM of the system by adding the momenta of the constituents.

---

[2]If it were PK pitching, the ball might reach home plate sooner if it were sent by parcel post!

[3]Recall that a particle, by definition, experiences only translational motion.

TOTAL MOMENTUM   Suppose that the system under consideration is comprised of $\mathcal{N}$ point(-like) particle constituents. The masses and velocities of the individual particles are denoted by M_j and $\vec{v}_j$ respectively, with the index, j, running from $1 \to \mathcal{N}$. [Each particle is assigned a distinct value.] The linear momentum of each constituent is

$$\vec{p}_j = M_j \, \vec{v}_j \, .$$

The aggregate, or TOTAL, momentum of the system is

$$\vec{P}_{\text{Total}} = \sum_{j=1}^{\mathcal{N}} \vec{p}_j = \vec{p}_1 + \vec{p}_2 + \vec{p}_3 + \ldots + \vec{p}_{\mathcal{N}} \, .$$

It is vitally important to realise that **this is a vector sum over the particles constituting the system.** The index j labels the particles, and each particle's momentum vector has one, two, or three components depending on the constraints upon its motion.

This is the very first time [in these notes] that we have dared to add together dynamical constructs applying to *different* particles.

> ASIDE: In discussing forces and dynamics, we have been quite dogmatic about distinguishing the object upon which each particular force acts and carefully summing only those forces which act on the same[4] object. We end this chapter with what appears to be an astonishing *volte face*, claiming that it is not only possible, but even desirable to construct the total momentum of a system of particles.
>
> *We'll see why over the course of the next few chapters.*

---

[4]Recall the care taken to avoid spurious and contradictory misapplications of N3 in Chapter 10.

Chapter 30

Systems of Particles and Centre of Mass

Until the last chapter, we exclusively focussed our studies on the kinematics and dynamics of [single, isolated, point-like] particles or very simple systems of particles. The study of more general systems of particles requires the introduction of various high-powered concepts and techniques. The first of these is the CENTRE OF MASS, henceforth denoted "CofM." There are two archetypical situations to examine:

∴   A system of discrete point(-like) particles

□   A rigid extended body, *i.e.*, a continuous distribution of matter

A priori these seem radically distinct. However, their deep similarities outweigh their superficial differences.

CENTRE of MASS   The centre of mass of a $\left\{ \begin{array}{c} \text{system of point(-like) particles} \\ \text{rigid extended body} \end{array} \right\}$ is the

mass-weighted average position of the $\left\{ \begin{array}{c} \text{system} \\ \text{body} \end{array} \right\}$.

∴   For a system of $\mathcal{N}$ point(-like) particles, the CofM is located at

$$\vec{R}_{\text{CofM}} = \frac{1}{M_{\text{Total}}} \sum_{i=1}^{\mathcal{N}} M_i \, \vec{r}_i \,,$$

where M_i and $\vec{r}_i$ denote the mass and position, respectively, of the ith point(-like) particle. The position of each particle is *weighted* by the particle's mass. The weighted sum is divided by the total mass to yield the average position.

[We will more fully explicate mass-weighted averaging later in this chapter.]

□   For a [rigid extended] body, the CofM position is

$$\vec{R}_{\text{CofM}} = \frac{1}{M_{\text{Total}}} \int dm \, \vec{r} \,,$$

where the integral is taken over the mass elements, dm, comprising the body. In order to make progress in this computation, the mass distribution must be expressed in terms of spatial quantities. The mass density function,[1] $\rho(\vec{r})$, effects the conversion of the integral over mass elements to an equivalent integral over the coordinate parameters describing the size and shape of the extended body,

$$dm = \rho(\vec{r}) \, d[\text{Volume}] \,.$$

---

[1] The density is not actually a function, but rather a **distribution**.

Q: How are these two archetypical situations—point(-like) and continuous—related, when they appear to be so different?

A: $\cdots \longleftarrow \square$

The sum over discrete particles arises as a special case of the integration over mass elements when the mass density is exactly ZERO everywhere except at the $\mathcal{N}$ discrete points corresponding to the locations of the point(-like) particles. Integration over such a distribution of mass leads naturally to the weighted sum.

A: $\cdots \longrightarrow \square$

To see that the treatment of the rigid extended body case follows from the discrete point-like mass formalism, we need only recollect the mathematical definition of the integral in terms of the limit of discrete sums over increasingly fine partitions.

> ASIDE: Imagine partitioning the continuous, extended, rigid body into a collection of $\mathcal{N}$ chunks. Provided that the chunks are small, we can model each of them as a [point(-like)^2] mass, ΔM_i, concentrated at a particular location,^{3} $\vec{r}_i$. The approximate location of the centre of mass is computed by the discrete sum over the $\mathcal{N}$–partition. A better approximation is obtained by REFINING the partition. This amounts to increasing the number of chunks to $\widetilde{\mathcal{N}} > \mathcal{N}$, with all $\widetilde{\Delta M_i} \leq \Delta M_i$. In the simultaneous limits $\mathcal{N} \to \infty$ and all $\Delta M_i \to 0$, the SUM formally becomes the INTEGRAL and converges^4 at the unique position identified as the location of the CofM of the extended distribution of matter.

A reason for emphasising this iterative approximation scheme is that one frequently encounters cases of extended bodies for which the integral(s) are not easily solvable in closed form. For such cases, one must settle for an approximate numerical evaluation of the location of the CofM obtained from a finite $\mathcal{N}$-partition of the body.

The [approximate] location of the CofM can be computed by application of the above formulae, and exploitation, whenever possible, of symmetries to render the computations simpler. We shall see numerous examples of these approaches in the remainder of this chapter.

> ASIDE: Experimentally, the approximate position of the CofM of an extended object can be crudely determined using a hook, some rope, and a plumb bob with a chalk line. The strategy is: suspend the body from [at least] three distinct and widely separated points in succession. Hang the plumb bob from the same point from which the body is suspended, and snap the chalk line. The chalk markings will provide three [or more] lines marking planes whose geometric intersection is at the unique centre of mass of the object.

....

EXAMPLE [*Various Centres of Mass Which are Amenable to Symmetry Analysis*]

A billiard ball   The CofM of a billiard ball is located at its geometric centre. [It is implicitly assumed that the mass distribution of the material composing the ball is spherically symmetric. The ball need not be homogeneous.^{5}]

---

^{2}We employ the term "chunks" to expressly avoid pathological situations in which the partitions are leaf-like or string-like.

^{3}Any sort of reasonable convention for associating the particular location to the chunk will suffice, including: the geometric centre of the chunk, its upper-left corner, its south-western-most part, *etc.*

^{4}Let's not consider the mathematical pathologies which might interfere with smooth convergence.

^{5}A substance or system is **homogeneous** if, and only if, its local properties are independent of position, *i.e.*, it is uniform in composition and structure.

A homogeneous cube   The CofM of such a cube is located at its geometric centre.

A rectangular sheet of plywood   The CofM is located at the geometric centre of the plane rectangle corresponding to the shape of the plywood.

> ASIDE: The plywood sheet is a 3-d solid slab-like object. It is usually safe to assume that it is uniform in the direction of its thickness, and just concern ourselves with the location of the CofM in the length and width directions, effectively reducing the problem to 2-d.

Two identical billiard balls   As noted above, each of the balls acts like a point mass located at its geometric centre. Thus, the CofM of the two equal-mass ball system lies halfway along the line which joins the centres of the two balls.

A perfect doughnut [mathematically a *torus*]   By symmetry, the CofM of the doughnut is located in the middle of the "hole."

[These last two centres of mass occurred at points in space where no actual mass resides.]

EXAMPLE [*Centre of Mass of Two Distinct Point-like Objects*]

We seek to determine the CofM for two balls of unequal mass under the assumptions that

- Each ball acts like a point mass located at its respective CofM.
- The CofM of each ball is coincident with its geometric centre.

Geometry and intuition together suggest that the CofM of the system lies along the line joining the centres, between the two balls and nearer to the heavier one.

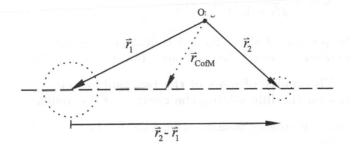

FIGURE 30.1   The Location of the Centre of Mass of Two Balls

Q: Can one prove these claims? Can one ascertain the CofM position unambiguously?

A: Yes and yes! For two point(-like) objects bearing masses $M_{1,2}$ and residing at $\vec{r}_{1,2}$ as illustrated in Figure 30.1, the position of the CofM is obtained by appropriate specialisation of the generic formula:

$$\vec{R}_{\text{CofM}} = \frac{1}{M_{\text{Total}}} \sum_{i=1}^{2} M_i \vec{r}_i = \frac{1}{M_1 + M_2} \left(M_1 \vec{r}_1 + M_2 \vec{r}_2 \right).$$

The line joining the two centres of the objects is generated by the difference in the two position vectors. Thus the locus of points which constitute the line can be described by the following formula:

$$\vec{L}(t) = \vec{L_0} + \left(\vec{r}_2 - \vec{r}_1\right) t\,,$$

where $\vec{L_0}$ is a vector to a particular point on the line, and t is a parameter, $t \in (-\infty, \infty)$. This prescription is not unique, because:

  * Any point on the line may be chosen to be the starting point, $\vec{L_0}$ [corresponding to the parametric origin $t = 0$].

  * The generating vector [or the parameter, t] may be rescaled.

WLOG, one may choose $\vec{L_0} = \vec{r}_1$, in which case $\vec{L}(t) = \vec{r}_1 + \left(\vec{r}_2 - \vec{r}_1\right) t$. It is easy to see that

$$\vec{L}(0) = \vec{r}_1 \qquad \text{and} \qquad \vec{L}(1) = \vec{r}_1 + \left(\vec{r}_2 - \vec{r}_1\right) = \vec{r}_2\,,$$

verifying that the locus of points, $\vec{L}(t)$, includes the centres of both balls.

Q: Where precisely along this line does the CofM lie?

A: Let's distribute the factor of $1/M_{\text{Total}}$ through the sum over particles in the discrete point-like expression for the CofM position. With creative rearrangement of factors and terms, we obtain

$$\vec{R}_{\text{CofM}} = \frac{M_1}{M_1 + M_2}\,\vec{r}_1 + \frac{M_2}{M_1 + M_2}\,\vec{r}_2$$

$$= \left[1 + \frac{M_1}{M_1 + M_2} - 1\right]\vec{r}_1 + \frac{M_2}{M_1 + M_2}\,\vec{r}_2$$

$$\vdots \qquad\qquad \text{noting that} \qquad \frac{M_1}{M_1+M_2} - 1 = -\frac{M_2}{M_1+M_2}$$

$$= \left[1 - \frac{M_2}{M_1 + M_2}\right]\vec{r}_1 + \frac{M_2}{M_1 + M_2}\,\vec{r}_2$$

$$= \vec{r}_1 + \frac{M_2}{M_1 + M_2}\left(\vec{r}_2 - \vec{r}_1\right).$$

The formula for the position of the CofM is identical in structure to $\vec{L}(t)$, the generic equation for the line which intersects the two points, thus proving the following claim.

**The centre of mass of the two ball system
lies on the line joining the centres of the balls.**

The mass of each ball is necessarily positive. Therefore, the factor which plays the rôle of the parameter t satisfies

$$0 < \frac{M_2}{M_1 + M_2} < 1\,,$$

where, as we saw earlier, $t = 0$ corresponds to $\vec{r}_1$, and $t = 1$ corresponds to $\vec{r}_2$.

**By the monotonicity of linear functions, $\vec{R}_{\text{CofM}}$ must lie between the two
centres of the balls. Furthermore, $t \equiv M_2/(M_1 + M_2)$ ensures that the CofM
lies nearer to the weightier particle.**

IF the two balls have equal mass, $M_1 = M_2$, THEN the CofM lies at the midpoint of the line between the centres of the balls, exactly as expected from symmetry.

The CofM of a Body with Continuous Distribution of Substance

Suppose that we wish to determine the location of the centre of mass of an irregular lump of material substance. The total mass of the lump and the position of its CofM are formally defined via

$$M_{\text{Total}} = \int dm\,, \quad \text{and} \quad M_{\text{Total}}\, \vec{R}_{\text{CofM}} = \int dm\, \vec{r}\,.$$

Q: What does it mean to integrate over dm?

A: Exactly what one would think! *I.e.,* to integrate over the distribution of mass, giving weight in proportion to the amount of mass present in the small neighbourhood of each point. The lump resides in space, and its extent [size and shape] and internal structure may be characterised in terms of spatial coordinates.

> ASIDE: This may well qualify as THE MOST completely obvious statement appearing in these chapters, but one is obliged sometimes to call attention to assumptions or preconceptions hidden in plain sight!

The mass density bridges the gap between the mass distribution and the spatial description of the lump. The density function may be [formally] extended to all of space by defining it to be identically ZERO everywhere outside of the lump. The mass density typically varies throughout the distribution; however, in the special case in which the density is constant [wherever it is non-zero], the distribution is deemed to be UNIFORM.

1-d In an effectively one-dimensional system,

$$dm = \lambda\, d[\,\text{length}\,]\,,$$

where λ is the **lineal mass density** [*a.k.a.* the mass per unit length]. An effectively 1-d system might be a long and very thin object which can be approximated by a line, or a system with sufficient symmetry that only one component of the position of the CofM requires calculation.

2-d In an effectively two-dimensional system,

$$dm = \sigma\, d[\,\text{area}\,]\,,$$

where σ is the **areal mass density** [*a.k.a.* the mass per unit area]. An effectively 2-d system might be a very thin flat object which can be approximated by a plane, or a system with sufficient symmetry that only two components of the position of the CofM need be calculated.

3-d In a fully three-dimensional system,

$$dm = \rho\, d[\,\text{volume}\,]\,,$$

where ρ is the **volume mass density** [*a.k.a.* the mass per unit volume]. This is the generic case, accommodating any sort of regular or irregular distribution of matter.

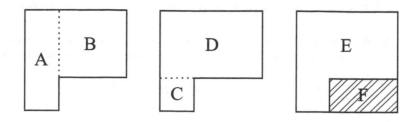

FIGURE 30.2   Various Partitions of a Piece of Notched Plywood

EXAMPLE [*Centres of Mass for Irregular Objects*]

A rectangular sheet of plywood has a rectangular notch cut from one corner as shown in Figure 30.2. The CofM of the notched plywood can be computed in a variety of ways. Three methods which effectively convert analysis of the continuous sheet into consideration of an equivalent two point-particle system are presented below.

AB   Partition the figure into two rectangles, **A** and **B**. Compute the mass of each rectangular constituent:

$$M_A = \sigma \, \text{Area}(A) = \sigma \, \big(\text{length}(A) \times \text{width}(A) \big),$$
$$M_B = \sigma \, \text{Area}(B) = \sigma \, \big(\text{length}(B) \times \text{width}(B) \big),$$

where σ is the areal mass density associated with this particular grade of plywood. The centres of mass of parts **A** and **B** lie at their respective geometric centres. The locations of these, $\vec{r}_A$ and $\vec{r}_B$, can be expressed in terms of a common coordinate system. The combined CofM of the two rectangles [treated as point-like objects] is

$$
\begin{aligned}
\vec{R}_{\text{CofM}} &= \frac{1}{M_{\text{Total}}} \sum_{i=\{A,B\}} M_i \vec{r}_i \\
&= \frac{1}{M_A + M_B} \big(M_A \vec{r}_A + M_B \vec{r}_B \big) \\
&= \frac{\text{Area}(A) \, \vec{r}_A + \text{Area}(B) \, \vec{r}_B}{\text{Area}(A) + \text{Area}(B)} \\
&= \frac{\text{Area}(A) \, \vec{r}_A + \text{Area}(B) \, \vec{r}_B}{\text{Total Area}}.
\end{aligned}
$$

The second-to-last equality arose as a consequence of the uniformity of σ, the areal mass density.

CD   Partition the figure into two rectangles, **C** and **D**. The mass of each rectangular constituent is, with σ constant,

$$M_C = \sigma \, \text{Area}(C) = \sigma \, \big(\text{length}(C) \times \text{width}(C) \big),$$
$$M_D = \sigma \, \text{Area}(D) = \sigma \, \big(\text{length}(D) \times \text{width}(D) \big).$$

The centres of mass of parts **C** and **D** reside at their geometric centres, $\vec{r}_C$ and

$\vec{r}_D$. The combined CofM of the two rectangles is

$$\vec{R}_{\text{CofM}} = \frac{1}{M_{\text{Total}}} \sum_{i=\{C,D\}} M_i\,\vec{r}_i = \frac{1}{M_C + M_D}\left(M_C\,\vec{r}_C + M_D\,\vec{r}_D\right)$$

$$= \frac{\text{Area}(C)\,\vec{r}_C + \text{Area}(D)\,\vec{r}_D}{\text{Area}(C) + \text{Area}(D)}$$

$$= \frac{\text{Area}(C)\,\vec{r}_C + \text{Area}(D)\,\vec{r}_D}{\text{Total Area}}\,.$$

Once again, the constant areal density, σ, has cancelled out in the computation.

EF   [*This method works, but handle with care!*]

Partition the figure into two rectangles, **E** and **F**, where **E** is the **E**ntire sheet of unnotched plywood and **F** is the **F**orgotten notch. The mass of each rectangular "constituent" is

$$M_E = \sigma\,\text{Area}(E) = \sigma\left(\text{length}(E) \times \text{width}(E)\right),$$

$$M_F = -\sigma\,\text{Area}(F) = -\sigma\left(\text{length}(F) \times \text{width}(F)\right).$$

Here M_E is the mass of an unnotched sheet, while $|M_F|$ $[M_F < 0]$ is the mass of the missing part. The centres of mass of **E** and **F** are at their respective centres. The total mass of the notched board is, once again, obtained by *adding*[6] the constituent masses, $M_{\text{Total}} = M_E + M_F$. In the same manner, the CofM of the notched figure is

$$\vec{R}_{\text{CofM}} = \frac{1}{M_{\text{Total}}} \sum_{i=\{E,F\}} M_i\,\vec{r}_i = \frac{1}{M_E + M_F}\left(M_E\,\vec{r}_E + M_F\,\vec{r}_F\right)$$

$$= \frac{\text{Area}(E)\,\vec{r}_E - \text{Area}(F)\,\vec{r}_F}{\text{Area}(E) - \text{Area}(F)}$$

$$= \frac{\text{Area}(E)\,\vec{r}_E - \text{Area}(F)\,\vec{r}_F}{\text{Total Area}}\,.$$

In effect, we have added a **F**ictitious piece bearing *negative mass* to an unnotched sheet!

EXAMPLE [*Centre of Mass of a Thin Uniform Rod*]

A thin rod of length L and constant lineal mass density λ_0 is illustrated in Figure 30.3.

Q: What is the mass of the rod? Where does its centre of mass lie?

A: Let's see!

The total mass of the rod is

$$M_{\text{Total}} = \int dm = \int_0^L \left[\lambda_0\,dx\right] = \lambda_0\,x\Big|_0^L = \lambda_0\,(L - 0) = \lambda_0\,L\,.$$

Unsurprisingly, the mass of the uniform rod is equal to its constant mass per unit length times its length!

---

[6]The factor of -1 incorporated into the mass of the missing section ensures that its addition is tantamount to subtraction.

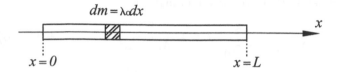

FIGURE 30.3 A Thin Rod with Uniform Density

The location of the CofM of the rod is determinable from

$$M_{\text{Total}}\, R_{\text{CofM},x} = \int dm\, x = \int_0^L x\,\big[\lambda_0\,dx\big] = \lambda_0 \left[\frac{1}{2}x^2\right]_0^L = \frac{1}{2}\,\lambda_0\,(L^2 - 0^2) = \frac{1}{2}\,\lambda_0\,L^2\,.$$

Dividing through by the value of the total mass calculated above, we obtain

$$R_{\text{CofM},x} = \frac{\frac{1}{2}\lambda_0\,L^2}{\lambda_0\,L} = \frac{1}{2}\,L\,.$$

Also unsurprising is the discovery that the centre of mass of the uniform rod is located at its geometric centre!

CENTRE of MASS   The physical significance of the centre of mass is that when a net force acts on a rigid extended body so as to produce only translational motion, the body responds as though all of its mass were concentrated at the location of its centre of mass.

This provides the *a posteriori* justification for treating particles as though they were point-like, as we have been doing since the beginning of this VOLUME!

ADDENDUM: The General Notion of Averaging

Six physics students wrote an exam and obtained the following set of scores.[7]

Student	Score
Zoe	80
Yvan	90
Xavier	90
Walter	90
Vera	90
Ursula	100

Exam Scores for a Particular Class of Physics Students

Q: What is the average score?

A: Let's determine the answer via three equivalent methods.

METHOD ONE: [*The Standard Averaging Technique*]

$$\text{Average Score} = \frac{1}{\# \text{ of students}} \sum_{i \in \{Z,Y,X,W,V,U\}} (i\text{th score})$$

$$= \frac{1}{6}\left(80 + 90 + 90 + 90 + 90 + 100\right) = \frac{540}{6}$$

$$= 90\,.$$

METHOD TWO: [*Scores Weighted by Number of Students*]
To reduce the number of computational steps [and concomitantly, the likelihood of arithmetic error] it is preferable to instead sum over the scores, weighted by the number of students.

Score	Students	Number
80	Zoe	1
90	Yvan, Xavier, Walter, Vera	4
100	Ursula	1

Exam Scores for a Particular Class of Physics Students (Keyed by Score)

$$\text{Average Score} = \frac{1}{\# \text{ of students}} \sum_{i \in \{80,90,100\}} (i) \times (\text{number of students with this score})$$

$$= \frac{1}{6}\Big((80)(1) + (90)(4) + (100)(1)\Big) = \frac{540}{6}$$

$$= 90\,.$$

---

[7] An educational psychologist might view this example with approbation. Gender balance and a variety of ethnicities are reflected in the first names. In addition, there is the subliminal confidence builder arising from the awarding of the highest score to Ursula, *i.e.*, "U" which is homophonically "you" [and also "ewe," he said sheepishly]. Regrettably, this subliminal trick fails on this occasion for two reasons. The first is that I am exposing it, so it is no longer subliminal, and the second is that the exam was out of 1000!

METHOD THREE: [*Scores Weighted by Fraction of Students*]
This really is just a refinement of the second method. This sum over the scores is weighted
by the fraction of students with that particular score.

Score	Students	Number	Fraction
80	Zoe	1	1/6
90	Yvan, Xavier, Walter, Vera	4	2/3
100	Ursula	1	1/6

Exam Scores for a Particular Class of Physics Students (Keyed by Score)

$$\text{Average Score} = \sum_{i \in \{80,90,100\}} (i) \times (\text{fraction of students with this score})$$

$$= (80)\left(\frac{1}{6}\right) + (90)\left(\frac{2}{3}\right) + (100)\left(\frac{1}{6}\right) = 13\frac{1}{3} + 60 + 16\frac{2}{3}$$

$$= 90 \, .$$

The three methods presented above all yield the same result because they are precisely
equivalent means of computing the average. The conclusion to be drawn from this digression
is that

$$\vec{R}_{\text{CofM}} = \frac{1}{M_{\text{Total}}} \sum_{i=1}^{N} M_i \vec{r}_i = \sum_{i=1}^{N} \frac{M_i}{M_{\text{Total}}} \vec{r}_i \, ,$$

really is **the mass-weighted average position of the system of particles.**

Chapter 31

Seven Amazing Properties of the Centre of Mass

Having learned how to compute the CofM position of a system of particles [or a rigid extended body], we can entertain the possibility that this position might be changing in time. The time rate of change of the position of the centre of mass is its velocity,

$$\vec{V}_{\text{CofM}} = \frac{d\vec{R}_{\text{CofM}}}{dt}.$$

This is great, but let's see what it means in the case of $\mathcal{N}$ discrete point[-like] particles. As usual, the particles are labelled with the index i running from 1 to $\mathcal{N}$. Thus,

$$\vec{V}_{\text{CofM}} = \frac{d\vec{R}_{\text{CofM}}}{dt} = \frac{d}{dt}\left[\frac{1}{M_{\text{Total}}}\sum_{i=1}^{\mathcal{N}} M_i \vec{r}_i\right].$$

We shall assume that the M_i are all constant in time, and therefore M_{Total} is too. Hence,

$$\vec{V}_{\text{CofM}} = \frac{1}{M_{\text{Total}}}\sum_{i=1}^{\mathcal{N}} M_i \frac{d\vec{r}_i}{dt} = \frac{1}{M_{\text{Total}}}\sum_{i=1}^{\mathcal{N}} M_i \vec{v}_i,$$

where $\vec{v}_i = \frac{d\vec{r}_i}{dt}$ is the velocity of the ith particle. This constitutes our **first** amazing result.

VELOCITY of the CENTRE of MASS The velocity of the centre of mass of a system of particles is the mass-weighted average velocity of the particles in the system:

$$\vec{V}_{\text{CofM}} = \frac{1}{M_{\text{Total}}}\sum_{i=1}^{\mathcal{N}} M_i \vec{v}_i.$$

But there's more. Recall that the MOMENTUM of the ith particle is

$$\vec{p}_i = M_i \vec{v}_i,$$

and thus the velocity of the CofM of the collection of particles is expressible as

$$\vec{V}_{\text{CofM}} = \frac{1}{M_{\text{Total}}}\left[\sum_{i=1}^{\mathcal{N}} \vec{p}_i\right] = \frac{\vec{P}_{\text{Total}}}{M_{\text{Total}}}.$$

Our **second** amazing result is that **the total momentum of the system of particles is the product of the total mass and the velocity of the centre of mass**, *viz.*,

$$\vec{P}_{\text{Total}} \equiv M_{\text{Total}}\,\vec{V}_{\text{CofM}}.$$

Recall that we went out on a limb in Chapter 29, summing momenta of different particles to construct the total momentum of a system. Here we see that the aggregate momentum attains physical meaning through its association with the motion of the centre of mass.

Knowing how to compute the velocity of the CofM of a system of particles [or an extended body], we are able to consider how this velocity might be changing in time. The time rate of change of the velocity of the centre of mass is the acceleration of the centre of mass, *i.e.*,

$$\vec{A}_{\text{CofM}} = \frac{d\vec{V}_{\text{CofM}}}{dt} .$$

In the case of $\mathcal{N}$ discrete point-like particles, this acceleration becomes

$$\vec{A}_{\text{CofM}} = \frac{d\vec{V}_{\text{CofM}}}{dt} = \frac{d}{dt}\left[\frac{1}{M_{\text{Total}}}\sum_{i=1}^{\mathcal{N}} M_i\,\vec{v}_i\right] .$$

We are assuming, as before, that each of the M_i and M_{Total} are constant in time, and so,

$$\vec{A}_{\text{CofM}} = \frac{1}{M_{\text{Total}}}\sum_{i=1}^{\mathcal{N}} M_i\,\frac{d\vec{v}_i}{dt} = \frac{1}{M_{\text{Total}}}\sum_{i=1}^{\mathcal{N}} M_i\,\vec{a}_i ,$$

where $\vec{a}_i = \frac{d\vec{v}_i}{dt}$ is the acceleration of the ith particle. This is our **third** amazing result.

ACCELERATION of the CENTRE of MASS   For a system comprised of discrete particles [or a rigid body], the acceleration of the centre of mass,

$$\vec{A}_{\text{CofM}} = \frac{1}{M_{\text{Total}}}\sum_{i=1}^{\mathcal{N}} M_i\,\vec{a}_i ,$$

is the mass-weighted average acceleration of the particles.

But there's yet more. Newton's Second Law,

$$M_i\,\vec{a}_i = \vec{F}_{\text{NET},i} ,$$

relates the acceleration of the ith particle, $\vec{a}_i$, to the net force acting upon it. This net force is obtained by summing the forces which act on i (in the customary manner). Precise analogy with the [immediately prior] analysis of the CofM velocity militates for writing

$$M_{\text{Total}}\,\vec{A}_{\text{CofM}} = \sum_{i=1}^{N} \vec{F}_{\text{NET},i} = \vec{F}_{\text{Total}} ,$$

where $\vec{F}_{\text{Total}}$ is the sum[1] of ALL of the forces which act on ALL of the particles in the system. We next take note of an important subtlety. WLOG, the forces operative on each of the particles may be cast into two classes: **internal** and **external**.

INTERNAL   Internal forces arise due to interaction with other particles in the system, or parts of the rigid body.

---

[1] "*Vujà dé!*" we have never done this before! [However, it was foreshadowed at the end of Chapter 10.]

EXTERNAL   External forces arise from interactions with objects which lie outside of the system or body.

One can separately add forces within each class *en route* to obtaining the net force acting on a given particle:

$$\vec{F}_{\text{NET},i} = \vec{F}_{\text{NET, int'l},i} + \vec{F}_{\text{NET, ext'l},i} \, .$$

The internal force on the ith particle admits possible contributions from all other particles in the system:

$$\vec{F}_{\text{NET, int'l},i} = \sum_{\substack{j=1 \\ j \neq i}}^{N} \vec{F}_{\text{net},ji} \, .$$

We are obligated to write $\vec{F}_{\text{net},ji}$ for the interaction force exerted **by** particle j **upon** particle i to allow for multiple modes of interaction[2] between pairs of particles.

> ASIDE: In expressing the net internal force as a sum over the particles in the system, the possibility of self-interaction has been explicitly dismissed by the "$j \neq i$" restriction in the sum. The very idea of $\vec{F}_{\text{net},ii}$ is ambiguous,[3] and so we ensure that these terms do not appear in the sum. Alternatively, one may choose to define $\vec{F}_{\text{net},ii} \equiv 0$, and then sum over all particles, including the one which is being acted on.

The total force acting on the system is

$$\vec{F}_{\text{Total}} = \sum_{i=1}^{N} \vec{F}_{\text{NET},i} = \sum_{i=1}^{N} \left[\sum_{\substack{j=1 \\ j \neq i}}^{N} \vec{F}_{\text{net},ji} \right] + \sum_{i=1}^{N} \vec{F}_{\text{NET, ext'l},i} \, .$$

A double sum, $\{i, j\}$, over the particles in the system appears in the calculation of the TOTAL internal force. The interaction forces obey N3, and thus a perfect pairwise cancellation among these interaction forces will ensue when the sum is extended over all particles in the system. Hence, **the total *internal* force is exactly zero**—always!

The **fourth** amazing result is that **the total force acting upon the system is the net *external* force**, *viz.*, the sum of the external forces acting upon the entire set of particles in the system.

$$\vec{F}_{\text{Total}} = \vec{F}_{\text{NET ext'l}} \, .$$

> ASIDE: This provides *a posteriori* justification for having written
>
> $$M \, \vec{a} = \vec{F}_{\text{NET}} \, ,$$
>
> for blocks and other objects, earlier in these notes. Extended objects were treated as single particles located at the positions of their respective centres of mass. The dynamics of these bodies depended solely upon the forces that were imposed "from the outside," rather than internal forces (*i.e.*, those that kept them rigid).

---

[2]Perhaps pairs of particles are joined by two, or more, springs/ropes, or perhaps they exert both gravitational and electromagnetic forces upon each other.
[3]**Q:** Is it possible for one to lift oneself off the ground by pulling on one's own shoelaces?

In a system of particles, the second amazing result reads

$$\vec{P}_{\text{Total}} \equiv M_{\text{Total}}\,\vec{V}_{\text{CofM}}\,.$$

Since this is a *bona fide* [instantaneous] physical relation involving time-dependent quantities [at an unspecified time], the time derivative of this equation must be a valid physical relation. Maintaining the assumed constancy of the various M_i and M_{Total} while taking the derivative leads to

$$\frac{d}{dt}\,\text{LHS} = \frac{d\vec{P}_{\text{Total}}}{dt}\,, \qquad \text{AND} \qquad \frac{d}{dt}\,\text{RHS} = M_{\text{Total}}\,\vec{A}_{\text{CofM}} = \vec{F}_{\text{NET ext'l}}\,.$$

The **fifth** amazing result is that **the net external force is equal to the time rate of change of the total momentum.** Thus,

$$\frac{d\vec{P}_{\text{Total}}}{dt} = \vec{F}_{\text{NET ext'l}}\,,$$

generates the equations of motion governing the dynamics of particular systems of particles and extended bodies.

> ASIDE: **Q:** Wasn't this result obtained when momentum was first introduced?

> **A:** No, not really. In Chapter 29 we were still labouring under the restriction to single particles undergoing translational motion. This general result is applicable to systems of particles and extended bodies.

The **sixth** amazing property stems directly from the fifth, but has powerful implications in its own right. It is the LAW OF MOMENTUM CONSERVATION.

The LAW of CONSERVATION of MOMENTUM   WHEN the net external force acting on a system vanishes, THEN the total momentum of the system is conserved, *i.e.,* it is a constant of the motion.

$\vec{P}$ The law follows from the determination, above, that the net external force dictates the time rate of change of the total momentum.

$\vec{P}$ Note that the conserved quantity is the total momentum of the system. This does not prevent or impede the exchange of momentum among constituents of the system.

The CENTRE of MASS FRAME   The centre of mass frame is an IRF in which the CofM of the system under consideration is at rest, and hence the total momentum of the system vanishes,

$$\vec{P}_{\text{Total}}\Big|_{\text{CofM Frame}} = 0\,.$$

The CofM frame is also known as the Centre of Momentum frame. This particular frame is often advantageous for analysis of collisions between two particles.

The **seventh** wonderful feature is that the total kinetic energy of a system [as measured in any particular IRF] can be uniquely decomposed into the **bulk** kinetic energy ascribable to the whole system and the **relative** kinetic energy ascribable to the constituents.

BULK   The bulk kinetic energy depends only on the total mass of the system M_{Total}, and the centre of mass velocity $\vec{V}_{\text{CofM}}$.

RELATIVE   The relative kinetic energy [*a.k.a.* internal kinetic energy] arises from the relative motions of particles comprising the system.

PROOF of the BULK and RELATIVE DECOMPOSITION of the KINETIC ENERGY

For a system of $\mathcal{N}$ discrete particles with masses M_i, and velocities $\vec{v}_i$ [in some IRF],

$$M_{\text{Total}} = \sum_{i=1}^{\mathcal{N}} M_i \,, \qquad \text{and} \qquad \vec{V}_{\text{CofM}} = \frac{1}{M_{\text{Total}}} \sum_{i=1}^{\mathcal{N}} M_i \, \vec{v}_i \,,$$

respectively. The velocity of the ith particle, $\vec{v}_i$, may always be expressed as the [vector] sum of the CofM velocity and its relative velocity, *viz.*,

$$\vec{v}_i = \vec{V}_{\text{CofM}} + \vec{v}_{\text{rel},i} \,.$$

ASIDE: An observer in the centre of mass IRF sees each particle moving with its relative velocity. Here we eschew the customary practice of denoting the CofM velocities by $\vec{u}$.

The kinetic energy of the ith particle is

$$K_i = \frac{1}{2} M_i \left| \vec{v}_i \right|^2 = \frac{1}{2} M_i \left| \vec{V}_{\text{CofM}} + \vec{v}_{\text{rel},i} \right|^2 = \frac{1}{2} M_i \left| \vec{V}_{\text{CofM}} \right|^2 + M_i \, \vec{V}_{\text{CofM}} \cdot \vec{v}_{\text{rel},i} + \frac{1}{2} M_i \left| \vec{v}_{\text{rel},i} \right|^2 \,,$$

where the identity:

$$\left| \vec{a} + \vec{b} \right|^2 = (\vec{a} + \vec{b}) \cdot (\vec{a} + \vec{b}) = \left| \vec{a} \right|^2 + 2\, \vec{a} \cdot \vec{b} + \left| \vec{b} \right|^2 \,, \quad \text{for all pairs of vectors, } \{\vec{a}, \vec{b}\} \,,$$

has been invoked. The total kinetic energy of the system is the sum of the kinetic energies of the constituent particles. Performing the sum, substituting the expressions for each particle's kinetic energy, and recognising the DISTRIBUTIVITY of vector summation through the dot product, yields

$$K_{\text{Total}} = \sum_{i=1}^{\mathcal{N}} K_i = \frac{1}{2} \left[\sum_{i=1}^{\mathcal{N}} M_i \right] \left| \vec{V}_{\text{CofM}} \right|^2 + \left[\sum_{i=1}^{\mathcal{N}} M_i \, \vec{v}_{\text{rel},i} \right] \cdot \vec{V}_{\text{CofM}} + \left[\sum_{i=1}^{\mathcal{N}} \frac{1}{2} M_i \left| \vec{v}_{\text{rel},i} \right|^2 \right] \,.$$

FIRST TERM   The masses alone are summed, with equal weighting. This sum is multiplied by constant factors. The first term is the BULK kinetic energy,

$$K_{\text{CofM}} = \frac{1}{2} M_{\text{Total}} \left| \vec{V}_{\text{CofM}} \right|^2 \,.$$

SECOND TERM   The expression appearing within the brackets imbedded in the second term is the mass-weighted sum of the relative velocities.

Let's begin our analysis of this particular mass-weighted sum with a statistical digression. The amount by which each individual datum (in a data set) differs from the average value is that value's *deviation from the mean*. For example, recalling our dear friends and their test scores from the Addendum to Chapter 30, we can assign to each person his or her deviation from the mean (90), as in the table below.

Student	Score	Deviation
Zoe	80	-10
Yvan	90	0
Xavier	90	0
Walter	90	0
Vera	90	0
Ursula	100	+10

Exam Scores and Deviations for a Particular Class of Physics Students

Q: What happens when we sum the deviations over the set of students?

A: We get exactly zero, as can be easily verified.

$$\text{Total Deviation} = \sum_{i \in \{Z,Y,X,W,V,U\}} \left(\text{Deviation}_i \right) = -10 + 0 + 0 + 0 + 0 + 10 = 0 \,.$$

This is not an accident. Thinking precisely:

The average value is that number which yields equal amounts of positive and negative deviation amongst all of the data.

The expression in brackets must vanish,

$$\left[\sum_{i=1}^{\mathcal{N}} M_i \, \vec{v}_{\text{rel},i} \right] = \vec{0} \,,$$

because $\vec{v}_{\text{rel},i}$ corresponds to the deviation of the velocity of the ith particle from the system average value and the sum is over all of the particles. The dot product of any vector and $\vec{0}$ is always ZERO, nullifying the second term.

THIRD TERM   The third term consists of the sum of the kinetic energies possessed by the constituent particles by virtue of their relative motions as [consistently] viewed in the CofM IRF. This third term is, quite deservingly, called the RELATIVE kinetic energy of the system.

$$K_{\text{relative},i} = \frac{1}{2} M_i \left| \vec{v}_{\text{rel},i} \right|^2 ,$$

$$K_{\text{relative}} = \sum_{i=1}^{\mathcal{N}} K_{\text{relative},i} \,.$$

Recombining the three terms yields the mathematical statement of our **seventh** wonder:

$$K_{\text{Total}} \equiv K_{\text{CofM}} + K_{\text{relative}} \,.$$

In chapters to come, we shall exploit this decomposition property of the kinetic energy to acquire a better understanding of certain systems.

Chapter 32

Collisions

Consider a system comprised of two particles which interact by means of a contact force (*i.e.*, a collision). The LAW OF MOMENTUM CONSERVATION [promulgated in Chapter 31]:

IF the net external force on the particles vanishes,
THEN $\vec{P}_{\text{Total}}$ is constant in time,

is operative here.

The actual value of the total momentum is of secondary importance and depends on the IRF in which it is computed; what matters is that the force of interaction, being internal to the system, is unable to affect the total momentum.

> ASIDE: When external forces are present, all is not necessarily lost. If these external forces and time scales yield a net external impulse which is relatively small, then invoking the impulse approximation amounts to legislating "approximate" momentum conservation.

Q: Okay, momentum is conserved in a collision. How about mechanical energy?

A: Typically, mechanical energy is dissipated in a collision. A collision taking place in a single instant and at a particular location [as is the case in our model] cannot affect the potential energy of the system. Thus, the collision may only effect a change in the system's kinetic energy. Empirical observation and analysis reveal that a continuum of energy behaviour, bounded by two extreme cases, is possible. The phenomenological parameter dictating where along this continuum any particular collision lies is called the **elasticity** of the colliding particles.

Perfectly Elastic	Partially Elastic	Completely Inelastic
$K_f = K_i$	$K_f < K_i$	$K_f < K_i$, maximally
Conservation of Mechanical Energy	Some Mechanical Energy Lost	Maximal Loss of Mechanical Energy

We shall concern ourselves with the perfectly elastic case in the rest of this chapter, deal with the completely inelastic case in Chapter 33, and forego tackling partial elasticity.

COMPLETELY ELASTIC COLLISION OF TWO PARTICLES

Two particles collide in a perfectly elastic manner. We assume that before and after the collision, the particles move inertially. [No long-distance forces act between the particles, and external forces, if they exist, precisely cancel.] The first panel of Figure 32.1 shows a snapshot of the system before the collision. The particles have masses M_1 and M_2 respectively, and prior to the collision they have initial velocities $\vec{v}_{1,i}$ and $\vec{v}_{2,i}$. Knowledge of the masses and the initial velocities is enough to define the initial state of the system [at least to the extent required for our present purposes].

FIGURE 32.1   Two Particles Undergoing an Elastic Collision

ASIDE: The velocities of the particles are measured with respect to a particular IRF, associated with an observer at rest in that frame.

    The insistence that the particles move inertially prior to and following the collision renders the pre- and post-collision velocities constant. This additional constraint is relaxed in the example which ends this chapter.

THE IDEA: **Apply the conservation laws for momentum and kinetic energy in order to make predictions about the final state of the system.**

The initial momenta of the particles and the system are

$$\left.\begin{array}{l} \vec{p}_{1,i} = M_1\,\vec{v}_{1,i} \\[4pt] \vec{p}_{2,i} = M_2\,\vec{v}_{2,i} \end{array}\right\} \implies \vec{P}_{\text{Total},i} = \vec{p}_{1,i} + \vec{p}_{2,i} = M_1\,\vec{v}_{1,i} + M_2\,\vec{v}_{2,i}\,.$$

The initial kinetic energies of the particles and the system are

$$\left.\begin{array}{l} K_{1,i} = \tfrac{1}{2}\,M_1\,\bigl|\vec{v}_{1,i}\bigr|^2 \\[6pt] K_{2,i} = \tfrac{1}{2}\,M_2\,\bigl|\vec{v}_{2,i}\bigr|^2 \end{array}\right\} \implies K_{\text{Total},i} = K_{1,i} + K_{2,i} = \frac{1}{2}\,M_1\,v_{1,i}^2 + \frac{1}{2}\,M_2\,v_{2,i}^2\,.$$

ASIDE: It is possible to analyse collisions in terms of the momentum, mass, and kinetic energy of the colliding particles, by writing

$$K_{1,i} = \frac{\bigl|\vec{p}_{1,i}\bigr|^2}{2\,M_1} \quad\text{and}\quad K_{2,i} = \frac{\bigl|\vec{p}_{2,i}\bigr|^2}{2\,M_2}\,.$$

BANG! — the collision occurs.

We shall insist that the corporal integrity of the particles is preserved. In other words, no mass is transferred, exchanged, or otherwise lost by either particle, so

$$M_{1,i} = M_{1,f} \equiv M_1 \quad\text{and}\quad M_{2,i} = M_{2,f} \equiv M_2\,.$$

After the collision has taken place, the particles have velocities $\vec{v}_{1,f}$ and $\vec{v}_{2,f}$ respectively. While we may not know the values of these velocities, we can see that they exist. Formally, we can compute the final momenta and kinetic energies of the particles in precisely the same manner as the initial ones.

[The two sets of formulae should be equivalent under the exchange of labels $i \leftrightarrow f$.]

Proceeding as before, not merely switching the labels, we obtain the final values of the momenta and kinetic energies [and confirm our intuition in the process].

$$\left.\begin{array}{r}\vec{p}_{1,f} = M_1\,\vec{v}_{1,f} \\ \vec{p}_{2,f} = M_2\,\vec{v}_{2,f}\end{array}\right\} \implies \vec{P}_{\text{Total},f} = \vec{p}_{1,f} + \vec{p}_{2,f} = M_1\,\vec{v}_{1,f} + M_2\,\vec{v}_{2,f}$$

$$\left.\begin{array}{r}K_{1,f} = \tfrac{1}{2}\,M_1\,\left|\vec{v}_{1,f}\right|^2 \\ K_{2,f} = \tfrac{1}{2}\,M_2\,\left|\vec{v}_{2,f}\right|^2\end{array}\right\} \implies K_{\text{Total},f} = K_{1,f} + K_{2,f} = \frac{1}{2}\,M_1\,v_{1,f}^2 + \frac{1}{2}\,M_2\,v_{2,f}^2$$

ASIDE: One may work exclusively with kinetic energies and momenta [and mass] by writing

$$K_{1,f} = \frac{\left|\vec{p}_{1,f}\right|^2}{2\,M_1} \quad \text{and} \quad K_{2,f} = \frac{\left|\vec{p}_{2,f}\right|^2}{2\,M_2}\,.$$

If you have a sneaking suspicion that we have not actually done anything yet ... you are completely correct! The calculations of total momentum and kinetic energy in the pre- and post-collision states are completely generic and [nearly] devoid of meaning. The physics enters when we apply the momentum conservation law to the system and, for the completely elastic case, insist on kinetic energy conservation as well.

$$\vec{P}_{\text{Total},f} = \vec{P}_{\text{Total},i} \qquad\qquad K_{\text{Total},f} = K_{\text{Total},i}$$

$$\vec{p}_{1,f} + \vec{p}_{2,f} = \vec{p}_{1,i} + \vec{p}_{2,i} \quad \text{and} \quad K_{1,f} + K_{2,f} = K_{1,i} + K_{2,i}$$

$$M_1\,\vec{v}_{1,f} + M_2\,\vec{v}_{2,f} = M_1\,\vec{v}_{1,i} + M_2\,\vec{v}_{2,i} \qquad \tfrac{1}{2}\,M_1\,v_{1,f}^2 + \tfrac{1}{2}\,M_2\,v_{2,f}^2 = \tfrac{1}{2}\,M_1\,v_{1,i}^2 + \tfrac{1}{2}\,M_2\,v_{2,i}^2$$

This set of equations relates the final and initial states of motion for the system of two particles experiencing a perfectly elastic collision.

These dynamical equations consist of a vector equation [MOMENTUM] and a scalar equation [KINETIC ENERGY] and are valid in any number of dimensions. In what follows we presume complete knowledge of the initial state of motion of the system and attempt to predict the final state.

ASIDE: It is also possible to *retrodict* the initial state from measurements of the final state. This should not be surprising, since a bit of thought reveals that elastic collisions are symmetric under time reversal!

Let's separately investigate systems in 3-, 2-, and 1-d.

3-d In three dimensions, there are three momentum relations and one kinetic energy relation for a total of four equations. Each particle's velocity, or momentum, possesses three independent [spatial] components. Thus, in a two-particle system, there are six *a priori* unknown quantities.

Four equations and **six** unknowns $\implies$ an indeterminate system!

One cannot uniquely determine the final state of the system (in 3-d) without an additional two items of independent data.

> ASIDE: Two final-state data which are commonly specified are: the plane generated by the two particles' final velocities, and the open angle between them.

2-d   In two dimensions, there are two momentum relations and one kinetic energy relation for a total of three equations. Two independent components in each of the particle's velocities, or momenta, lead to there being four unknown quantities.

> **Three** equations and **four** unknowns $\implies$ an indeterminate system!

The final state of the system (in 2-d) cannot be determined without the provision of an additional datum.

1-d   In one dimension, one momentum relation and one kinetic energy relation yield a total of two equations. However, in one dimension, there is only one component in each particle's velocity [momentum], and hence a total of two unknown quantities.

> **Two** equations and **two** unknowns $\implies$ a determinate system! [Perhaps.[1]]

In 1-d, the final state of the system is uniquely determined by its initial conditions.

All of the above analyses took place in the IRF of a generic observer. We dub this IRF the "laboratory frame." We are free to re-analyse the collision from the perspective of any IRF. A sometimes advantageous perspective is that of an observer who is at rest with respect to [equivalently, "moving alongside of"] particle 2 in the initial state.

[This is reminiscent of the example in Chapter 3 involving PK, NK, and the police car.]

In this particular case, the initial momenta of the particles and of the system are

$$\left.\begin{array}{c} \vec{p}_{1,i} = M_1\,\vec{v}_{1,i} \\ \vec{p}_{2,i} = M_2\,\vec{v}_{2,i} = 0 \end{array}\right\} \implies \vec{P}_{\text{Total},i} = \vec{p}_{1,i} + \vec{p}_{2,i} = \vec{p}_{1,i} = M_1\,\vec{v}_{1,i}\,.$$

The initial kinetic energies of the particles and the system are

$$\left.\begin{array}{c} K_{1,i} = \frac{1}{2}\,M_1\,\left|\vec{v}_{1,i}\right|^2 \\ K_{2,i} = \frac{1}{2}\,M_2\,\left|\vec{v}_{2,i}\right|^2 = 0 \end{array}\right\} \implies K_{\text{Total},i} = K_{1,i} + K_{2,i} = \frac{1}{2}\,M_1\,v_{1,i}^2\,.$$

BANG! — the collision occurs.

---

[1]The mathematical pathologies which make this qualifier necessary are assumed to not occur.

After the collision has taken place, the particles have velocities $\vec{v}_{1,f}$ and $\vec{v}_{2,f}$ respectively. The expressions for the final momenta and kinetic energies are exactly the same as those obtained earlier, in the laboratory IRF.

$$\left.\begin{array}{l} \vec{p}_{1,f} = M_1\,\vec{v}_{1,f} \\ \vec{p}_{2,f} = M_2\,\vec{v}_{2,f} \end{array}\right\} \implies \vec{P}_{\text{Total},f} = \vec{p}_{1,f} + \vec{p}_{2,f} = M_1\,\vec{v}_{1,f} + M_2\,\vec{v}_{2,f}$$

$$\left.\begin{array}{l} K_{1,f} = \frac{1}{2}\,M_1\,\left|\vec{v}_{1,f}\right|^2 \\ K_{2,f} = \frac{1}{2}\,M_2\,\left|\vec{v}_{2,f}\right|^2 \end{array}\right\} \implies K_{\text{Total},f} = K_{1,f} + K_{2,f} = \frac{1}{2}\,M_1\,v_{1,f}^2 + \frac{1}{2}\,M_2\,v_{2,f}^2$$

Applying momentum and kinetic energy conservation provides the relation between initial and final quantities.

$$\vec{P}_{\text{Total},f} = \vec{P}_{\text{Total},i} \qquad\qquad K_{\text{Total},f} = K_{\text{Total},i}$$
$$\vec{p}_{1,f} + \vec{p}_{2,f} = \vec{p}_{1,i} \qquad \text{and} \qquad K_{1,f} + K_{2,f} = K_{1,i}$$
$$M_1\,\vec{v}_{1,f} + M_2\,\vec{v}_{2,f} = M_1\,\vec{v}_{1,i} \qquad\qquad \tfrac{1}{2}\,M_1\,v_{1,f}^2 + \tfrac{1}{2}\,M_2\,v_{2,f}^2 = \tfrac{1}{2}\,M_1\,v_{1,i}^2$$

These dynamical equations relate the final and initial states of the system as seen by an observer who is, initially, riding alongside particle 2.

This set of equations is no more than a specialisation of the [more general] laboratory IRF set. One makes use of the rest frame of particle 2 if it simplifies the setting-up and analysis of the problem. The essential physics is exactly the same for all observers. Those details which are observer dependent ... well, *they* can be different.

There is a particular frame of reference which is best suited [meaning algebraically simplest] for analysing collisions. This preferred frame is the Centre of Mass frame [*a.k.a.* Centre of Momentum frame] that was introduced in Chapter 31. Recall that, for the two colliding particles, the [initial] velocity of the centre of mass is

$$\vec{V}_{\text{CofM}} = \frac{1}{M_1 + M_2}\left(M_1\,\vec{v}_{1,i} + M_2\,\vec{v}_{2,i}\right).$$

In the CofM frame, $\vec{V}_{\text{CofM}} = 0$. There are three general comments to make before we proceed with the analysis.

CofM We have been assuming that the net external force vanishes, and, as a consequence, the velocity of the centre of mass remains unchanged through the collision. [This is why we did not bother to place an "*i*" subscript on the $\vec{V}_{\text{CofM}}$ in the above expression.]

CofM A particular system of interest may consist of more than two particles. The entire system will possess a [physically relevant and well-defined] centre of mass position and velocity. It is not the system-wide centre of mass that interests us in the analysis of a collision between two constituents, but only that of the two body sub-system consisting of the colliding particles. [The non-participating particles are often referred to as *spectators*.]

CofM An empirical and logical fact is that [virtually[2]] all collisions are between pairs of particles. When three or more particles collide, it is [almost[3]] always through a combination of separate two body collisions occurring at distinct times.

---

[2]This CAVEAT is to keep the lawyers at bay.
[3]Ditto.

Starting from observed velocities in some other (*i.e.*, laboratory) frame, we can compute the velocity of the CofM observed in this other frame, *viz.*,

$$\vec{V}_{\text{CofM}} = \frac{1}{M_1 + M_2} \left(M_1\, \vec{v}_{1,i} + M_2\, \vec{v}_{2,i} \right).$$

We can then effect the conversion to the CofM frame by subtracting away the bulk motion, leaving only the relative motion of the (two-particle) system:

$$\vec{u}_{1,i} = \vec{v}_{1,i} - \vec{V}_{\text{CofM}} \qquad \text{and} \qquad \vec{u}_{2,i} = \vec{v}_{2,i} - \vec{V}_{\text{CofM}}.$$

To remind ourselves that we are in the CofM frame, we denote the particle velocities by $\vec{u}$ rather than $\vec{v}$. The total initial momentum vanishes in the CofM frame (tautologically):

$$\vec{P}_{\text{Total},i} = \vec{p}_{1,i} + \vec{p}_{2,i} = M_1\, \vec{u}_{1,i} + M_2\, \vec{u}_{2,i} \equiv 0.$$

The initial kinetic energy of the system is just the relative kinetic energy of the particles, because the CofM is at rest.

$$K_{1,i} = \tfrac{1}{2}\, M_1 \left| \vec{u}_{1,i} \right|^2 = \tfrac{1}{2}\, M_1\, u_{1,i}^2$$

$$K_{2,i} = \tfrac{1}{2}\, M_2 \left| \vec{u}_{2,i} \right|^2 = \tfrac{1}{2}\, M_2\, u_{2,i}^2$$

$$\implies \quad K_{\text{Total},i} = K_{1,i} + K_{2,i} = K_{\text{relative},i} = \tfrac{1}{2}\, M_1\, u_{1,i}^2 + \tfrac{1}{2}\, M_2\, u_{2,i}^2.$$

BANG! — the collision occurs.

After the collision has taken place, the particles have velocities $\vec{u}_{1,f}$ and $\vec{u}_{2,f}$ respectively. The expressions for the final momenta and kinetic energies are not unlike those obtained earlier.

$$\left. \begin{aligned} \vec{p}_{1,f} &= M_1\, \vec{u}_{1,f} \\ \vec{p}_{2,f} &= M_2\, \vec{u}_{2,f} \end{aligned} \right\} \implies \vec{P}_{\text{Total},f} = \vec{p}_{1,f} + \vec{p}_{2,f} = M_1\, \vec{u}_{1,f} + M_2\, \vec{u}_{2,f}$$

$$\left. \begin{aligned} K_{1,f} &= \tfrac{1}{2}\, M_1 \left| \vec{u}_{1,f} \right|^2 \\ K_{2,f} &= \tfrac{1}{2}\, M_2 \left| \vec{u}_{2,f} \right|^2 \end{aligned} \right\} \implies K_{\text{Total},f} = K_{1,f} + K_{2,f} = K_{\text{relative},f} = \tfrac{1}{2}\, M_1\, u_{1,f}^2 + \tfrac{1}{2}\, M_2\, u_{2,f}^2$$

Momentum and kinetic energy conservation provide the relations between the final and initial states of motion.

$$\vec{P}_{\text{Total},f} \equiv 0 \equiv \vec{P}_{\text{Total},i} \qquad \text{and} \qquad K_{\text{Total},f} = K_{\text{relative},f} = K_{\text{relative},i} = K_{\text{Total},i}.$$

It is all too easy to think that we have just been pushing symbols around without actually computing anything. This misperception arises, in part, because we have been considering the general case, in which the collision occurs in *d*-dimensions. Now, let's explicitly consider the situation in 1-d, where the final state of the system is completely solvable in terms of initial data [without the need for additional information].

EXAMPLE [*Completely Elastic Collision of Two Particles in One Dimension*]

Let's undertake three independent analyses in: the lab frame, the rest frame of particle 2, and the centre of mass frame.

LAB FRAME   Conservation of momentum and kinetic energy in one dimension implies

$$M_1\, v_{1,f} + M_2\, v_{2,f} = M_1\, v_{1,i} + M_2\, v_{2,i}$$
$$\frac{1}{2}\, M_1\, v_{1,f}^2 + \frac{1}{2}\, M_2\, v_{2,f}^2 = \frac{1}{2}\, M_1\, v_{1,i}^2 + \frac{1}{2}\, M_2\, v_{2,i}^2 \, .$$

The general solution[4] of these equations,

$$v_{1,f} = \frac{M_1 - M_2}{M_1 + M_2}\, v_{1,i} + \frac{2\, M_2}{M_1 + M_2}\, v_{2,i}$$
$$v_{2,f} = \frac{2\, M_1}{M_1 + M_2}\, v_{1,i} - \frac{M_1 - M_2}{M_1 + M_2}\, v_{2,i} \, ,$$

expresses the final velocities of the particles in terms of their masses and initial velocities.

1 ↔ 2   The formulae are symmetric under the exchange of particle labels. [The physics should not depend on our choice of names for the particles!]

$M_1 = M_2$ A special case occurs when the particles have equal mass. Then, as is clear from the formulae,

$$v_{1,f} = v_{2,i} \qquad \text{and} \qquad v_{2,f} = v_{1,i} \, ,$$

and we say that the particles have "exchanged velocities."

REST FRAME   When particle 2 is initially at rest and the entire system is confined to one dimension, the expressions of conservation of momentum and kinetic energy are

$$M_1\, v_{1,f} + M_2\, v_{2,f} = M_1\, v_{1,i}$$
$$\frac{1}{2}\, M_1\, v_{1,f}^2 + \frac{1}{2}\, M_2\, v_{2,f}^2 = \frac{1}{2}\, M_1\, v_{1,i}^2 \, .$$

Solving[5] yields

$$v_{1,f} = \frac{M_1 - M_2}{M_1 + M_2}\, v_{1,i}$$
$$v_{2,f} = \frac{2\, M_1}{M_1 + M_2}\, v_{1,i} \, ,$$

as the expression of the final velocities of the particles in terms of their masses and the initial velocity of particle 1 [as particle 2 is initially at rest]. It is evident that this solution is a special case of the laboratory frame solution quoted above.

_____

[4]That this is a coupled set of non-linear equations (quadratic terms appear in the kinetic energy equation) does make the task of solving them harder, but not insurmountably so.

[5]Ditto!

1 ↮ 2   The formulae are not symmetric under exchange of particle labels, since the initial state clearly distinguishes between the particle which is moving, 1, and that which is at rest, 2.

$M_1 < M_2$ IF the incident particle is less massive than the struck particle, THEN the incident particle bounces back in the direction from whence it came, while the struck particle moves forward.

$M_1 = M_2$ IF the particles have equal mass, THEN the incident particle stops and the struck particle moves forward with the velocity of the incident particle. The particles, as expected, have "exchanged velocities."

$M_1 > M_2$ IF the incident particle is more massive than the struck particle, THEN both particles proceed in the direction of motion of the incident particle.

ASIDE: These types of collisions occur in American football hand-off plays. The fullback drives to the line and stutter-steps to momentarily arrest the [motion of] prospective tacklers on the other team.

- If the fullback runs into a defensive lineman [typically more massive than he], then the fullback bounces backward, while the lineman is rocked back toward his own goal.

- If the fullback runs into a linebacker [of comparable mass], then he comes nearly to a stop, while the linebacker is knocked toward his own goal at about the same speed as the fullback was travelling.

- If the fullback runs into a safety [less massive than he], then he is slowed in his forward progress (but not stopped), while the safety "gets popped" toward his own goal.

Those who instruct players in the *fine art* of tackling emphasise the importance of grasping the ball carrier, thereby maximising the inelasticity [to come, in Chapter 33].

CofM FRAME   When we view an elastic collision from the CofM frame in 1-d the conservation equations become

$$M_1 \, \vec{u}_{1,f} + M_2 \, \vec{u}_{2,f} = 0 = M_1 \, \vec{u}_{1,i} + M_2 \, \vec{u}_{2,i}$$
$$\frac{1}{2} M_1 u_{1,f}^2 + \frac{1}{2} M_2 u_{2,f}^2 = \frac{1}{2} M_1 u_{1,i}^2 + \frac{1}{2} M_2 u_{2,i}^2 .$$

The only solution of this set of equations is

$$u_{1,f} = -u_{1,i} \qquad \text{and} \qquad u_{2,f} = -u_{2,i} .$$

Particles undergoing collision are *reflected* when viewed in the CofM frame!

EXAMPLE [*An Elastic Newton's Pendulum*]

A "Newton's Pendulum" gizmo, often sold at spiffy science shoppes, consists of $\mathcal{N}$ steel balls separately hung as pendula, but in contact. When one ball is pulled back and released, it swings, and upon striking the others, stops, at which instant the ball on the other end swings upwards. Analogous behaviour is exhibited when one pulls back two, three, ..., or even $\mathcal{N} - 1$ balls, and is *sooooo coooool* that it inspires [otherwise normal] people to purchase these products!   [The little balls "click" ever so nicely as they collide, too.]

We make a physical/mathematical model of the system in order to address a common question. A number $n = 1$ or 2 of the $\mathcal{N} = 5$ bobs is (are) pulled back to a height h above the equilibrium point and held there, while the other $\mathcal{N} - n$ bobs remain at rest at the equilibrium height, as shown in Figure 32.2.

The strings are spaced along the common support with separations equal in size to the diameter of the bobs in order that the collisions take place at the equilibrium height, and that the bobs are moving horizontally at the instant they collide.

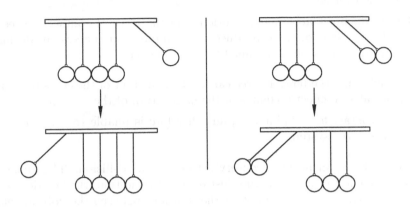

FIGURE 32.2 Newton's Pendulum

The raised bob(s) is (are) released, swing(s) and collide(s) elastically with the others. After the collision, the bob(s) at the other end swing(s) up to a maximum height above equilibrium H, as in Figure 32.2.

Q: How does the final height, H, compare with the initial height, h?

A: Everyone's first inclination is to *guess* that $H = h$, and we shall verify this guess by means of the following line of argument.

PRE-   Before the Collision

The forces which act are gravity and the tension in the string.

- Gravity is a conservative force.

- The tension force does no net work because everywhere along the circular arc of the path, the tension is directed perpendicular to the direction of motion.

Hence, the total mechanical energy of the system is conserved in the pre-collision phase, while the momentum is not. We can exploit the conservation of mechanical energy to compute v_b, the speed of the moving bob(s) just before striking the set at rest. The subscripted b means "before".

$$\Delta E = 0 \implies \Delta K = -\Delta U_{\mathrm{g}}$$

$$\Delta K = K_f - K_i = \frac{1}{2}\,(n\,M)\,v_b^2 - 0 = \frac{1}{2}\,n\,M\,v_b^2$$

$$\text{AND} \quad -\Delta U_{\mathrm{g}} = -(n\,M)\,g\,(-h) = n\,M\,g\,h$$

$$\implies \quad \frac{1}{2}\,n\,M\,v_b^2 = n\,M\,g\,h, \quad \text{and hence} \quad v_b = \sqrt{2\,g\,h}\,.$$

The speed of the moving bobs, just before the collision, does not depend on how many of them there are! In the instant before the collision the motion of the bob(s) is purely horizontal, so the total momentum of the system is

$$\vec{P}_{\text{Total},b} = n\,M\,v_b\,[\leftarrow]\,.$$

MID-   During the Collision

The forces which act are gravity, tension, and the force of contact between the colliding bobs. None of these produce a net impulse on the system during the exceedingly short time scale assumed for the collision.

- Gravity and the tension force cancel at the instant of the collision and thus are not able to effect a change in the momentum of the system.

- The contact force is internal, and therefore is unable to contribute a net impulse to the system.

Thus, the total momentum is conserved through the collision. The momentum just after is equal to the momentum just before. In addition, the kinetic energy is conserved, since we are assuming that the collisions between the bobs are elastic.

$$\vec{P}_{\text{Total},a} = \tilde{n}\,M\,\vec{v}_a \equiv n\,M\,v_b\,[\leftarrow] = \vec{P}_{\text{Total},b}\,,$$

$$K_{\text{Total},a} = \frac{1}{2}\,\tilde{n}\,M\,v_a^2 \equiv \frac{1}{2}\,n\,M\,v_b^2 = K_{\text{Total},b}\,,$$

where $\tilde{n}$ is the number of bobs in motion after the collision. The only consistent solution of this set of equations is[6]

$$\tilde{n} = n \qquad \text{and} \qquad \vec{v}_a = \vec{v}_b\,,$$

exactly as we have illustrated in Figure 32.2.

POST-   After the Collision

The forces which act are gravity and the tension in the string, just as it was before the collision. Requiring that energy be conserved relates the speed of the aggregate after the collision, v_a, to the maximum height that it attains, H. A retracing of the PRE- argument confirms that the n bobs on the "far side" indeed rise to the same height as that from which the n original bobs were released.

Of course, this isn't the end of the story. The struck bob(s) rise, stop, fall, collide with the stationary set of $\mathcal{N} - n$ bobs, causing the original n to return to the height from which they were originally released. The system *oscillates* [the motion repeats in time] in such a manner as is pleasing to behold.

---

[6]Think about the analysis of elastic collisions in the rest frame of one of the particles.

Chapter 33

Completely Inelastic Collisions

In a completely inelastic collision the colliding particles fuse [a.k.a. stick together].

COMPLETELY INELASTIC COLLISION OF TWO PARTICLES

Two particles collide in a perfectly inelastic manner. We assume that before the collision the particles move inertially. [No long-distance forces act between the particles, and external forces, if they exist, precisely cancel.] See the first panel of Figure 33.1 for a snapshot of the system before the collision. The incident particles have masses M_1 and M_2 and velocities $\vec{v}_{1,i}$ and $\vec{v}_{2,i}$, with respect to a particular ["laboratory"] IRF. Knowledge of the masses and the initial velocities is enough to define the initial state of the system [at least to the extent required for our present purposes]. After the collision, the **aggregate** formed by the two particles, now fused into one, is assumed to also move inertially.[1]

FIGURE 33.1 Two Particles Undergoing an Inelastic Collision

The initial momenta of the particles and the system are

$$\left.\begin{array}{l} \vec{p}_{1,i} = M_1\,\vec{v}_{1,i} \\ \vec{p}_{2,i} = M_2\,\vec{v}_{2,i} \end{array}\right\} \implies \vec{P}_{\text{Total},i} = \vec{p}_{1,i} + \vec{p}_{2,i} = M_1\,\vec{v}_{1,i} + M_2\,\vec{v}_{2,i}\,.$$

The initial kinetic energies of the particles and the system are

$$\left.\begin{array}{l} K_{1,i} = \frac{1}{2}\,M_1\,\left|\vec{v}_{1,i}\right|^2 \\ K_{2,i} = \frac{1}{2}\,M_2\,\left|\vec{v}_{2,i}\right|^2 \end{array}\right\} \implies K_{\text{Total},i} = K_{1,i} + K_{2,i} = \frac{1}{2}\,M_1\,v_{1,i}^2 + \frac{1}{2}\,M_2\,v_{2,i}^2\,.$$

ASIDE: It is possible to express the kinetic energies in terms of the momenta, writing

$$K_{1,i} = \frac{\left|\vec{p}_{1,i}\right|^2}{2\,M_1} \quad \text{and} \quad K_{2,i} = \frac{\left|\vec{p}_{2,i}\right|^2}{2\,M_2}\,.$$

---

[1]These assumptions merely aid in defining the initial and final velocities, as mentioned in Chapter 32.

BANG! — the collision occurs.

The corporal integrity of the two particles appearing in the initial state is **NOT** preserved. The two incident particles fuse into a single entity. No mass is lost in the formation of the aggregate, so

$$M_f \equiv M_1 + M_2.$$

After the collision has taken place, the aggregate proceeds with a single, unique, final velocity $\vec{v}_f$. The momentum and kinetic energy of the final-state particle are then [formally] computable.

$$\vec{P}_{\text{Total},f} = M_f\,\vec{v}_f = (M_1 + M_2)\,\vec{v}_f$$
$$K_{\text{Total},f} = \tfrac{1}{2}\,M_f\left|\vec{v}_f\right|^2 = \tfrac{1}{2}\,(M_1 + M_2)\left|\vec{v}_f\right|^2.$$

ASIDE: We can work exclusively with kinetic energy and momenta by writing

$$K_{\text{Total},f} = \frac{\left|\vec{P}_{\text{Total},f}\right|^2}{2\,M_f}.$$

For collisions in general, the total momentum is conserved, so

$$\vec{P}_{\text{Total},f} = \vec{P}_{\text{Total},i} \quad\Longrightarrow\quad \vec{p}_f = \vec{p}_{1,i} + \vec{p}_{2,i} \quad\Longrightarrow\quad M_f\,\vec{v}_f = M_1\,\vec{v}_{1,i} + M_2\,\vec{v}_{2,i}$$

provides the relation between the final and initial states of the system.

....

EXAMPLE [*Completely Inelastic Collision of Two Particles*]

Q: Does knowledge of the initial state of the system determine the final state?

A: You betcha!   [From Conservation of Momentum alone! And in any number of dimensions!]

LAB FRAME   The final velocity of the aggregate is completely determined via

$$M_f\,\vec{v}_f = M_1\,\vec{v}_{1,i} + M_2\,\vec{v}_{2,i} \quad\Longrightarrow\quad \vec{v}_f = \frac{M_1\,\vec{v}_{1,i} + M_2\,\vec{v}_{2,i}}{M_1 + M_2},$$

which is the velocity of the centre of mass of the system!

> ASIDE: A moment's thought convinces one of the necessity of this result. Since the aggregate is but a single particle, its velocity is the velocity of its centre of mass [by definition!]. As the collision forces are internal to the system, the velocity of the centre of mass must be unchanged from that found in the initial state.

REST FRAME   As in Chapter 32, we can re-analyse from the point of view of an observer riding alongside particle 2 in the initial state.

The initial momenta of the particles and the system are

$$\left.\begin{array}{l} \vec{p}_{1,i} = M_1\,\vec{v}_{1,i} \\ \vec{p}_{2,i} = M_2\,\vec{v}_{2,i} = 0 \end{array}\right\} \quad\Longrightarrow\quad \vec{P}_{\text{Total},i} = \vec{p}_{1,i} + \vec{p}_{2,i} = \vec{p}_{1,i} = M_1\,\vec{v}_{1,i}.$$

BANG! — the collision occurs.

After the collision has taken place, the aggregate has velocity $\vec{v}_f$. The expression for the final momentum is

$$\vec{P}_{\text{Total},f} = M_f\,\vec{v}_f = (M_1 + M_2)\,\vec{v}_f\,.$$

Applying momentum conservation provides the final velocity in terms of M_1, M_2, and the initial velocity of particle 1.

$$\vec{P}_{\text{Total},f} = \vec{P}_{\text{Total},i} \quad \Longrightarrow \quad \vec{v}_f = \frac{M_1}{M_1 + M_2}\,\vec{v}_{1,i}\,.$$

En route to repeating the analysis of the collision in the CofM frame, we remind ourselves that in an inelastic collision some of the kinetic energy is dissipated.

Q: So, where does this energy go?

A: Some goes into producing the mechanical deformation of the particles when they fuse together. This energy eventually appears as heat[2] or some form of chemical energy. Also, some kinetic energy is converted into the sound, and perhaps light, energy that emanates from the collision.

CRUNCH!!

Q: How much kinetic energy is lost?

A: We'll answer this question anon.

Since the type of collision discussed above involves a loss of kinetic energy, it does not possess the property of time-reversal invariance. Nonetheless, it is possible for a single particle in the initial state to break up into two, or more, final state pieces which then move apart. This, however, requires the addition of mechanical energy to the system. The standard scenario is an explosion[3] causing the initial particle to separate into two or more fragments. The velocities of these fragments are partially constrained by conservation of momentum and overall energy [mechanical plus all other relevant forms].

---

[2]If you knead a doughy substance or repeatedly flex a rod or sheet of metal, it will become warm to the touch. The warmth arises from the work done in deforming the substance.

[3]By explosion, we mean any sudden conversion of [chemical or spring] energy to mechanical energy.

CofM
FRAME   In the centre of mass frame, the post-collision velocity of the aggregate is ZERO [tautologically] since the CofM is at rest [motionless] with respect to an observer in this frame. The observed velocities in a generic (*i.e.*, laboratory) frame enable the computation of the velocity of the centre of mass (in the lab frame):

$$\vec{V}_{\text{CofM}} = \frac{M_1 \vec{v}_{1,i} + M_2 \vec{v}_{2,i}}{M_1 + M_2}.$$

The particle velocities observed in the CofM frame are the relative velocities,

$$\vec{u}_{1,i} = \vec{v}_{1,i} - \vec{V}_{\text{CofM}} \quad \text{and} \quad \vec{u}_{2,i} = \vec{v}_{2,i} - \vec{V}_{\text{CofM}}.$$

The total initial momentum vanishes and the initial kinetic energy of the system is just the relative kinetic energy of the particles, because the CofM is at rest in this frame.

$$\vec{P}_{\text{Total},i} = \vec{p}_{1,i} + \vec{p}_{2,i} = M_1 \vec{u}_{1,i} + M_2 \vec{u}_{2,i} \equiv 0$$
$$K_{\text{Total},i} = K_{1,i} + K_{2,i} = K_{\text{relative},i} = \tfrac{1}{2} M_1 u_{1,i}^2 + \tfrac{1}{2} M_2 u_{2,i}^2.$$

BANG! — the collision occurs.

After the collision has taken place, the aggregate has velocity $\vec{u}_f$. The expressions for the final momenta and kinetic energies are

$$\vec{P}_{\text{Total},f} = M_f \vec{u}_f = (M_1 + M_2) \vec{u}_f$$
$$K_{\text{Total},f} = K_{\text{relative},f} = \tfrac{1}{2} M_f u_f^2 = \tfrac{1}{2} (M_1 + M_2) u_f^2.$$

Momentum conservation provides the relation between initial and final states and fixes the final velocity of the aggregate to be exactly zero.

$$\vec{P}_{\text{Total},f} \equiv 0 \quad \Longrightarrow \quad \vec{u}_f \equiv 0 \quad \Longrightarrow \quad K_{\text{Total},f} \equiv 0.$$

Now we have obtained the answer to the question posed above.

A: All of the RELATIVE kinetic energy, *i.e.*, that associated with the internal motion of the system. The bulk kinetic energy is associated with the motion of the centre of mass of the system and can only be changed by the action of an external force, and not by a collision.

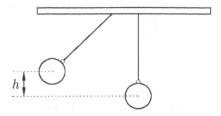

FIGURE 33.2   Initial State of Inelastic Newton's Pendulum

EXAMPLE [*An Inelastic Newton's Pendulum*]

Consider a Newton's Pendulum system comprised of a pair of equal-mass bobs attached to strings of length l, hung from a common height, as in the initial configuration shown in Figure 33.2. One bob rests at the equilibrium position, the other has been moved (along a circular arc) to a height h above the equilibrium point and is held there, at rest, until released. It swings and collides, inelastically, with the lower bob. [We make the approximation that the bobs are "small" with respect to the string length, l, and that the collision occurs when the incident bob reaches the bottom of its swing and is moving horizontally.] After the collision, the two bobs swing together, up to a maximum height above equilibrium, H, as in Figure 33.3.

Q: How does H compare with h?

A: One's intuition prompts the guess that $H = \frac{h}{2}$, but, in fact, the correct answer is $H = \frac{h}{4}$ as we expose by the following line of argument.

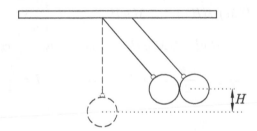

FIGURE 33.3   Final State of Inelastic Newton's Pendulum

PRE-   Before the Collision

The forces which act are gravity (which is conservative) and the tension in the string (which does no net work). Hence, the total mechanical energy of the system is conserved while the momentum is not. We can exploit the conservation of mechanical energy to compute the speed of the moving bob, v_b, just before it strikes the bob at rest. Again, the subscripted b denotes "before".

$$\Delta E = 0 \quad \Longrightarrow \quad \Delta K = -\Delta U_g$$

$$\Delta K = K_f - K_i = \frac{1}{2} M v_b^2 - 0 = \frac{1}{2} M v_b^2 \quad \text{and} \quad -\Delta U_g = -M g\,(-h) = M g h$$

$$\Longrightarrow \quad \frac{1}{2} M v_b^2 = M g h, \quad \text{and hence} \quad v_b = \sqrt{2 g h}.$$

We are assuming that the swinging ball is moving horizontally in the instant before the collision, so its momentum [the total momentum of the system] is

$$\vec{P}_{\text{Total},b} = M\,v_b\,[\rightarrow]\,.$$

MID-   **During the Collision**

The forces which act are gravity, tension, and the force of contact between the bobs. None of these produce a net impulse on the system during the exceedingly short time scale assumed for the collision. Thus, the total momentum just after the collision is equal to the momentum immediately before, *viz.*,

$$\vec{P}_{\text{Total},a} = (2M)\,\vec{v}_a \equiv \vec{P}_{\text{Total},b} = M\,v_b\,[\rightarrow] \quad \Longrightarrow \quad \vec{v}_a = \frac{1}{2}\,v_b\,[\rightarrow]\,.$$

Conservation of momentum during the collision has enabled us to determine the velocity of the two-bob aggregate immediately after the collision.

POST-   **After the Collision**

Once again, the forces which act are gravity and the tension in the string. This is just like the situation before the collision. Conservation of energy yields the relation between the speed of the aggregate after the collision, v_a, and the maximum height, H, that it attains.

$$\Delta E = 0 \quad \Longrightarrow \quad \Delta K = -\Delta U_{\text{g}}$$

$$\Delta K = K_f - K_i = 0 - \frac{1}{2}\,(2M)\,v_a^2 = -M\,v_a^2$$

$$-\Delta U_{\text{g}} = -(2M)\,g\,(H) = -2\,M\,g\,H$$

$$\Longrightarrow \quad M\,v_a^2 = 2\,M\,g\,H \qquad \text{and} \qquad v_a = \frac{1}{2}\,v_b = \sqrt{\frac{g\,h}{2}}$$

$$\Longrightarrow \quad \frac{g\,h}{2} = 2\,g\,H \quad \Longrightarrow \quad H = \frac{h}{4}\,.$$

Cool, eh?

ADDENDUM: Collisions Reexamined

Here we explicitly examine both elastic and inelastic [1-d] collisions of two bodies in various frames. The mass of particle 1 is taken[4] to be 4 kg, while that of particle 2 is 1 kg. All three of the observers discussed below have agreed to synchronise their watches to a common time, t, and have deemed that the collision occurred at the instant $t = 2\,\text{s}$. Furthermore, the point of collision is labelled 5 m in each of their respective frames. Notwithstanding the explicit use of metres and seconds, the behaviour of a system of colliding bodies scales with the choice of units. In the chart below, the units are elided. The trajectory of particle 1 is illustrated with a solid line, and that of particle 2 by a dash-dot line. The CofM moves along the narrower dashed line, while the aggregate, in the inelastic case, follows the thicker dashed line.

---

[4]Only the ratio of the particle masses is truly significant.

ELASTIC COLLISION			
FRAME	LAB	#1 at rest	CofM
Initial Conditions	$v_{1i} = -2 \quad v_{2i} = 3$	$v_{1i} = 0 \quad v_{2i} = 5$	$v_{1i} = -1 \quad v_{2i} = 4$
$V_{\text{CofM},i}$	-1	1	0
$K_{\text{Total},i}$	$12\frac{1}{2}$	$12\frac{1}{2}$	10
Trajectories			
Final Conditions	$v_{1f} = 0 \quad v_{2f} = -5$	$v_{1f} = 2 \quad v_{2f} = -3$	$v_{1f} = 1 \quad v_{2f} = -4$
$V_{\text{CofM},f}$	-1	1	0
$K_{\text{Total},f}$	$12\frac{1}{2}$	$12\frac{1}{2}$	10
ΔK_{Total}	0	0	0

Several facts are noteworthy.

•→ ←•   The observers regard the same collision from distinct viewpoints.

•→ ←•   One of the observers moves along with particle 1, seeing it initially at rest. Since the labels are interchangeable, this is not unlike the "rest frame of particle 2" examined earlier in Chapters 32 and 33.

•→ ←•   The time-reversal symmetry of Newton's Laws carries over to collisions. That is, each of these spacetime diagrams, illustrating multi-particle trajectories, read from left [PAST] to right [FUTURE], may also be read from right to left, yielding another set of multi-particle trajectories which also correspond to valid collisions. To make this point emphatically, in this particular elastic collision, the lab- and rest-frame observers explicitly verify this symmetry. The time-reversed form of an inelastic collision is better termed an explosion, as the system begins with but one particle in the initial states and passes to a final state consisting of two particles along with additional kinetic energy. This additional energy is obtained by conversion from some other form, *e.g.*, chemical, electromagnetic, or potential.

•→ ←•   The total momentum determined by each observer is conserved in both elastic and inelastic collisions. The particular value of the conserved momentum is observer-dependent. The total amount of kinetic energy is conserved in elastic collisions. It is a consequence of time-reversal symmetry that the kinetic energies determined in the lab and rest frames happen to be equal in value. The total kinetic energy is not conserved for inelastic collisions. However, all observers agree on the amount of energy lost ["gained" in the time-reversed cases] as such energy transfers must comport with the Law of Energy Conservation.

INELASTIC COLLISION			
FRAME	**LAB**	**#1 at rest**	**CofM**
Initial Conditions	$v_{1i} = -2 \quad v_{2i} = 3$	$v_{1i} = 0 \quad v_{2i} = 5$	$v_{1i} = -1 \quad v_{2i} = 4$
$V_{\text{CofM},i}$	-1	1	0
$K_{\text{Total},i}$	$12\frac{1}{2}$	$12\frac{1}{2}$	10
Trajectories			
Final Conditions	$v_f = -1$	$v_f = 1$	$v_f = 0$
$V_{\text{CofM},f}$	-1	1	0
$K_{\text{Total},f}$	$2\frac{1}{2}$	$2\frac{1}{2}$	0
ΔK_{Total}	-10	-10	-10

Chapter 34

Rotation

Thus far we have devoted ourselves to the study of systems of particles and rigid extended bodies undergoing translational motion.

Q: Is that it? Are we done, then?

A: Nope. Particles and rigid extended bodies can exhibit rotational motion, too.

RIGID ROTATION ABOUT A FIXED AXIS   In rigid rotation about a fixed axis:

- The direction singled out by the rotational axis is unchanging.

  [This excludes the PRECESSIONAL MOTION of a tilted spinning top, in which the top's axis is itself rotating about the vertical direction.]

- The axis is not translating through space.

  [In ROLLING MOTION, the axis moves with the object.]

- All points on the rotating body remain at the same distance from the axis of rotation.

  [When one plays with a hula hoop, parts of the hoop migrate in-and-out as the hoop goes round-and-round. A SPIROGRAPH™ toy *illustrates* the effects of a moving axis and relative motion with respect to the axis.]

Imagine a *generic* non-spherical object rotating about a fixed axis.

Q: What?! You can't do this?

A: That's okay. I can't either. So instead imagine a wine bottle[1] which is somehow firmly attached to an axis, about which it is free to rotate, as pictured in Figure 34.1. In the sketch, the axis about which the body rotates comes straight out of the page. A line orthogonal to and intersecting the axis of rotation serves to define a reference direction from which angles are measured.

The rotating system displayed in Figure 34.1 has its axis of rotation directed perpendicular both to the page and to the symmetry-axis of the bottle. This axis alignment simplifies the drawing by obviating the need for perspective. There is no physical significance to this choice; it merely makes it easier to visualise the rotating system.

Kinematic analysis of the rotating object commences with the choice of two points:

$\mathcal{A}$ a point on the axis of rotation, $\mathcal{A}$, designated as *the point* about which the rotation occurs, and

---

[1] *Mouton Cadet—Rouge ou Blanc* varietals—is a charming French Bordeaux.

FIGURE 34.1   A Sketch Showing the Salient Features of the Spinning Wine Bottle

$\mathcal{P}$ a reference point, $\mathcal{P}$, located on or within the rotating body, but not on the axis of rotation.

While it is advantageous to have $\mathcal{A}$ and $\mathcal{P}$ lie in a common plane perpendicular to the axis, this need not be assumed.

There exists a [UNIQUE, though time-dependent] vector, $\overrightarrow{\mathcal{AP}}$, extending from $\mathcal{A}$ to $\mathcal{P}$. This vector has a projection onto the plane which is perpendicular to the rotation axis and contains the ray indicating the reference direction. The angle θ, lying between the projected vector and the reference direction, characterises the angular position of the body. The fact that the body is rotating requires that the angular position be a function of time, $\theta(t)$.

Our finely honed[2] physics instincts lead us to inquire: "How is it changing in time?"

In order to precisely frame these and other questions, we now investigate angular kinematics, as we once began with rectilinear kinematics long ago in Chapter 2.

ANGULAR DISPLACEMENT   Angular displacement is the net difference in the angular positions, θ_i and θ_f, of the reference point on the body [with respect to the reference direction] occurring through a time interval Δt which commenced at t_i and ended at t_f. That is,

$$\Delta\theta = \theta(t_f) - \theta(t_i) = \theta_f - \theta_i.$$

There are [at least] eight noteworthy items to expound upon.

○   The SI unit associated with angular displacement is the radian. The radian is a dimensionless unit.[3] There are 2π radians, equivalent to 360 degrees, in a [planar Euclidean[4]] circle. Conversion between radians and degrees is readily effected by

---

[2]The snarky among us might rephrase this as *"finally* honed."
[3]Although this sounds oxymoronic, it's not!
[4]In non-Euclidean geometries, circles subtend more or less than 2π radians.

means of the following factors.

$$360° = 2\,\pi\,\text{rad} \quad \Longrightarrow \quad \begin{cases} 1\,\text{rad} = \dfrac{180°}{\pi} \simeq 57.2958° \simeq 57.3° \\[2mm] 1° = \dfrac{\pi}{180}\,\text{rad} \simeq 0.01745\,\text{rad} \end{cases}$$

○ IF the rotating object's progress $\theta_i \to \theta_f$ throughout the time interval Δt is not monotonic, THEN the angular distance traversed by the rotating object will exceed the angular displacement $\Delta\theta$.

○ The angular displacement may be positive, negative, or zero. POSITIVE displacement corresponds to net anti-clockwise rotation, while NEGATIVE displacement corresponds to net clockwise rotation. ZERO net displacement occurs when $\theta_f = \theta_i$, *i.e.*, when there is no net change in angular position.

○ Winding matters crucially in the determination of angular displacement.

IF the bottle rotates anti-clockwise by exactly $\mathcal{N}$ full turns, THEN the angular displacement is

$$\Delta\theta = \mathcal{N} \times (2\,\pi) = +2\mathcal{N}\pi \neq 0\,,$$

despite the fact that all points on the body have undergone no translational displacement whatsoever. *I.e.*, at the end of the time interval each occupies exactly the same point in space as it did at the beginning.

Similarly, IF the bottle rotates clockwise by exactly $\mathcal{M}$ full turns, THEN

$$\Delta\theta = -\mathcal{M} \times (2\,\pi) = -2\mathcal{M}\pi \neq 0\,.$$

Concisely: angular displacements are not "congruent modulo $2\,\pi$."

○ INCORRECT REASONING

One who finds these subtleties dizzying might attempt to spin *falsely* thus:

> Since shifting one's point of view from above the page to below the page has the effect of exchanging the senses of clockwise and anti-clockwise, the POSITIVE/anti-clockwise *vs.* NEGATIVE/clockwise distinction made above is meaningless.

The ineluctable flaw in the above argument is its failure to account for the particular choice made to direct the axis of rotation UP out of the page. [This fact is made evident in Figure 34.1 and explicitly described in the surrounding text.] This choice has fixed an orientation which then unambiguously distinguishes clockwise from anti-clockwise [and hence negative and positive angular displacements]. Crawling under the page in no way *undermines* the convention.

○ **Q:** Did we err in calling this angular displacement [rather than DIFFERENCE]?

A: Nope. It's displacement; the net angle through which the object rotates.

Q: Rectilinear displacement[5] is a vector quantity. How can $\Delta\theta$ have direction?

A: The angular displacement is directed perpendicular to the plane in which the rotation occurs, *i.e.*, along the axis of rotation with the orientation determined by a NEW RHR.

---

[5] Recall the definitions and discussions in Chapters 2 and 6.

RHR   Position your right hand with palm orthogonal to rays extending perpendicular from the axis and fingers curving along the forward direction of spatial paths taken by bundles of neighbouring points. Your right thumb indicates the sense along the axis in which the angular displacement is directed. By convention (and unless we have a good reason for doing otherwise) we assign[6] this to be the positive sense of increase of the angle.

The angular displacement vector is, therefore, unambiguously determined to be

$$\Delta \vec{\theta} = \vec{\theta}_f - \vec{\theta}_i = (\theta_f - \theta_i) \, [\text{RHR}].$$

The angles themselves have vectorial nature: their magnitudes are directed along the [fixed] rotation axis, with orientation assigned according to the RHR.

○   The assumed rigidity of the rotating body makes the angular displacement unique and independent of the choice of reference point.

[Non-rigid objects may experience **shear**[7] during rotation.]

○   In many instances, our goal shall be to determine the angular position of the rotating object as a function of time, *viz.*, $\theta(t)$.

AVERAGE ANGULAR VELOCITY   The average angular velocity is equal to the angular displacement [occurring throughout the time interval] divided by the duration of the time interval, Δt.

$$\vec{\omega}_{\text{av}} = \frac{\Delta \vec{\theta}}{\Delta t} = \frac{\vec{\theta}_f - \vec{\theta}_i}{t_f - t_i} = \frac{\vec{\theta}(t_f) - \vec{\theta}(t_i)}{t_f - t_i} \,.$$

As only situations in which the axis is fixed are considered here, the vector symbols are often omitted, letting the direction be implicit, and the orientation be specified by the sign.

INSTANTANEOUS ANGULAR VELOCITY   The instantaneous angular velocity is obtained by consideration of the average angular velocity throughout an infinitesimally small time interval, *i.e.*, $\Delta t \to 0$.

$$\vec{\omega} = \lim_{\Delta t \to 0} \vec{\omega}_{\text{av}} = \lim_{\Delta t \to 0} \frac{\Delta \vec{\theta}}{\Delta t} = \frac{d\vec{\theta}}{dt} \,.$$

ANGULAR SPEED   The instantaneous angular speed[8] is the magnitude of the angular velocity,

$$\omega = \left| \frac{d\vec{\theta}}{dt} \right| \,.$$

The angular velocity has magnitude ω and is directed along the axis of rotation, with orientation consistent with the RHR.

---

[6]Similar considerations are operative in the choice of orientation in 1-d rectilinear motion.
[7]Shear is discussed in VOLUME II.
[8]Angular speed was introduced, in the context of particles undergoing circular motion, in Chapter 8.

ASIDE: *Caveat lector!* It is common practice to drop the vector notation and hence elide the distinction between angular speed and velocity. Thus the sign of

$$\omega = \frac{d\theta}{dt}$$

carries all of the information needed to specify the vector velocity. Precisely the same convention was adopted for one-d rectilinear kinematics throughout Chapters 2–4.

The units associated with angular velocity are

$$[\omega] = \text{radians per second}.$$

These are [almost] the same units as cycle frequency, since radians are dimensionless. Cycle frequency, denoted by ν (the Greek letter "nu," is pronounced "new"),[9] is the rotation rate expressed in terms of revolutions per unit time. The SI unit of cycle frequency is

$$[\nu] = \text{cycles per second} = \frac{1}{\text{s}} = \text{Hz} = \text{hertz},$$

so-named in honour of Heinrich Hertz, a nineteenth century German physicist.

The conversion from angular speed [*a.k.a.* angular velocity or angular frequency] to frequency is accomplished by

$$\nu = \frac{\omega}{2\pi},$$

since there are 2π radians in a full cycle [circle].

As there is no reason whatsoever, physical, mathematical, or otherwise, to presume that the instantaneous angular velocity is constant, the inevitability of angular acceleration must be accepted.

AVERAGE ANGULAR ACCELERATION   Average angular acceleration is the change in the instantaneous angular velocity [occurring through the time interval Δt] divided by the duration of the time interval.

$$\vec{\alpha}_{\text{av}} = \frac{\Delta\vec{\omega}}{\Delta t} = \frac{\vec{\omega}_f - \vec{\omega}_i}{t_f - t_i}.$$

INSTANTANEOUS ANGULAR ACCELERATION   The instantaneous angular acceleration is obtained via consideration of the limit of the average angular acceleration throughout Δt as the duration of the time interval vanishes:

$$\vec{\alpha} = \lim_{\Delta t \to 0} \vec{\alpha}_{\text{av}} = \frac{d\vec{\omega}}{dt}.$$

When restricted to a fixed axis of rotation, as is the case here, the angular acceleration vector must align with the rotation axis. The orientation can again be specified by the sign, enabling the consistent re-expression of the kinematical formulae without explicit vector notation:

$$\alpha = \frac{d\omega}{dt} = \frac{d^2\theta}{dt^2}.$$

---

[9]Some authors represent cycle frequency by "*f*" for *fathomable* reasons.

It is not too surprising that:

Inversion of the definitions of the kinematical quantities [via integration] enables determination of the velocity from the acceleration and of the position from the velocity.

The case in which the angular acceleration is constant in magnitude [the direction is constant when the axis is fixed] deserves particular consideration, for the integrations are readily performed. Here,

$$\alpha = \bar{\alpha} = \text{constant} \quad \Longrightarrow \quad \omega = \omega_0 + \bar{\alpha}\,t \quad \Longrightarrow \quad \theta = \theta_0 + \omega_0\,t + \frac{1}{2}\,\bar{\alpha}\,t^2\,,$$

where ω_0 and θ_0 are constants of integration.[10] Furthermore, the angular velocity equation is linear in time and hence is easily inverted to express t in terms of the change in angular velocity and the constant angular acceleration:

$$t = \frac{\omega - \omega_0}{\bar{\alpha}}\,.$$

Substitution of this into the equation for the angle yields the **third** constant angular acceleration kinematical formula:

$$\omega^2 - \omega_0^2 = 2\,\bar{\alpha}\big(\Delta\theta\big)\,.$$

This last bit cements a [thus far] exact analogy between rotation about a fixed axis and rectilinear motion in 1-d, *viz.*,

$$\theta \sim x \qquad\qquad \omega \sim v \qquad\qquad \alpha \sim a.$$

Eventually [in Chapter 36], this analogy shall be extended so as to provide a correspondence between rotational dynamics and rectilinear dynamics. In the interim, Chapter 35, we shall undertake some angular kinematical analysis, and shall introduce and examine the direct correspondence between the rotational motion of a rigid extended object and the instantaneous translational motions of its constituent parts.

[Clear correspondence bests analogical association, eh?]

---

[10]Recall the discussions of the physical import of such kinematical constants in Chapters 2 and 6.

Chapter 35

Rotation and Translation

The discussion of fixed-axis rotational kinematics in the preceding chapter yielded definitions of angular position, displacement, velocity, and acceleration, *viz.*,

$$\theta(t), \qquad \Delta\theta = \theta(t_f) - \theta(t_i), \qquad \omega(t) = \frac{d\theta}{dt}, \qquad \alpha(t) = \frac{d\omega}{dt} = \frac{d^2\theta}{dt^2},$$

respectively. Under the assumption that the axis of rotation is fixed, the vector natures of these quantities may remain implicit.

In the following pair of examples, we shall explore and exploit the analogy between rotation about a fixed axis and rectilinear motion in 1-d, *viz.*,

$$\theta \sim x \qquad\qquad \omega \sim v \qquad\qquad \alpha \sim a.$$

Afterward, we will recast the motion of a rotating object in terms of particular motions of points on the surface of or within the bulk of the body.

EXAMPLE [*Fixed-Axis Constant Angular Acceleration Kinematics*]

A CD player spins up a disk from rest to 480 RPM in four seconds flat.

Qα: What is the average angular acceleration of the disk while it is spun up?
Qθ: Through what total angle does the disk rotate while being spun up?

In order to answer these questions we shall effectively assume that the CD experiences uniform[1] angular acceleration while it is spun up. The constant angular acceleration kinematical formulae read

$$\alpha = \bar{\alpha} = \text{constant} \quad\Longrightarrow\quad \begin{cases} \omega = \omega_0 + \bar{\alpha}\,t \\ \theta = \theta_0 + \omega_0\,t + \frac{1}{2}\bar{\alpha}\,t^2\,, \\ \omega^2 = \omega_0^2 + 2\,\bar{\alpha}(\Delta\theta) \end{cases}$$

where ω_0 and θ_0 are constants of integration determined by the initial conditions particular to the problem at hand.

Aα: The constant average angular acceleration is equal to the change of the angular velocity divided by the duration of the time interval in which the change occurred. The initial angular velocity is zero, since the disk was at rest at the start of the four second spin up. The final angular velocity must be computed from the cycle frequency:

$$480 \text{ RPM} = \frac{480}{60} \text{ Hz} = \frac{480 \times 2\,\pi}{60} \text{ rad/s} = 16\,\pi \text{ radians per second.}$$

---

[1]The average value of a constant [acceleration] function is equal to the value of the constant.

Thus, the average angular acceleration is determined to be

$$\bar{\alpha} = \frac{\Delta\omega}{\Delta t} = \frac{16\,\pi - 0}{4} = 4\,\pi \ \text{rad/s}^2.$$

This answers the first question.

Aθ: Determining the total angle through which the disk rotates may be accomplished in a variety of ways. We'll explore two of these means below.

METHOD ONE: [$\alpha \to \theta$ *via* t]
The angular displacement at time $t = 4\,\text{s}$ may be obtained by simply substituting this time into the constant acceleration kinematical formula for the angular position, now that the acceleration is known.

$$\Delta\theta = \theta(4) - \theta_0 = \left[\omega_0\,t_f + \frac{1}{2}\,\bar{\alpha}\,t_f^2\right]\Bigg|_{t_f=4} = 0 \times 4 + \frac{1}{2}\,(4\,\pi)\,4^2 = 32\,\pi \ \text{rad}.$$

This angular displacement, $32\,\pi\,\text{radians}$, corresponds to 16 revolutions of the disk.

METHOD TWO: [*Using the third kinematical relation*]
The angular displacement may be determined, without explicit reference to the final time, by solving

$$\omega_f^2 = \omega_0^2 + 2\,\bar{\alpha}\,(\Delta\theta),$$

where $\omega_f = 16\,\pi$, $\omega_0 = 0$, and $\bar{\alpha} = 4\,\pi$. The result, so obtained,

$$\Delta\theta = \frac{(16\,\pi)^2 - 0}{2 \times 4\,\pi} = 32\,\pi,$$

cannot fail to agree with the result obtained via METHOD ONE.

EXAMPLE [*Fixed-Axis Rotational Kinematics*]

Suppose that [somehow] it is known that the angular position of the reference point on the bottle of *Mouton Cadet*, valid in a [continuous] temporal interval including times $t = 1\,\text{s}$ and $t = 2\,\text{s}$, is expressed by

$$\theta(t) = \beta\,t^3,$$

for some constant β, with units of $[\beta] = \text{radians/s}^3$. We seek to determine the angular accelerations, velocities and positions of the bottle at the two times $t = 1\,\text{s}$ and $t = 2\,\text{s}$.

[Compare this example to *Determining Velocity From a Known Trajectory* in Chapter 2.]

θ   Angular Position

The angular positions are obtained directly. At $t = 1$ second, the angular position of the reference point is $\theta(1) = \beta \times (1\,\text{s})^3 = \beta\,\text{rad}$, while at $t = 2$ seconds, $\theta(2) = \beta \times (2\,\text{s})^3 = 8\,\beta\,\text{rad}$.

ω   Angular Velocity

The angular velocity is the derivative of the angular position function

$$\omega(t) = \frac{d\theta(t)}{dt} = \frac{d}{dt}\,(\beta\,t^3) = 3\,\beta\,t^2.$$

The angular velocity is $3\,\beta\,\text{rad/s}$ at time $t = 1\,\text{s}$, and $12\,\beta\,\text{rad/s}$ at $t = 2\,\text{s}$.

α Angular Acceleration

The acceleration is the instantaneous time rate of change of the velocity, so

$$\alpha(t) = \frac{d\omega(t)}{dt} = \frac{d}{dt}\left(3\,\beta\,t^2\right) = 6\,\beta\,t\,.$$

The angular acceleration is $6\,\beta\,\text{rad/s}^2$ at $t = 1\,\text{s}$ and $12\,\beta\,\text{rad/s}^2$ at $t = 2\,\text{s}$.

EXACT CORRESPONDENCE BETWEEN ANGULAR AND RECTILINEAR KINEMATICS

The description of angular motion is predicated on the spatial position of a reference point on the rotating body acting as a proxy for the BULK rotation of the entire body.

Q: Precisely what happens to the reference point as the body rotates?

A: The reference point moves along a circular trajectory in a plane perpendicular to the axis of rotation.

Here are the details. Let r_p denote the perpendicular distance from the axis of rotation to the reference point $\mathcal{P}$. The locus of points constituting the trajectory of the reference point is a circle with radius r_p. During the short time interval $\Delta t = t_f - t_i$, the body has rotated and the reference point has moved from an initial to a final point in space. See Figure 35.1 for a sketch of the situation. The rotation of the bottle responsible for the movement of the reference point is characterised by the angle, $\Delta\theta$, lying between the rays to the reference point at the final and initial times,

$$\Delta\theta = \theta_f - \theta_i = \theta(t_f) - \theta(t_i) = \theta(t_i + \Delta t) - \theta(t_i)\,.$$

The distance travelled by the reference point is [the arc length of the circular segment],

$$\Delta S = r_p\,\Delta\theta\,.$$

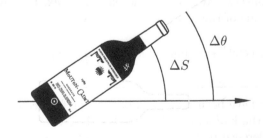

FIGURE 35.1 Angular Motion of a Rigid Extended Body and the Rectilinear Motion of a Reference Point

The average tangential speed of the reference point throughout Δt is[2]

$$v_{\mathrm{av},t} = \frac{\Delta S}{\Delta t} = r_{\!P}\,\frac{\Delta\theta}{\Delta t} = r_{\!P}\,\omega_{\mathrm{av}}\,.$$

The instantaneous tangential speed of the reference point is obtained, as usual, by taking the limit of the average tangential speed as the time interval shrinks to zero.

$$v_t = \lim_{\Delta t\to 0} v_{\mathrm{av},t} = r_{\!P}\,\lim_{\Delta t\to 0}\omega_{\mathrm{av}} = r_{\!P}\,\omega\,,$$

where, in the last equality, ω is the contemporaneous value of the angular velocity.

The average tangential acceleration throughout the time interval $\Delta t = t_f - t_i$, in which the tangential speed changes by an amount $\Delta v_t = v_{t,f} - v_{t,i} = r_{\!P}\,\omega_f - r_{\!P}\,\omega_i = r_{\!P}\,\Delta\omega$, is

$$a_{\mathrm{av},t} = \frac{\Delta v_t}{\Delta t} = r_{\!P}\,\frac{\Delta\omega}{\Delta t} = r_{\!P}\,\alpha_{\mathrm{av}}\,.$$

The instantaneous tangential acceleration is

$$a_t = \lim_{\Delta t\to 0} a_{\mathrm{av},t} = r_{\!P}\,\lim_{\Delta t\to 0}\alpha_{\mathrm{av}} = r_{\!P}\,\alpha\,.$$

Thus, the tangential velocity and acceleration of a point on the rotating body [located a perpendicular distance $r_{\!P}$ from the axis] are related to their rotational counterparts by:

$$v_t = \omega\,r_{\!P} \qquad \text{and} \qquad a_t = \alpha\,r_{\!P}\,.$$

IF $\alpha = 0$ is constant, THEN ω is constant, and hence the reference point undergoes UNIFORM CIRCULAR MOTION. We recall that this motion is not inertial, as discussed in Chapter 8, where centripetal acceleration,

$$a_c = \frac{v^2}{R}\,,$$

was first introduced. Here, the radius is $R = r_{\!P}$, the [tangential] speed is $v = v_t = \omega\,r_{\!P}$, and thus

$$a_c = \omega^2\,r_{\!P}\,.$$

This expression for the magnitude of the centripetal acceleration, in terms of angular quantities, is completely general and in particular does not depend on the assumption that the tangential acceleration vanishes. When both accelerations exist, they are orthogonal by definition, and the magnitude of the total acceleration is

$$|\vec{a}| = \left| a_c[\,\text{centripetal}\,] + a_t[\,\text{tangential}\,] \right| = \sqrt{a_c^2 + a_t^2} = r_{\!P}\,\sqrt{\omega^4 + \alpha^2}\,.$$

This said, we note that we have been speaking about the reference point as though it were a particle. Indeed, we can think of a little chunk of the rigidly rotating object with its CofM located at [or near] the reference point as being a particle! We'll do just this in the next chapter on an excursion to uncover the rotational analogue of mass.

---

[2]The symbol "t" does double duty here, signifying *tangential* as well as denoting the time parameter.

Chapter 36

Introduction to Rotational Dynamics

Let's start our investigation of rotational dynamics by computing the kinetic energy of a rigid body rotating about a fixed axis with angular speed/velocity ω.

Q: Wait! How in the world are we to do this?

A: Here's how:

PARTITION   the body into little chunks.

COMPUTE   the translational kinetic energy of each chunk.

SUM   the kinetic energy contributions from each and every chunk, and *voilà*!

We will have computed the kinetic energy of the body.

This might look like legerdemain, for it was our intent to compute the rotational kinetic energy of the entire body, while the proposed method sums the translational kinetic energies of chunks constituting a partition of the body. In fact, this method is efficacious precisely because kinetic energy **is** energy of motion. It does not matter whether the motion is [characterised as[1]] rotation or translation – all that matters is that *it moves*.

With this plan in hand, we can compute the rotational kinetic energy and, serendipitously, we'll uncover the rotational analogue of mass!

PARTITION   Partition[2] the object into $\mathcal{N}$ chunks, as illustrated in Figure 36.1.

The mass of the ith chunk is Δm_i. [The use of the symbol Δm presages the passage to the limit of an infinite partition.] The selection of a single reference point internal to each chunk uniquely specifies its distance from the axis of rotation, d_i, and its velocity, $\vec{v}_i$.

COMPUTE   Compute the translational kinetic energy of each chunk.

The kinetic energy of the ith chunk is

$$\Delta K_i = \frac{1}{2} \Delta m_i \left| \vec{v}_i \right|^2 .$$

[The kinetic energy contributions of the chunks are marked with "Δ's," as well.] For rigid

---

[1]These weasel words are made necessary by the freedom to choose a single, overall IRF, or a different one at each instant. We encountered (*glossed over* is more accurate) this idea of a collection of instantaneous reference frames when we chose radial and tangential coordinates instantaneously aligned with the roller coaster carriage in the Loop-the-Loop example in Chapter 19. We'll see (or rather *gloss over*) something similar when we consider rolling in Chapter 43.

[2]One must be on guard against pathologies which may arise if the chunks are **filamentous** [line-like] or **foliated** [sheet-like].

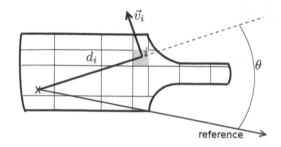

FIGURE 36.1   An $\mathcal{N}$-partition of the Spinning Wine Bottle

rotation, the tangential speed of the chunk, expressed in terms of the angular speed of the rotating body and the distance from the axis to the chunk, is

$$\left|\vec{v}_i\right| = \omega\, \mathrm{d}_i \, ,$$

as was shown in Chapters 8 and 35. Hence, the kinetic energy of the ith chunk is

$$\Delta K_i = \frac{1}{2}\, \Delta m_i\, \mathrm{d}_i^2\, \omega^2 \, .$$

SUM   Sum the energies of the chunks to obtain the total kinetic energy of the body.

Summing over the chunks [*a.k.a.* the mass elements in the partition] yields

$$K_{\text{Total}} = \sum_{i=1}^{\mathcal{N}} \Delta K_i = \sum_{i=1}^{\mathcal{N}} \left\{ \frac{1}{2}\, \Delta m_i\, \mathrm{d}_i^2\, \omega^2 \right\} = \frac{1}{2} \left(\sum_{i=1}^{\mathcal{N}} \Delta m_i\, \mathrm{d}_i^2 \right) \omega^2 \, .$$

In the last equality, the constants $1/2$ and ω^2 factored out of the expression in parentheses.

The form of the result for the total amount of rotational kinetic energy is reminiscent of that for translational motion, $K_{\text{translational}} = \frac{1}{2}\, m\, v^2$, and we are thus inspired to write

$$K_{\text{rotational}} = \frac{1}{2}\, I\, \omega^2 \, .$$

The factor I, acting as the rotational analogue of mass, is called the MOMENT OF INERTIA.

MOMENT of INERTIA   The moment of inertia, I, of a system of $\mathcal{N}$ discrete point-like particles rotating about a fixed axis is

$$I = \sum_{i=1}^{\mathcal{N}} \Delta m_i\, \mathrm{d}_i^2 \, ,$$

where Δm_i and d_i denote the mass and the perpendicular distance from the axis, respectively, pertinent to each of the $\mathcal{N}$ particles.

A continuous distribution of matter undergoing rigid rotation about a fixed axis can be formally partitioned into a finite collection of $\mathcal{N}$ discrete point-like chunks. This partition may be infinitely refined by taking the simultaneous limits, $\{\mathcal{N} \to \infty \text{ AND } \Delta m_i \to 0, \forall i\}$, providing passage from the discrete sum over point(-like) particles to an integral over the mass distribution.

$$I = \int dm \, \mathrm{d}^2 ,$$

where d denotes the perpendicular distance of each mass element, dm, from the axis of rotation. **The moment of inertia of a body is its *mass-weighted squared distance from the axis of rotation.***

As usual, there are several additional comments to be made.

- The dimensions and SI units associated with the moment of inertia are

$$[I] = [\,\text{Mass}\,] \cdot [\,\text{Length}\,]^2 = \mathbf{kg} \cdot \mathbf{m}^2 .$$

- **The moment of inertia is the rotational analogue of mass.**

  Nonetheless, there is an essential distinction. Mass is [taken to be[3]] an intrinsic quantity, whereas moments of inertia depend on how mass is distributed with respect to an axis of rotation.

- **The moment of inertia for a system is equal to the sum of the moments of inertia of its constituents, computed with respect to the same axis.** That is, the PRINCIPLE OF LINEAR SUPERPOSITION applies to moments of inertia.

EXAMPLE [*Moment of Inertia of a Baton Twirled about Its Centre*]

A baton may be crudely modelled as a massless rod, of length $2\,a$, and two point(-like) endpieces, each of mass M. Suppose that the baton is spinning about an axis perpendicular to the rod and through its centre.

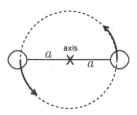

FIGURE 36.2 A Simple Model of a Baton Spinning about Its Centre

The baton's mass-weighted squared distance from the axis [*i.e.,* its moment of inertia] is

$$I_{\text{baton, centre}} = I_{\text{rod, centre}} + I_{\text{endpieces, centre}} ,$$

---

[3]Pardon the weasel words. The age-old philosophical debates about the nature of mass continue to this day. In Chapter 10, the [empirically consistent] claim was made [for modelling purposes] that mass is intrinsic. In *reality*™, mass may be predicated on the existence of the rest of the universe (Mach's Principle), or on the properties of the putative Higgs field. In either of these possible cases, mass becomes a relational quantity.

where $I_{\text{rod, centre}} = 0$, because it is massless, and

$$I_{\text{endpieces, centre}} = \sum_{i=1}^{2} m_i \, d_i^2 = M \, a^2 + M \, a^2 = 2 \, M \, a^2 \, .$$

Therefore, the total moment of inertia of this baton, while it rotates about its centre, is

$$I_{\text{baton, centre}} = 2 \, M \, a^2 \, .$$

Let's ensure that this result agrees with our expectations. This moment of inertia grows

M linearly with the mass of the endpieces, and

a quadratically with increasing length of the rod.

EXAMPLE [*Moment of Inertia of a Baton Twirled about an Endpoint*]

Consider the same baton as in the example above, spinning about an axis perpendicular to the rod and passing through one end.

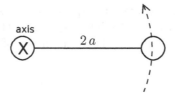

FIGURE 36.3   A Simple Model of a Baton Spinning about Its End

The mass-weighted squared distance from the axis for the baton is

$$I_{\text{baton, end}} = I_{\text{rod, end}} + I_{\text{endpieces, end}} \, ,$$

where $I_{\text{rod, end}} = 0$, again, because the rod is assumed to be massless, and

$$I_{\text{endpieces, end}} = \sum_{i=1}^{2} m_i \, d_i^2 = 0 + M \, (2 \, a)^2 = 4 \, M \, a^2 \, .$$

Therefore, the total moment of inertia of the baton, while it rotates about an endpoint, is

$$I_{\text{baton, end}} = 4 \, M \, a^2 \, .$$

This moment of inertia also behaves in accord with our physical intuitions.

M The (point-)mass located on the axis makes a vanishing contribution to the moment of inertia of the baton. The moment of inertia depends linearly on the mass of the off-axis endpiece.

a The moment of inertia grows quadratically with increasing length of the rod.

AXIS $I_{\text{baton, end}} > I_{\text{baton, centre}}$.

EXAMPLE [*Moment of Inertia of a Rod Spinning about an Endpoint*]

A thin uniform rod of total mass M and length L rotates about one of its endpoints. The rod acts like a 1-d filament with a constant lineal mass density, $\lambda_0 = \frac{M}{L}$.

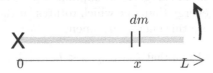

FIGURE 36.4 A Thin Uniform Rod Spinning about an Endpoint

Let us characterise the position of each mass element in the rod by a coordinate x with origin on the axis of rotation and increasing to L at the other endpoint. Then each mass element, $dm = \lambda_0\,dx$, is at distance $\mathrm{d} = x$, from the axis of rotation. With this propitious parameterisation, the integral expression for the moment of inertia becomes

$$I_{\text{rod, end}} = \int dm\,\mathrm{d}^2 = \int_0^L [\lambda_0\,dx]\,x^2 = \lambda_0 \int_0^L x^2\,dx = \lambda_0 \left[\frac{1}{3}x^3\right]_0^L = \frac{1}{3}\lambda_0\,L^3\,.$$

While this formula for I is accurate, it is still somewhat opaque. Invoking the definition of the lineal mass density, $\lambda_0 = M/L$, allows the re-expression of the moment of inertia as

$$I_{\text{rod, end}} = \frac{1}{3}M\,L^2\,.$$

This is much more evidently a mass-weighted squared distance.

EXAMPLE [*Moment of Inertia of a Rod Spinning about Its Midpoint*]

Consider the same rod as in the example above, spinning about an axis perpendicular to the rod and passing through its midpoint.

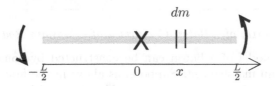

FIGURE 36.5 A Thin Uniform Rod Spinning about Its Midpoint

Each mass element, $dm = \lambda_0\,dx$, occupies a unique position in the rod expressed by $x \in [-\frac{L}{2}, \frac{L}{2}]$. The origin of coordinates has been chosen to lie on the axis of rotation. The computation of I proceeds as follows:

$$I_{\text{rod, mid}} = \int dm\,\mathrm{d}^2 = \int_{-\frac{L}{2}}^{\frac{L}{2}} x^2\,[\lambda_0\,dx] = \lambda_0 \left[\frac{1}{3}x^3\right]_{-\frac{L}{2}}^{\frac{L}{2}} = \frac{1}{3}\lambda_0 \left[\frac{L^3}{8} - \frac{-L^3}{8}\right] = \frac{1}{12}\lambda_0\,L^3\,.$$

Re-expression in terms of the total mass provides a nicer looking result:

$$I_{\text{rod, mid}} = \frac{1}{12}M\,L^2\,.$$

That

$$I_{\text{rod, midpoint}} < I_{\text{rod, endpoint}}$$

seems eminently reasonable based upon our experiences and developed intuition. Furthermore, it can be apprehended from an energetic standpoint, too. The rod rotating about an endpoint possesses more kinetic energy than one which rotates [at the same angular frequency] about its midpoint. Let us explore this energy argument.

Energetic Proof that $I_{\text{rod, midpoint}} < I_{\text{rod, endpoint}}$

For every mass element along one arm of the rod rotating about its midpoint, there is a corresponding mass element on the rod experiencing rotation about its endpoint which has exactly the same motion, and thus exactly the same kinetic energy.

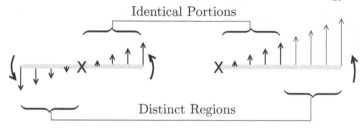

FIGURE 36.6   Motion of Mass Elements along Rods Rotating about Mid and End Points

Every other mass element along the rod rotating about its endpoint is farther from the axis of rotation, and hence moving with greater speed, than ALL mass elements on the other arm of the rod which is rotating about its midpoint.

The system with more kinetic energy at fixed rotational speed must have a greater moment of inertia.

EXAMPLE [*Moment of Inertia of a Baton Comprised of a Massive Rod and Endcaps*]

A slightly more realistic model of a baton can be constructed by considering two point masses at the ends of a thin massive rod. Suppose, as above in the first example, that the rod has length $2a$, and the endcaps each have mass M_C. The rod has mass M_R uniformly distributed along its length. The total moment of inertia of a baton rotating about an axis perpendicular to its rod and through its midpoint is

$$I_{\text{baton}} = I_{\text{endcaps}} + I_{\text{rod}} = 2\, M_C\, a^2 + \frac{1}{12}\, M_R\, (2\, a)^2 = \left(2\, M_C + \frac{M_R}{3} \right) a^2\,.$$

Rather than do any new calculations, we have employed linear superposition and adapted results from earlier examples. Addition of the moments of inertia of the constituents of the baton is meaningful because each has been correctly computed with respect to the same axis of rotation.

[Next up: more moments of inertia, and then two pertinent theorems!]

Chapter 37

Mo' Moments of Inertia

In the last chapter, it was realised that the moment of inertia, the mass-weighted squared distance from the axis of rotation, fulfilled the rôle in rotational dynamics analogous to that of mass in rectilinear mechanics. Here we shall compute four distinctly *hoopy*[1] moments of inertia. A regularity occurring in these [seemingly uncorrelated] results shall be revealed when we develop a particular theorem in Chapter 38.

EXAMPLE [*Moment of Inertia of a Thin Circular Hoop about Its Symmetry Axis*]

Consider a thin circular hoop of mass M and radius R spinning about its axis of symmetry. The "thin" descriptor allows us to model it as a filament with a lineal mass density λ.

[This hoop is not assumed to be uniform. Its density may vary with position.]

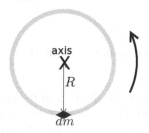

FIGURE 37.1   A Thin Hoop Spinning about Its Axis of Symmetry

As the hoop consists of a continuous distribution of matter,

$$I_{\text{hoop, sym-axis}} = \int dm \; \mathrm{d}^2 \,.$$

In this particular instance, all of the matter in the hoop [the collection of mass elements] happens to be at the same distance from the axis, $\mathrm{d} = R$, and hence

$$I_{\text{hoop, sym-axis}} = R^2 \int dm \,.$$

The integral of dm over the mass distribution yields the total mass of the hoop, M, irrespective of how the mass is distributed. Therefore, the moment of inertia of a thin circular hoop rotating about its axis of symmetry is

$$I_{\text{hoop, sym-axis}} = M R^2 \,.$$

This is the same as the moment of inertia of a point mass M circling the axis at distance R, because in both cases ALL of the mass is at the same distance from the axis.

---

[1]Here is yet another gratuitous reference to *The Hitchhiker's Guide to the Galaxy*.

EXAMPLE [*Moment of Inertia of an Annular Cylinder about Its Axis of Symmetry*]

An annular cylinder with length L and outer and inner radii R_o and R_i, respectively, is composed of a material having uniform mass density ρ_0. The cylinder rotates about its symmetry axis as pictured in Figure 37.2.

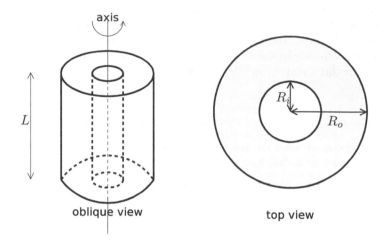

FIGURE 37.2   An Annular Cylinder Spinning about Its Axis of Symmetry

The first task is to parameterise the mass distribution in terms of coordinates which encode the geometry [size and shape] of the cylinder, and its relation to the axis of rotation. The most natural description employs cylindrical polar coordinates, (r, θ, z), for the radial, azimuthal, and axial dimensions, respectively. The region of space in which the matter distribution [the cylinder] exists is quantified by specifying the coordinate ranges:

$$r \in [R_i, R_o], \qquad \theta \in [0, 2\pi), \qquad z \in [0, L].$$

The simplicity of these ranges is proof that these coordinates are well-suited.

The second task is to obtain an expression for the differential mass elements comprising the annular cylinder. The [uniform] mass density relates the distribution of mass to the geometry of the body,

$$dm = \rho_0 \, d[\text{Volume}].$$

In cylindrical polar coordinates, the volume element is

$$d[\text{Volume}] \equiv [dz][dr][r \, d\theta] = r \, dr \, d\theta \, dz.$$

The third task, now that we have the parametric expression for the mass elements and the parameter ranges for the cylinder, is to compute the mass [*a.k.a.* the translational inertia] of the annular cylinder.

$$M_{\text{a-cyl}} = \int dm$$

$$= \int \rho_0 \, d[\text{Volume}]$$

$$= \int_{\{R_i,0,0\}}^{\{R_o,2\pi,L\}} \rho_0 \, r \, dr \, d\theta \, dz$$

$$= \rho_0 \left[\int_{R_i}^{R_o} r \, dr \right] \left[\int_0^{2\pi} d\theta \right] \left[\int_0^L dz \right].$$

The assumption of uniform density has enabled factorisation of the three-dimensional integration into a product of three independent one-dimensional integrals.

[Such factorisation may not occur in circumstances where the density varies.]

In the case at hand, each of the integrals can be evaluated:

$$\int_{R_i}^{R_o} r \, dr = \frac{1}{2}\left(R_o^2 - R_i^2\right), \qquad \int_0^{2\pi} d\theta = 2\pi - 0 = 2\pi, \qquad \text{and} \qquad \int_0^L dz = L - 0 = L.$$

Thus, the total mass is

$$M_{\text{a-cyl}} = \rho_0 \, \pi \left(R_o^2 - R_i^2\right) L.$$

ASIDE: This result is certainly correct since it conforms to the tautology

Total Mass ≡ (mass per unit volume) × (volume).

The volume of the annular cylinder is the area of the top surface, $\pi\left(R_o^2 - R_i^2\right)$, multiplied by the height, L.

The fourth task is the computation of the moment of inertia:

$$I = \int dm \, \text{d}^2.$$

The cylinder rotates about its axis of symmetry, identified with the z-axis in the chosen parameterisation. There is a unique distance from the axis of rotation to the location of each and every mass element, $\text{d} = r$, throughout the distribution. As before, $dm = \rho_0 \, r \, dr \, d\theta \, dz$, and thus

$$I = \int_{\{R_i,0,0\}}^{\{R_o,2\pi,L\}} \rho_0 \, r^3 \, dr \, d\theta \, dz$$

$$= \rho_0 \left[\int_{R_i}^{R_o} r^3 \, dr \right] \left[\int_0^{2\pi} d\theta \right] \left[\int_0^L dz \right].$$

The expression for I has factorised into three separate 1-d integrations:

$$\int_{R_i}^{R_o} r^3 \, dr = \frac{1}{4}\left(R_o^4 - R_i^4\right), \qquad \int_0^{2\pi} d\theta = 2\pi - 0 = 2\pi, \qquad \text{and} \qquad \int_0^L dz = L - 0 = L,$$

which combine to yield the moment of inertia of the annular cylinder about its symmetry axis,

$$I_{\text{a-cyl, axis}} = \rho_0 \, \frac{\pi}{2} \left(R_o^4 - R_i^4\right) L.$$

This result is made more intelligible by expressing the difference of quartics in terms of the sum and difference of squares:

$$\left(R_o^4 - R_i^4\right) = \left(R_o^2 + R_i^2\right)\left(R_o^2 - R_i^2\right).$$

One finally obtains, for the moment of inertia of the annular cylinder, the result,

$$I_{\text{a-cyl, axis}} = \frac{1}{2}\left[\rho_0\,\pi\left(R_o^2 - R_i^2\right)L\right]\left(R_o^2 + R_i^2\right) = \frac{1}{2}M_{\text{a-cyl}}\left(R_o^2 + R_i^2\right).$$

While this is certainly a handsome expression, we must still ensure that it is in accord with our reasoned expectations. As usual, we consider particular limits.

○   Cylindrical Shell

The annular cylinder reduces to a thin shell of radius R, in the double limit where $R_i \to R$ AND $R_o \to R$. Substituting into our final derived result,[2] the moment of inertia becomes

$$I_{\text{shell, axis}} = M\,R^2,$$

which is exactly what one should expect based on the observation that **all of the mass of the shell is located a distance R from the axis of rotation**.

> ASIDE: This limiting case devolves to a thin circular hoop rotating about its axis of symmetry, as was studied in an example found in Chapter 36. The present computation corroborates the earlier result.

●   Solid Cylinder

The cylinder is solid, rather than annular, when the limits $R_o = R$ AND $R_i \to 0$ are taken simultaneously. Thus, for a solid cylinder of radius R,

$$M_{\text{s-cyl}} = \rho_0\,\pi\,R^2\,L \qquad \text{and} \qquad I_{\text{s-cyl, axis}} = \frac{1}{2}M_{\text{s-cyl}}\,R^2.$$

....

EXAMPLE [*Moment of Inertia of a Thin Hoop about a Diameter*]

A thin uniform hoop, of total mass M and radius R, spins about an axis which lies in the plane of the hoop and passes through the centre. Pithily, we call this "spinning about a diameter of the hoop." As the hoop is uniform, one expects that its lineal mass density is

$$\lambda_0 = \frac{\text{Total Mass}}{\text{Circumference}} = \frac{M}{2\,\pi\,R}.$$

ASIDE: Just for grins, let's prove it.

Assuming that the mass is distributed uniformly along the arc length of the hoop,

$$dm = \lambda_0\,ds = \lambda_0\,R\,d\theta,$$

and therefore

$$M = \int dm = \lambda_0\,R\int_0^{2\pi} d\theta = \lambda_0\,(2\,\pi\,R).$$

---

[2]CAVEAT: Naive substitution leads one to a NULL result owing to the [spurious] vanishing of the mass of the cylinder in this particular limit. Realisation that the density becomes singular to preserve the magnitude of the mass resolves seeming contradictions.

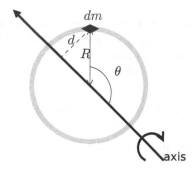

FIGURE 37.3 A Thin Hoop Spinning about a Diameter

The hoop's mass-weighted squared distance from the axis is

$$I = \int dm \, \mathrm{d}^2 \, .$$

The differential mass element dm expands as above, while the perpendicular distance to the axis varies with position along the hoop. According to the geometry of the system, and the choice of the angle θ [with reference to the negative direction along the axis of rotation] for the purpose of labelling the mass elements,

$$\mathrm{d} = R \sin(\theta) \, .$$

[This is correct for $0 \le \theta \le \pi$. For $\pi < \theta < 2\pi$, the distance is $d = -R\sin(\theta)$. Squaring makes this distinction moot.] Thus,

$$I_{\text{hoop, diam}} = \int_0^{2\pi} R^2 \sin^2(\theta) \, [\lambda_0 \, R \, d\theta] = \lambda_0 \, R^3 \left[\int_0^{2\pi} \sin^2(\theta) \, d\theta \right] \, .$$

Everything boils down to determining the value of the integral within the brackets. We discuss three ways in which this integral can be performed.

METHOD ONE: [*Standard Calculus Trick*]
The identity

$$\sin^2(x) = \frac{1 - \cos(2x)}{2} \, ,$$

enables re-expression of the integrand in a transparently integrable form, *viz.*,

$$\int_0^{2\pi} \sin^2(\theta) \, d\theta = \frac{1}{2} \int_0^{2\pi} d\theta - \frac{1}{2} \int_0^{2\pi} \cos(2\theta) \, d\theta = \frac{1}{2} \Big[\theta \Big]_0^{2\pi} - \frac{1}{2} \left[\frac{\sin(2\theta)}{2} \right]_0^{2\pi}$$

$$= \pi - 0 = \pi \, .$$

METHOD TWO: [*Look-up Tables*]
The second method is to use tables, or Maple™ [or any other sufficiently powerful computer algebra system], to get the result: π.

METHOD THREE: [*Inspiration*]

A third approach is to exploit the similarities of the **sine** and **cosine** functions over a complete cycle $[0 \to 2\pi]$. They differ only by a relative phase shift of $\pi/2$ **rad**. Squaring preserves the similarity and the phase shift, and thus the areas under the respective curves for an entire oscillation [*i.e.,* the integrals] are equal:

$$\int_0^{2\pi} \sin^2(\theta)\, d\theta \equiv \int_0^{2\pi} \cos^2(\theta)\, d\theta\,.$$

Thus, we can creatively rewrite the integrand as a sum of terms which then collapse by virtue of a trigonometric identity:

$$\int_0^{2\pi} \sin^2(\theta)\, d\theta = \int_0^{2\pi} \left[\frac{\sin^2(\theta) + \cos^2(\theta)}{2} \right] d\theta = \int_0^{2\pi} \left[\frac{1}{2} \right] d\theta = \frac{1}{2}\,(2\pi) = \pi\,.$$

Whichever method is employed for the purpose of the computation, one invariably obtains

$$I_{\text{hoop, diam}} = \pi\,\lambda_0\,R^3\,.$$

The beauty that one might have hoped for in this result is not at all evident, until the expression is recast in terms of the total mass of the hoop:

$$I_{\text{hoop, diam}} = \frac{1}{2}\,M\,R^2\,.$$

EXAMPLE [*Moment of Inertia of a Thin Hoop about a Tangent*]

The same hoop as employed in the above example spins about an axis lying in the plane of the hoop and tangent to it at a single point. As the total mass of the hoop has already been determined, we can start right away on the moment of inertia.

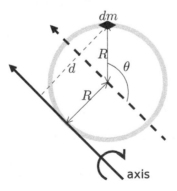

FIGURE 37.4   A Thin Uniform Hoop Spinning about a Tangent

The differential mass element, once again, is

$$dm = \lambda_0\, ds = \lambda_0\, R\, d\theta\,,$$

while its distance from the axis of rotation is now

$$d = R + R\sin(\theta) = R\left(1 + \sin(\theta)\right),$$

where we are keeping the same convention for measuring the angle as we did in the previous example, even though the axis of rotation is different. Thus,

$$
\begin{aligned}
I_{\text{hoop, tang}} &= \int dm \; \mathrm{d}^2 \\
&= \int_0^{2\pi} \left[\lambda_0 \, R \, d\theta\right] R^2 \left(1 + \sin(\theta)\right)^2 \\
&= \lambda_0 \, R^3 \left[\int_0^{2\pi} \left\{1 + 2\sin(\theta) + \sin^2(\theta)\right\} d\theta\right].
\end{aligned}
$$

The three terms comprising the integrand can be separately integrated.

1st   The first term integrates to 2π.

2nd   The second term is ZERO [by symmetric integration or direct calculation].

3rd   The third term yields π.

Taken together, these three terms sum to the coefficient of the moment of inertia of the uniform hoop rotating about a tangent:

$$I_{\text{hoop, tang}} = 3\pi\lambda_0 R^3 = \frac{3}{2} M R^2.$$

In the next chapter, we shall describe in detail the PARALLEL AXIS THEOREM. Application of this theorem provides an algebraic means of computing moments of inertia of a given object from knowledge of its moment about a particular axis. In addition, we shall encounter the PERPENDICULAR AXIS THEOREM. Although it has much less utility, it is nonetheless a cool theorem with an inordinately cute proof.

ADDENDUM: An Alternative Approach to the Annular Cylinder

One of the computational techniques employed in the centres of mass for irregular objects example [found in Chapter 30] considered an **E**ntire [regular] object plus a **F**orgotten piece. This was shown to be equivalent to taking the **E**ntire object and adding a **F**ictitious portion with NEGATIVE mass density. This technique may be [judiciously] employed for moments of inertia, too.

An annular cylinder with length L, and inner and outer radii R_i and R_o, comprised of material with uniform density ρ_0, was the subject of the second example in this chapter. In this instance, the **E**ntire object is a solid cylinder with length L and radius R_o, while the **F**ictitious piece is a cylinder of length L with radius R_i.

The volume of the entire piece is $V_E = \pi\, R_o^2\, L$. Since the material is uniform, its mass and moment of inertia are

$$M_E = \rho_0\, V_E = \rho_0\, \pi\, R_o^2\, L \quad \text{and} \quad I_E = \tfrac{1}{2}\, M_E\, R_o^2 = \tfrac{1}{2}\, \rho_0\, \pi\, R_o^4\, L\,.$$

In similar fashion, the volume, mass, and moment of inertia of the fictitious piece are:

$$V_F = \pi\, R_i^2\, L\,, \qquad M_F = -\rho_0\, V_F = -\rho_0\, \pi\, R_i^2\, L\,,$$
$$\text{and} \qquad I_F = \tfrac{1}{2}\, M_F\, R_i^2 = -\tfrac{1}{2}\, \rho_0\, \pi\, R_i^4\, L\,.$$

The volume, mass, and moment of inertia of the annular cylinder are obtained by careful additions of the corresponding fictitious contributions to the entire quantities.

$V = V_E - V_F$	$M = M_E + M_F$
$\quad = \pi\, \left[R_o^2 - R_i^2 \right]\, L$	$\quad = \rho_0\, \pi\, \left[R_o^2 - R_i^2 \right]\, L$

$$\begin{aligned} I &= I_E + I_F \\ &= \tfrac{1}{2}\, \rho_0\, \pi\, \left(R_o^4 - R_i^4 \right)\, L \\ &= \tfrac{1}{2}\, \rho_0\, \pi\, \left[R_o^2 - R_i^2 \right] \left(R_o^2 + R_i^2 \right)\, L \\ &= \tfrac{1}{2}\, M\, \left(R_o^2 + R_i^2 \right) \end{aligned}$$

These are precisely the same results as were obtained earlier.

Chapter 38

Moment of Inertia Theorems

It so happens that several of the moments of inertia computed in the examples in Chapters 36 and 37 are related via the PARALLEL AXIS THEOREM.

PARALLEL AXIS THEOREM   The parallel axis theorem expresses the moment of inertia of a body about a particular axis in terms of its moment of inertia about another axis, parallel to the first and passing through the CofM of the body,

$$I \equiv I_{\text{CofM}} + M D^2 .$$

The added bit consists of a [dimensionally appropriate] combination of: the total mass of the body, M, and the [unique minimal] distance between the parallel axes, D.

Before constructing an incontrovertible proof of this theorem, it behooves us to first verify it in the special cases which we have already examined.

EXAMPLE [*Parallel Axis Theorem: Uniform Rods*]
In Chapter 36, the moments of inertia for a uniform rod rotating about parallel axes through one endpoint and the midpoint of the rod were separately computed to be

$$I_{\text{rod, end}} = \frac{1}{3} M L^2 \qquad \text{and} \qquad I_{\text{rod, mid}} = \frac{1}{12} M L^2 ,$$

respectively. The CofM of the rod is at its midpoint. The distance between the axes is one-half of the length of the rod (*i.e., $L/2$*). Incorporation of these specific features into the RHS of the statement of the parallel axis theorem yields

$$I_{\text{rod, mid}} + M \left(\frac{L}{2} \right)^2 = \left(\frac{1}{12} + \frac{1}{4} \right) M L^2 = \frac{1}{3} M L^2 \equiv I_{\text{rod, end}} ,$$

confirming the veracity of the theorem in this case.

EXAMPLE [*Parallel Axis Theorem: Uniform Hoops*]
In Chapter 37, the moments of inertia for a uniform thin hoop rotating about a diameter and a tangent were computed to be

$$I_{\text{hoop, diam}} = \frac{1}{2} M R^2 \qquad \text{and} \qquad I_{\text{hoop, tang}} = \frac{3}{2} M R^2 .$$

The CofM of the hoop is at its centre, so $I_{\text{CofM}} = I_{\text{hoop, diam}}$. The distance between these parallel axes is equal to R, the radius of the hoop. Therefore,

$$I_{\text{hoop, diam}} + M R^2 = \left(\frac{1}{2} + 1 \right) M R^2 = \frac{3}{2} M R^2 = I_{\text{hoop, tang}}$$

holds, in accord with the theorem.

PROOF of the Parallel Axis Theorem

Setting the Stage for the Proof

WLOG we'll employ Cartesian coordinates, (x, y, z), such that the z-axis is aligned with the parallel axes of rotation. The 2-d planes perpendicular to the axes of rotation are all parallel to the xy-plane and distinguished by differing constant values of z.

The centre of mass lies at a particular point in space,

$$\vec{R}_{\text{CofM}} = (X_{\text{CofM}}, Y_{\text{CofM}}, Z_{\text{CofM}}).$$

In the plane perpendicular to the axes and through the CofM, there exists a unique point, $\vec{P}$, which lies on the other axis. The coordinates of $\vec{P}$ can be expressed in terms of an offset $(a, b, 0)$ from the CofM as follows:

$$\vec{P} = (P_x, P_y, P_z) = (X_{\text{CofM}} + a, Y_{\text{CofM}} + b, Z_{\text{CofM}}).$$

The MINIMUM distance between the axes, D, is

$$D = \sqrt{a^2 + b^2}.$$

Figure 38.1 shows this cross-section, with the locations of the two axes of rotation [one through the CofM and the other through $\vec{P}$] marked.

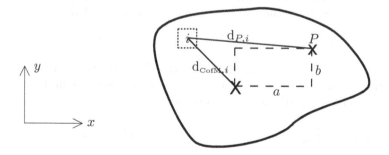

FIGURE 38.1   A Cross-Section of the Rotating Object Perpendicular to the Axes and through the CofM

The Heart of the Proof

The first step is to construct an $\mathcal{N}$-partition of the body. The partition shall be employed for the [parallel] computations of the moments of inertia about the two parallel axes.

The second step is to compute the contributions to the moments of inertia associated with individual mass elements in the partition invoking the point-like particle approximation. The ith chunk, possessing mass Δm_i, resides at (x_i, y_i, z_i). The perpendicular distance from the axis through the CofM to this chunk, $\text{d}_{\text{C},i}$, is

$$\text{d}_{\text{C},i}^2 = \left(x_i - X_{\text{CofM}}\right)^2 + \left(y_i - Y_{\text{CofM}}\right)^2.$$

The perpendicular distance to the other axis (the one passing through the point P), $d_{P,i}$, can be similarly determined.

$$
\begin{aligned}
d_{P,i}^2 &= \left(x_i - P_x\right)\right)^2 + \left(y_i - P_y\right)\right)^2 \\
&= \left(x_i - \left(X_{\text{CofM}} + a\right)\right)^2 + \left(y_i - \left(Y_{\text{CofM}} + b\right)\right)^2 \\
&= \left(\left(x_i - X_{\text{CofM}}\right) - a\right)^2 + \left(\left(y_i - Y_{\text{CofM}}\right) - b\right)^2 \\
&= \left(x_i - X_{\text{CofM}}\right)^2 - 2\,a\left(x_i - X_{\text{CofM}}\right) + a^2 + \left(y_i - Y_{\text{CofM}}\right)^2 - 2\,b\left(y_i - Y_{\text{CofM}}\right) + b^2 \,.
\end{aligned}
$$

The terms in the last expression can be recast as

$$
d_{P,i}^2 = d_{C,i}^2 - 2\,a\left(x_i - X_{\text{CofM}}\right) - 2\,b\left(y_i - Y_{\text{CofM}}\right) + D^2 \,.
$$

The contributions of the ith chunk to the total moments of inertia are

$$
\Delta I_{C,i} = \Delta m_i\, d_{C,i}^2 \qquad \text{and} \qquad \Delta I_{P,i} = \Delta m_i\, d_{P,i}^2 \,,
$$

respectively. These two moments of inertia, pertaining to the particular chunk bearing the label i, are expected to differ since they are with respect to distinct [albeit parallel] axes.

The third step is to sum over the mass elements in the partition to obtain the total moments of inertia of the body about the two axes.

$$
I_{\text{CofM}} \equiv \sum_{i=1}^{N} \Delta I_{C,i} = \sum_{i=1}^{N} \Delta m_i\, d_{C,i}^2 \qquad \text{and} \qquad I_P \equiv \sum_{i=1}^{N} \Delta I_{P,i} = \sum_{i=1}^{N} \Delta m_i\, d_{P,i}^2 \,.
$$

Truth be told, there is no content, beyond definitions, in the above expressions for the moments of inertia, until we incorporate the relations between $d_{P,i}$ and $d_{C,i}$. Proceeding to do just this, the summand for $I_{P,i}$ breaks up into four terms, and hence so does the sum.

$$
\begin{aligned}
I_P &= \sum_{i=1}^{N} \Delta m_i\left[d_{C,i}^2 - 2\,a\left(x_i - X_{\text{CofM}}\right) - 2\,b\left(y_i - Y_{\text{CofM}}\right) + D^2\right] \\
&= \sum_{i=1}^{N} \Delta m_i\, d_{C,i}^2 - 2\,a \sum_{i=1}^{N} \Delta m_i\left(x_i - X_{\text{CofM}}\right) - 2\,b \sum_{i=1}^{N} \Delta m_i\left(y_i - Y_{\text{CofM}}\right) + \sum_{i=1}^{N} \Delta m_i\, D^2
\end{aligned}
$$

Each separate sum in this expression is evaluated below.

1st The first term is I_{CofM}.

2nd The second term is zero because it amounts to a computation of the mass-weighted total deviation of the values of the x-components of the elements in the partition from the x-component of the CofM. By construction, this total deviation must vanish.

3rd The argument employed for the second term, applied to the y-component, ensures that the third term must be equal to zero as well.

4th The fourth term becomes

$$
\left(\sum_{i=1}^{N} \Delta m_i\right) D^2 = M\,D^2
$$

on performing the sum.

The fourth, and final, step is to gather the above results and conclude that

$$I_P = I_{\text{CofM}} + M D^2 \,.$$

ASIDE: One might be concerned that this proof is somehow less than complete because only that foliation of the body in the plane perpendicular to the axes and passing through the CofM was considered. These doubts are assuaged by the realisation that for fixed axes the z-coordinate value never reappears in the analysis after its introduction.

EXAMPLE [*Application of the Parallel Axis Theorem*]

A thin circular hoop of total mass M [uniformly distributed], and radius R rotates about an axis perpendicular to the plane of the hoop, and passing through one of its points.

Q: What is the moment of inertia of the uniform hoop about this axis?

A: Let's find out. First, we'll employ the parallel axis theorem to obtain the result, and then we'll verify it by direct computation.

axis
X

FIGURE 38.2   A Thin Hoop Rotating about an Axis through Its Rim

METHOD ONE: [*Applying the Parallel Axis Theorem*]

In Chapter 37, it was determined that the moment of inertia of a uniform thin hoop rotating about its axis of symmetry [passing through the CofM] is $I_{\text{CofM}} = M R^2$.

[This is because all of the hoop's mass is the same distance away from the axis of rotation.]

Applying the parallel axis theorem to this particular case yields

$$I_{\text{rim}} = I_{\text{CofM}} + M R^2 = 2 M R^2 \,.$$

METHOD TWO: [*Direct Computation*]

The moment of inertia,

$$I_{\text{hoop, edge}} = \int dm \, \mathrm{d}^2 \,,$$

can be computed in the usual manner. Once again, we shall employ polar coordinates. Since the object is lineal and uniform, the mass elements are parameterisable as

$$dm = \lambda_0 \, R \, d\theta \,,$$

where $\lambda_0 = M/(2\pi R)$ is the constant lineal mass density, and the angle, θ, varies from 0 to 2π in order to encompass the entire hoop.

The squared distance of the mass element from the axis of rotation at first seems well-nigh impossible to determine, until one realises that it arises directly from the LAW OF COSINES. Employing the conventions illustrated in Figure 38.3,

$$d^2 = 2R^2\left(1 - \cos(\theta)\right).$$

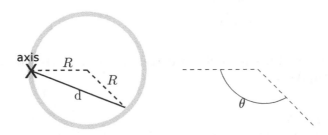

FIGURE 38.3 The Law of Cosines and the Squared Distance from the Axis

Thus,

$$I_{\text{hoop, edge}} = 2\lambda_0 R^3 \int_0^{2\pi} \left(1 - \cos(\theta)\right) d\theta = 2\lambda_0 R^3 \left[\int_0^{2\pi} 1\, d\theta - \int_0^{2\pi} \cos(\theta)\, d\theta\right]$$

$$= 2\lambda_0 R^3 \left[\theta\big|_0^{2\pi} - \sin(\theta)\big|_0^{2\pi}\right] = 2\lambda_0 R^3 \left[(2\pi - 0) - (0 - 0)\right] = 4\pi\lambda_0 R^3.$$

Invoking the expression for the total mass of the thin hoop, $M = \lambda_0\, 2\pi R$, the moment of inertia becomes

$$I_{\text{hoop, edge}} = 2MR^2,$$

exactly as was obtained [much more easily] via METHOD ONE.

There is another result, promoted to the august stature of a theorem, which holds for plane (2-d) figures only. It's called the PERPENDICULAR AXIS THEOREM, and while it has less practical utility than the parallel axis theorem, it is still *pretty cool.*

PERPENDICULAR AXIS THEOREM   Suppose that a planar object can be rotated about either of two mutually perpendicular axes of rotation which lie in the same plane as the object. Call these axes A_1 and A_2. Further suppose that the moments of inertia of the object about A_1 and A_2 are known to be I_1 and I_2 respectively. The two axes must intersect at a point Q. Consider a third axis, A_3, which is perpendicular to both A_1 and A_2, and also passes through Q. The moment of inertia of the planar object about A_3 is equal to the sum of the moments of inertia about the other two axes. *I.e.,*

$$I_3 = I_1 + I_2.$$

PROOF of the Perpendicular Axis Theorem

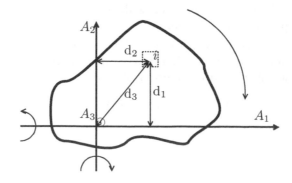

FIGURE 38.4 A Body with Two Coplanar Perpendicular Axes, A_1 and A_2, and a Third [Mutually Perpendicular] Axis, A_3

Setting the Stage for the Proof

The stage is mostly set in the extended description included in the statement of the theorem and illustrated in Figure 38.4. The distinct distances to the respective axes for a particular mass element constituent of the planar body are shown.

The Heart of the Proof

We can readily arrive at a proof of the theorem by means of the following argument.

AXIS 1   The moment of inertia about axis A_1 is $I_1 = \int dm\ \mathrm{d}_1^2$, where d_1 is the perpendicular distance from the axis A_1 to the mass element dm.

AXIS 2   The moment of inertia about axis A_2 is $I_2 = \int dm\ \mathrm{d}_2^2$, where d_2 is the distance from the axis A_2 to the dm.

AXIS 3   The moment of inertia about A_3 is given by an expression of the same form as those above, $I_3 = \int dm\ \mathrm{d}_3^2$. However, $\mathrm{d}_3^2 = \mathrm{d}_1^2 + \mathrm{d}_2^2$, and hence

$$I_3 = \int dm\ \mathrm{d}_3^2 = \int dm\ (\mathrm{d}_1^2 + \mathrm{d}_2^2) = \int dm\ \mathrm{d}_1^2 + \int dm\ \mathrm{d}_2^2 = I_1 + I_2\,.$$

Therefore, subject to the conditions explicitly mentioned in its statement, the perpendicular axis theorem is a valid relation among moments of inertia for a planar body.

Chapter 39

Torque

A rigid body is free to rotate about a fixed axis. A force, $\vec{F}$, with magnitude F and direction $\hat{F}$, acts on the body at the point labelled $\mathcal{P}$, as illustrated in Figure 39.1.

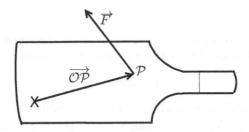

FIGURE 39.1 A Force Applied at a Point on a Rigid Rotating Body

The vector $\overrightarrow{O\mathcal{P}}$ resides in a plane perpendicular to the axis of rotation and extends from the axis to the point $\mathcal{P}$. The applied force has a component parallel to $\overrightarrow{O\mathcal{P}}$, dubbed $F_{\parallel}$. The remaining part(s) of the force vector, having vanishing projection onto $\overrightarrow{O\mathcal{P}}$, have magnitude $F_{\perp}$ and lie entirely in the plane perpendicular to $\overrightarrow{O\mathcal{P}}$.

> ASIDE: The decomposition of a vector into its component in a certain direction plus "the rest" is always possible,[1] with "the rest" self-consistently defined by $\vec{F_{\perp}} \equiv \vec{F} - F_{\parallel}\left[\widehat{\overrightarrow{O\mathcal{P}}}\right]$.

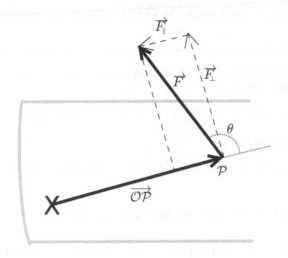

FIGURE 39.2 Vector Decomposition of a Force Acting at a Point

---

[1]Indeed, we have done this several times already. See Chapters 6 and 8, for early instances.

Theta, the angle between $\overrightarrow{OP}$ and $\vec{F}$ [when placed tail-to-tail[2]], is employed to express the parallel and perpendicular components of the force:

$$F_{\parallel} = F\,\cos(\theta)\,, \qquad \text{and} \qquad F_{\perp} = F\,\sin(\theta)\,.$$

Consideration of Figure 39.2 reveals the operative distinction between these components.

| $\parallel$ | PARALLEL | $\vec{F_{\parallel}}$ does NOT act to produce rotation about the axis. |
| $\perp$ | PERPENDICULAR | $\vec{F_{\perp}}$ acts to produce rotation about the axis. |

These observations lead us to define TORQUE.

TORQUE   Torque is a measure of the ability of a force to cause the object upon which it acts to rotate about an axis.

While accurate, this seems too nebulous. In order to study the effects of torque, we'll have to quantify it. As a first attempt, let's set the torque, τ [the Greek letter "tau"], equal to

$$\tau = r\,F\,\sin(\theta)\,,$$

where r is the perpendicular distance from the axis to the point of application of the force, *i.e.*, the magnitude of $\overrightarrow{OP}$; F denotes the magnitude of the force; and θ is the angle lying between the extension of $\overrightarrow{OP}$ and the force vector, as mentioned above. The salutary aspect of this putative formula is that it captures several essential aspects of torque.

$r = 0$ IF the force acts on the axis, THEN the torque vanishes.

> [A force applied directly on an axis cannot produce rotation about that axis.[3]]

$F = 0$ IF the force vanishes, THEN so too, necessarily, does the torque.

$\theta = \pi, 0$ IF the force acts toward, or away from, the axis, THEN the torque vanishes.

> [A force parallel or anti-parallel to $\overrightarrow{OP}$ cannot produce rotation about that axis.[4]]

$\theta = \frac{\pi}{2}$ IF the force acts at right angles to $\overrightarrow{OP}$, THEN the torque is maximised.

According to the proposed formula the dimensions of torque are [distance] × [force], and thus the SI units of torque must be

$$[\tau] = [r][F] = [F][r] = \text{newton}\cdot\text{metres} = \text{N}\cdot\text{m}\,.$$

Two comments must be made.

- The units of work are $\text{N}\cdot\text{m} = \text{J}$, joules. However, this should not be imbued with any significance:

  Torques are not energies and therefore are not expressed in joules.

- The Imperial unit for torque is the `pound-foot` (sometimes known as the `foot-pound`).

---

[2]These vectors actually reside in different spaces.
[3]Think about this the next time that you walk through a revolving door.
[4]Ditto.

There are two distinct parsings of $\tau = r\,F\,\sin(\theta)$ to consider.

- The force and angle terms may be combined:

$$\tau = r\,(F\,\sin(\theta)) = r\,F_\perp\,.$$

In this view, the torque is the product of the magnitude of the perpendicular component of the force and the distance from the axis to the point of application of the force.

- Alternatively, the angle may be combined with the distance:

$$\tau = (r\,\sin(\theta))\,F = r_\perp\,F\,.$$

The **moment arm**, $r_\perp$, is the minimum distance between the axis of rotation and the **line of action**[5] of $\vec{F}$ applied at $\mathcal{P}$. From this perspective, the torque is the product of the magnitude of the force and its moment arm with respect to the axis of rotation.

> The moment arm has the virtue of being a rather intuitive concept. If you doubt this, go ask an auto mechanic or plumber how he/she thinks of torque.

TORQUES OBEY A PRINCIPLE OF LINEAR SUPERPOSITION

Consider a body, free to rotate about an axis, subjected to three separate forces each acting at a distinct location, as illustrated in Figure 39.3.

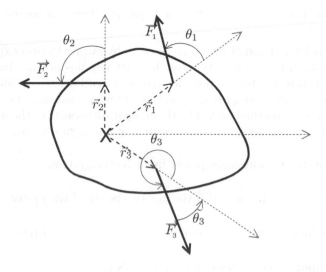

FIGURE 39.3 Linear Superposition for Torques

---

[5]The line of action passes through the specified point, extending [both ways] in the direction [parallel to that] of the particular vector.

The torques produced by $\vec{F_1}$ and $\vec{F_2}$, applied at $\mathcal{P}_1$ and $\mathcal{P}_2$, respectively, are

$$\tau_1 = r_1 \, F_1 \sin(\theta_1) \qquad \text{and} \qquad \tau_2 = r_2 \, F_2 \sin(\theta_2) \, .$$

Attempting computation of the torque provided by $\vec{F_3}$, we encounter a subtlety in that, to maintain consistency, we must use the very obtuse θ_3, rather than the acute angle $\widetilde{\theta}_3$.

$$\tau_3 = r_3 \, F_3 \sin\left(\theta_3\right) = -r_3 \, F_3 \sin\left(\widetilde{\theta}_3\right) \, .$$

ASIDE: One appreciates that forces $\vec{F_1}$ and $\vec{F_2}$ are both acting so as to cause rotation of the body to occur in an anti-clockwise sense about the axis, whereas $\vec{F_3}$ acting alone would give rise to a clockwise rotation. Therefore, that these torque contributions should appear with opposite signs, is not at all unreasonable. For the time being,[6] we shall just accept that **the angles are to be measured anti-clockwise from the extension of $\vec{r}_i$ to the force vector.** Alternatively, one can always employ the smaller angle between the force and position vectors and assign an overall factor of (± 1) depending on whether the force, acting alone, would give rise to anti-clockwise or clockwise rotation.

The net torque is the sum of the three contributing torques

$$
\begin{aligned}
\tau_{\text{net}} &= \tau_1 + \tau_2 + \tau_3 \\
&= r_1 \, F_1 \sin(\theta_1) + r_2 \, F_2 \sin(\theta_2) + r_3 \, F_3 \sin\left(\theta_3\right) \, . \\
&= r_1 \, F_1 \sin(\theta_1) + r_2 \, F_2 \sin(\theta_2) - r_3 \, F_3 \sin\left(\widetilde{\theta}_3\right)
\end{aligned}
$$

That torques adhere to the PRINCIPLE OF LINEAR SUPERPOSITION is consistent with, analogous to, and emergent from the fact that the forces themselves obey the same principle.

Q: What could possibly intervene to prevent us from immediately deposing our provisional expression for torque in favour of one which is generally correct?

A: An explication of the effects of torque and a discussion of energetics.

CLAIM: The existence of a net torque acting on a system affects the angular acceleration of the system.

This claim is broadly bolstered and clearly corroborated by our real world experiences with revolving doors, turntables, batons, and the like. In what remains of this chapter, we shall derive the relation between net torque and angular acceleration by consideration of the mechanical power injected into a rigid body, rotating about a fixed axis, by a collection of forces, each acting at a point on the body. The three separate aspects of the analysis which, when combined, constitute the complete and incontrovertible argument are given below.

POWER   Compute the instantaneous power input to the system.

ENERGY   Determine the rate at which the kinetic energy of the system is increasing.

CONS.   Invoke the [instantaneous version of the] Work–Energy Theorem:

$$\text{Input Power} = \text{Rate of Change of Kinetic Energy} \, .$$

---

[6]That is, until torque is redefined as a vector in Chapter 40.

INSTANTANEOUS POWER INPUT

The total amount of mechanical work done by the collection of forces acting on the body as it rotates through an angle $\Delta\phi$ during a time interval Δt can be computed. Dividing the work obtained by Δt will yield the average mechanical power input to the system, and taking the limit as the interval shrinks to zero provides the instantaneous power input. Let's do all of this in six steps.

i   Partition[7] the body into $\mathcal{N}$ chunks, such that the mass of the jth chunk is ΔM_j, while its [representative] position is $\vec{R}_j$. Let r_j denote the perpendicular distance of the jth chunk from the axis of rotation.

ii   Enumerate and characterise all of the forces which act on each chunk comprising the body. Sum the forces acting on the jth chunk to obtain the net[8] force, $\vec{F}_{\text{NET},j}$, applied to this particular chunk.

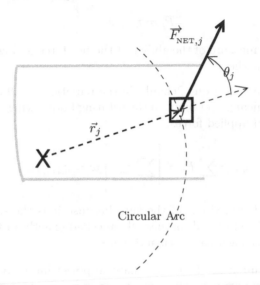

FIGURE 39.4   The Net Force Acting upon the jth Chunk of a Rotating Body

iii   During the elapse of a short interval of time, Δt, the entire body will have rotated through a small angle $\Delta\phi$. The jth chunk will have moved along a section of circular arc whose length is $\Delta s_j = r_j\,\Delta\phi$. Provided the time interval is sufficiently short, the vector displacement of the jth chunk, $\Delta\vec{s}_j$, has magnitude [approximately[9]] equal to the arc length and direction tangent to the path. Thus, the vector displacement is [approximately] orthogonal to the line extending from the axis to the chunk.

iv   The work done by the net force acting upon the jth chunk as the body rotates through the angle $\Delta\phi$ can be computed.

$$\Delta W_j = \vec{F}_{\text{NET},j} \cdot \Delta\vec{s}_j = F_{\text{NET},j}\,\Delta s_j\,\cos\big(\text{angle between force and displacement}\big).$$

---

[7]Choosing the partition so as to avoid pathological situations, of course.

[8]Anticipating a future result, one need only determine the net external force.

[9]The chunk's trajectory is along a circular arc, and thus the displacement is the corresponding secant for finite $\{\Delta\phi, \Delta t\}$. Realising that we shall eventually take the limit that $\Delta t \to 0$ emboldens us to make these approximations.

Close attention to the sketch in Figure 39.4, along with the tautology that the tangential direction is orthogonal to the ray from the axis to the reference point, reveals that the angle which lies between the net force and the tangential displacement is equal to $\frac{\pi}{2} - \theta_j$. Furthermore, $\cos\left(\frac{\pi}{2} - \theta_j\right) \equiv \sin\left(\theta_j\right)$, and thus

$$\Delta W_j = F_{\text{NET},j}\, r_j\, \Delta\phi\, \sin(\theta_j) = r_j\, F_{\text{NET},j}\, \sin(\theta_j)\, \Delta\phi\,.$$

v   Hence the instantaneous power input to this chunk of the rotating body is

$$P_j = \lim_{\Delta t \to 0} \frac{\Delta W_j}{\Delta t} = \lim_{\Delta t \to 0}\left\{ r_j\, F_{\text{NET},j}\, \sin(\theta_j)\, \frac{\Delta\phi}{\Delta t} \right\} = r_j\, F_{\text{NET},j}\, \sin(\theta_j)\, \omega\,,$$

where ω is the instantaneous angular speed of the entire rigid body. The expression for power can be recast in terms of the net torque acting on the chunk:

$$P_j = \tau_{\text{net},j}\, \omega\,.$$

The net torque is a measure of the ability of the net force acting on the chunk to cause rotation about the axis.

vi   Summing over the collection of chunks in the partition results in an expression for the total mechanical power input to the rotating body owing to the application of all of the external applied forces:

$$P_{\text{Net}} = \sum_{j=1}^{\mathcal{N}} P_j = \left[\sum_{j=1}^{\mathcal{N}} \tau_{\text{net},j} \right] \omega = \tau_{\text{net ext'l}}\, \omega\,.$$

The angular speed factored out of the sum because it is the same for all parts of the rigid body. The internal torques all cancelled exactly in the sum over the chunks due to N3 [as might have been anticipated].

This ends the first part: computing the net mechanical power input to the rotating body.

INSTANTANEOUS RATE OF INCREASE OF KINETIC ENERGY

The energy analysis of Chapter 36 informed us that the kinetic energy of a body undergoing rotation is

$$K = \frac{1}{2} I \omega^2\,.$$

Q: How is this kinetic energy changing in time?
A: We'll work this out three different ways below.

METHOD ONE: [*Most Straightforward Approach*]
Let's take the time derivative to get the time rate of change of the kinetic energy, assuming that $\frac{dI}{dt} = 0$, since, after all, the body is rigid and the axis is fixed. Thus,

$$\frac{dK}{dt} = \frac{1}{2} I \left(2\omega\, \frac{d\omega}{dt} \right) = I\, \omega\, \alpha\,.$$

METHOD TWO: [*Vive Les (Finite) Differences*]
Let's proceed in parallel with the analysis for power input, again making explicit reference to the situation pictured in Figure 39.4.

During the time interval Δt in which the the body rotates through an angle $\Delta\phi$, the total kinetic energy of the rotating body changes from its initial value, $K_i = \frac{1}{2} I \omega_i^2$, to its final value $K_f = \frac{1}{2} I \omega_f^2$. The assumed constancy of the moment of inertia means that the change in the kinetic energy ensues from the change in the angular speed of the body. Hence,

$$\Delta K = K_f - K_i = \frac{1}{2} I \left(\omega_f^2 - \omega_i^2 \right) .$$

IF the time interval is sufficiently small, THEN the instantaneous angular accelerations at times during the interval can be well-approximated by $\bar{\alpha}$, the average angular acceleration throughout the interval. The analysis of constant angular acceleration about a fixed axis undertaken in Chapter 34 exposed the kinematical relation

$$\omega_f^2 - \omega_i^2 = 2\,\bar{\alpha}\,\Delta\phi .$$

Substituting this result, now an approximation, into the expression for the change in the kinetic energy of the rotating body yields

$$\Delta K = \frac{1}{2} I \left(\omega_f^2 - \omega_i^2 \right) \simeq \frac{1}{2} I \left(2\,\bar{\alpha}\,\Delta\phi \right) = I\,\bar{\alpha}\,\Delta\phi .$$

Thus, the instantaneous time rate of change of the kinetic energy is

$$\frac{dK}{dt} = \lim_{\Delta t \to 0} \frac{\Delta K}{\Delta t} = I \lim_{\Delta t \to 0} \left\{ \bar{\alpha}\,\frac{\Delta\phi}{\Delta t} \right\} = I\,\alpha\,\omega .$$

In the last equality, we've invoked the definition of average angular speed. Also, we've realised that the instantaneous value of a time-dependent quantity is equal to the limit of its average value throughout a time interval whose duration approaches zero.

METHOD THREE: [*Vive Les (Finite) Differences – Encore*]
Let's revisit METHOD TWO, except this time eschewing the invocation of the "third" constant angular acceleration kinematical relation from Chapter 34.

For sufficiently short time scales, Δt, the linear approximation to changes in angular speed,

$$\omega_f \simeq \omega_i + \alpha\,\Delta t ,$$

is valid, and hence

$$\omega_f^2 - \omega_i^2 \simeq 2 \left(\alpha\,\Delta t \right) \omega_i + \left(\alpha\,\Delta t \right)^2 .$$

Therefore,

$$\Delta K = \frac{1}{2} I \left(\omega_f^2 - \omega_i^2 \right) \simeq I\,\alpha\,\omega_i\,\Delta t + \frac{1}{2} I \left(\alpha\,\Delta t \right)^2 ,$$

approximates the change in the kinetic energy from the start to the end of the short time interval. The average time rate of change of the kinetic energy throughout Δt is

$$\frac{\Delta K}{\Delta t} = I\,\alpha\,\omega_i + \frac{1}{2} I\,\alpha^2\,\Delta t .$$

The instantaneous time rate of change of the kinetic energy is the limit of its average value, as $\Delta t \to 0$,

$$\frac{dK}{dt} = \lim_{\Delta t \to 0} \frac{\Delta K}{\Delta t} = I\,\alpha\,\omega ,$$

where we have dropped the [now irrelevant] subscript "i," denoting "initial," on the ω.

Whichever of the above methods we find most to our liking, the inescapable result is that

$$\frac{dK}{dt} = I\,\alpha\,\omega\,.$$

POWER INPUT EQUALS RATE OF CHANGE OF KINETIC ENERGY

The Work–Energy Theorem states that the net mechanical work on a system manifests itself as a change in the total energy of the system. In the present situation, the change in the total energy is entirely accounted for on the kinetic energy side of the ledger. Thus, the instantaneous change in the kinetic energy of the rotating object is precisely equal to the input mechanical power!

$$P_{\text{Net}} = \frac{dK}{dt} \qquad \Longrightarrow \qquad \tau_{\text{net ext'l}}\,\omega = I\,\alpha\,\omega\,,$$

and thus

$$I\,\alpha = \tau_{\text{net ext'l}}\,.$$

This is the equation of motion relating the net external torque acting on the rigid object, its moment of inertia, and the angular acceleration that it experiences. It does not take much squinting to see that this dynamical relation is the rotational analogue of N2 [expressed for purely translational motion of particles and systems in Chapters 10 and 31]. We will subsequently refer to $I\,\alpha = \tau_{\text{net ext'l}}$ as

"the rotational version of N2,"

or "the angular expression of N2,"

or simply "N2."

Chapter 40

Torque-y Topics

Let's take the results of the last chapter and turn them about. That is, we'll start with

$$I\alpha = \tau_{\text{net ext'l}} \, ,$$

the angular expression of N2, and develop an energetic formulation of rotational dynamics. Recall that the mechanical work performed by a force acting through a displacement was defined in Chapter 21. It does no violence to, and indeed is consonant with, our analogy between fixed-axis rigid-body rotation and translational motion in 1-d to propose the following definition of mechanical work associated with a particular torque.

MECHANICAL WORK (provisional)   The mechanical work performed by a torque [with respect to a fixed axis] is its integral through an angular displacement [about the same fixed axis] from an initial angle, θ_i, to a final angle, θ_f. I.e.,

$$W_{if}[\tau] = \int_{\theta_i}^{\theta_f} \tau \, d\theta \, .$$

Two comments help to explicate this definition of work.

UNITS   **Q:** As work is a form of energy, the units had better be joules, hadn't they?

A: Consider the [dimensions and] units of the terms appearing on the RHS,

$$[\tau] = \texttt{N} \cdot \texttt{m} \qquad \text{and} \qquad [\theta] = \texttt{radians} \, .$$

The radian is a dimensionless unit, and therefore the units associated with the expression for work are $\texttt{N} \cdot \texttt{m} \cdot \texttt{rad} = \texttt{N} \cdot \texttt{m} = $ joules. This is not merely a semantic trick. Torque is not a form of energy, whereas the application of a torque through an angle θ is. This distinction is maintained in our assignment of units [*cf.* the comment made in Chapter 39].

PATH   The specification of the path [through angle-space] has been glossed over, because fixed axis rotational motion is essentially one-dimensional.

This definition of rotational work inspires us to propose a WORK–ENERGY THEOREM FOR ROTATION.

WORK–ENERGY THEOREM for ROTATION   The work done by the net external torque acting on a [rigid, fixed-axis] rotating body is equal to the change in its rotational kinetic energy.

PROOF of the Rotational Version of the Work–Energy Theorem

The arguments developed in our first analyses of energetics [Chapter 21 *et seq.*] carry over here, and militate for the identification of the net work done by all of the torques acting on the system with that performed by the net torque:

$$\text{Net Work} = \sum_n W_{if}[\tau_n] = W_{if}\left[\sum_n \tau_n\right] = W_{if}[\tau_{\text{net}}] = W_{if}[\tau_{\text{net ext'l}}] .$$

ASIDE: That the net work done on the system is equal to the work done by the net torque follows from both torques and integrals respecting the Principle of Linear Superposition. Newton's Third Law ensures that the net internal torque vanishes.

The proof of the theorem proceeds [just as in the translational case] from the expression for work, augmented by N2 and the definition of angular acceleration. The inverse chain rule again supplies the crucial step.

The work done by the net external torque is

$$W_{if}[\tau_{\text{net ext'l}}] = \int_{\theta_i}^{\theta_f} \tau_{\text{net ext'l}} \, d\theta .$$

The rotational version of N2 equates the net external torque [the CAUSE] with the product of the moment of inertia of the body and its angular acceleration [the EFFECT]. Thus, the integrand on the RHS may be rewritten, leading to

$$W_{if}[\tau_{\text{net ext'l}}] = \int_{\theta_i}^{\theta_f} I\alpha \, d\theta = I \int_{\theta_i}^{\theta_f} \alpha \, d\theta .$$

The assumption of rigidity ensures the constancy of the moment of inertia, enabling it to be factored out of the integrand in the last equality. The angular acceleration is

$$\alpha = \frac{d\omega}{dt} = \frac{d\omega}{d\theta}\frac{d\theta}{dt} = \frac{d\omega}{d\theta}\,\omega = \frac{1}{2}\frac{d\omega^2}{d\theta} ,$$

where the time derivative has been consistently converted to a derivative with respect to angle via the chain rule. Inserting this expression into the integral, and carefully changing the variable of integration from angle, θ [with limits $\theta_i \to \theta_f$], to squared angular speed, ω^2 [with corresponding limits $\omega_i^2 \to \omega_f^2$], the net work becomes

$$W_{if}[\tau_{\text{net ext'l}}] = \int_{\omega_i^2}^{\omega_f^2} \frac{1}{2}I \, d(\omega^2) = \frac{1}{2}I\left(\omega_f^2 - \omega_i^2\right) .$$

We are then led to propose a self-consistent[1] definition of rotational kinetic energy.

ROTATIONAL KINETIC ENERGY   The rotational kinetic energy ascribable to a body possessing moment of inertia I with respect to the [fixed] axis about which it is rotating at angular speed ω is

$$K = \frac{1}{2}\,I\,\omega^2 .$$

Thus, in light of the above argument and the definition of rotational kinetic energy,

$$W_{if}[\tau_{\text{net ext'l}}] = K_f - K_i = \Delta K .$$

---

[1]It is globally consistent too, since it agrees exactly with the expression for kinetic energy developed in Chapter 36.

This theorem is the rotational analogue of the translational Work–Energy Theorem presented in Chapter 23.

Deducing these energetic results has been great fun, yet there remains the somewhat unsettling fact that they are predicated upon the provisional expression for torque introduced in Chapter 39. Let's patch up the formula for torque, and eliminate the need for the *ad hoc* convention for measuring angles, by finally admitting that

TORQUE IS A VECTOR QUANTITY.

Recall that earlier we strenuously argued for the vector nature of the angular displacement and, more primitively, that the angular position itself is a vector quantity.

> ASIDE: The magnitude of the angular position vector, $\vec{\phi}$, is the size of the angle, $|\vec{\phi}| = \phi$, while its direction, $\hat{\phi}$, is parallel to the axis of rotation with orientation determined by means of a RHR. [The use of ϕ, rather than θ, to denote the angular position is no more significant than labelling a position (in 1-d) y rather than x.]
>
> RHR: Arrange your right hand so that outward rays from the axis are perpendicular to your palm. Align your fingers in the direction in which the angle, ϕ, is increasing. [They should curl naturally along with the trajectories of bundles of reference points on the body.] Your thumb then indicates the sense of positive angle along the axis of rotation.
>
> If the angular displacement is a vector, then so too are the angular velocity and acceleration:
>
> $$\vec{\omega} = \frac{d\vec{\phi}}{dt} = \frac{d\phi}{dt}\,\hat{\phi}\,, \qquad \text{and} \qquad \vec{\alpha} = \frac{d\vec{\omega}}{dt} = \frac{d^2\vec{\phi}}{dt^2} = \frac{d^2\phi}{dt^2}\,\hat{\phi}\,.$$
>
> It must be noted that these time derivatives simplified a great deal on account of the constancy of $\hat{\phi}$, *i.e.*, the assumption that the axis is fixed.

Plainly, the torque inherits a vector nature through its identification with $I\vec{\alpha}$ in N2.

Torque's definition as **a measure of the ability of a force to cause rotation,**

$$\tau = r\,F\,\sin(\theta)\,,$$

appears incommensurate with the idea that it is a vector. Casting our minds all the way back to Chapter 5 in search of inspiration, we recall the introduction of the vector cross product, and the promise that it would eventually find its uses. This is one of them. All of the features of torque that we have developed thus far are retained, and augmented with a directional aspect, by the redefinition of vector torque in terms of the cross product.

VECTOR TORQUE The vector torque,

$$\vec{\tau}_F = \vec{r} \times \vec{F}\,,$$

quantifies the ability of the force $\vec{F}$, applied at the position $\vec{r}$ relative to a POINT OF ROTATION,[2] to cause rotation about that particular point.

From this moment on, this shall be our understanding of torque. In practice, however, we will often not make explicit its vector nature, and instead treat it implicitly.

---

[2] Until now, we have considered rotation about a fixed axis only. Generally, rotation occurs about a point, and the definition of vector torque accommodates this. For rotation about an axis, think of the vector $\vec{r}$ as extending perpendicular from the axis to the point at which the force is applied.

In light of the newfound vectorial attribute of torque, it behooves us to redefine the mechanical work associated with the action of a torque through a particular angular displacement.

MECHANICAL WORK   The mechanical work performed by a torque acting through an angular displacement from an initial angle, $\vec{\theta}_i$, to a final angle, $\vec{\theta}_f$, is

$$W_{if}[\vec{\tau}] = \int_{\vec{\theta}_i}^{\vec{\theta}_f} \vec{\tau} \cdot d\vec{\theta}.$$

There is one remaining aspect of energetics to consider.

INSTANTANEOUS POWER   The instantaneous mechanical power input to the rotating system is equal to the dot product of the torque with the angular velocity:

$$P = \vec{\tau} \cdot \vec{\omega}.$$

This corresponds exactly to the formula for mechanical power in systems undergoing translation, *viz.*, $P = \vec{F} \cdot \vec{v}$, discerned in Chapter 28.

Chapter 41

Pulleys with Rotational Inertia

The mathematical models thus far employed to describe the dynamics of physical systems which include pulleys have been predicated upon the inertialessness of the ideal pulley [introduced in Chapter 11]. The goal here is to refine our analysis by incorporation of the inertia effects of real pulleys.

EXAMPLE [*A Pulley with Inertia Driven by the Weight of a Block*]

A disk-shaped pulley is mounted centrally on a fixed horizontal axle. The mass of the pulley is $M_\mathbb{P}$, its radius is $R_\mathbb{P}$, and its moment of inertia is $I_\mathbb{P}$.

> ASIDE: The radius, $R_\mathbb{P}$, is the distance from the centre of the axle to the channel in which the rope lies. No assumptions are here made about the thickness of the axle [except that it is less than twice the radius of the pulley].

A length of ideal rope is wound around the pulley and the free end is affixed to a block of mass m, as shown in Figure 41.1. We assume that the pulley spins on frictionless bearings, that no drag forces are operative, and that the acceleration due to gravity is constant.

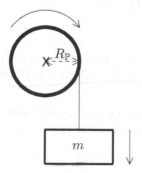

FIGURE 41.1 A Pulley with Inertia Driven by the Weight of a Block

Q: What is the acceleration of the block?

A: Let's follow the recipe, now incorporating the inertia of the pulley.

1. Sketch—see Figure 41.1.

2. There are two dynamical constituents: the block and the pulley.

   Block

   Weight ∘ $\vec{W} = m\,g$ [↓].

   Tension ∘ $\vec{T} = T$ [↑].

Pulley

Weight ∘   The weight of the pulley is $\vec{W_\mathbb{P}} = M_\mathbb{P}\, g \,[\downarrow]$.

Tension ∘   Tension in the rope, $\vec{T''} = T'$ [vertically down] $= T'\,[\downarrow]$ acts on the pulley.

Normal ∘   A support force, $\vec{P}$, holds the pulley in place.

With the above enumeration of the forces on each dynamical constituent, we can draw the two separate FBDs illustrated in Figure 41.2.

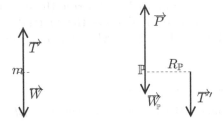

FIGURE 41.2   FBDs for the Block and Pulley

3. Coordinates: Let y be downward positive for the block, and upward positive for the pulley. Let clockwise be the positive sense of rotation.

4. $m\,\vec{a} = \vec{F}_{\text{NET}}$ and $I\,\vec{\alpha} = \vec{\tau}_{\text{net}}$

Block

Vertical: $m\,a_y = F_{\text{NET}y} = m\,g - T$.

Pulley

Vertical: The pulley experiences zero translational acceleration because the support force cancels the combined effects of the weight of the pulley and the tension in the rope.

Rotational: Each of the three forces which act on the pulley may contribute to the net torque that it experiences [about the centre of its axle], and so,

$$I\,\alpha = \tau_{\text{net}} = \tau_W + \tau_{T'} + \tau_P\,.$$

Let's examine each of these torques in turn.

τ_W The weight of the pulley is deemed to act at its CofM, which is at the centre of the axis of rotation. Therefore, the moment arm associated with the weight force is zero, and thus $\tau_W = 0$. [The weight of the symmetric pulley is unable to cause rotation about its axle.]

$\tau_{T'}$ The tension in the rope, T', is applied at the place where the rope unwinds from the pulley, which is a distance $R_\mathbb{P}$ from the axis. The tension is tangential to the rim of the pulley [at $\pi/2$ with respect to the moment arm], and hence

$$\tau_{T'} = T'\, R_\mathbb{P} \sin\left(\frac{\pi}{2}\right) = T'\, R_\mathbb{P}\,.$$

This particular torque is exerted in the clockwise [positive] sense.

τ_P The support force acts on the pulley from the axle, and is unable to produce rotation of the pulley about the axis: $\tau_P = 0$.

Hence, $\tau_{\text{net}} = T' R_{\mathbb{P}}$.

5. Now we must link the dynamical constituents.

Ideal Rope Ideal rope is a perfect transmitter of force ($T = T'$) AND is extensionless (the linear acceleration characteristic of the system has the same value, a, everywhere along the rope).

Pulley IF the rope unwinds from the [rigid] pulley without sliding, THEN the tangential speed of points on the rim of the pulley exactly matches the speed of the rope, AND the tangential acceleration of points on the rim must also equal a, the acceleration of both the rope and block. Hence the magnitude of the [clockwise] angular acceleration must be

$$\alpha = \frac{a}{R_{\mathbb{P}}}.$$

These linkages enable re-expression of N2 for the pulley in terms of the acceleration of the system and the tension in the ideal rope, *viz.*,

$$I_{\mathbb{P}} \alpha = \tau_{\text{net}} \quad \Longrightarrow \quad I_{\mathbb{P}} \frac{a}{R_{\mathbb{P}}} = T R_{\mathbb{P}} \quad \Longrightarrow \quad \frac{I_{\mathbb{P}}}{R_{\mathbb{P}}^2} a = T.$$

The dynamical equations governing this block–pulley system can be written as

$$\left\{ \begin{array}{l} m\,a = m\,g - T \\ \dfrac{I_{\mathbb{P}}}{R_{\mathbb{P}}^2}\,a = T \end{array} \right\}, \quad \text{two equations for unknowns } \{a, T\}.$$

The solutions of these equations are

$$a = \frac{1}{1 + \frac{I_{\mathbb{P}}}{m\,R_{\mathbb{P}}^2}}\,g, \quad \text{and} \quad T = \frac{I_{\mathbb{P}}}{R_{\mathbb{P}}^2} \frac{1}{1 + \frac{I_{\mathbb{P}}}{m\,R_{\mathbb{P}}^2}}\,g = \frac{m\,g}{1 + \frac{m\,R_{\mathbb{P}}^2}{I_{\mathbb{P}}}}.$$

_____

These results apply for the pulley of radius $R_{\mathbb{P}}$, mass $M_{\mathbb{P}}$, and moment of inertia $I_{\mathbb{P}}$. In any particular case, these three quantities are related. For instance:

IF the pulley is hoop-like, THEN $I_{\mathbb{P}} \simeq M_{\mathbb{P}}\,R_{\mathbb{P}}^2$,

IF the pulley is disk-like, THEN $I_{\mathbb{P}} \simeq \frac{1}{2}\,M_{\mathbb{P}}\,R_{\mathbb{P}}^2$.

In the remainder of this example, we shall suppose that the pulley behaves exactly as a uniform disk, in which case,

$$a = \frac{1}{1 + \frac{1}{2}\frac{M_{\mathbb{P}}}{m}}\,g \quad \text{and} \quad T = \frac{M_{\mathbb{P}}}{2\,m + M_{\mathbb{P}}}\,m\,g.$$

6. Let's argue for the physical consistency of our solutions for the acceleration and the tension via two physically relevant limits.

$M_{\mathbb{P}} \ll m$ Imagine that the disk-like pulley is much less massive than the block.

⋄   Our guess is that the system will have a relatively large acceleration. The general formula for the acceleration reveals

$$\lim_{M_{\mathbb{P}} \ll m} a = \lim_{M_{\mathbb{P}} \ll m} \frac{1}{1 + \frac{1}{2}\frac{M_{\mathbb{P}}}{m}} \, g \to g^{-} \, ,$$

exactly as we expect in this limit.

⋄   Our guess is that the tension in the rope is small[1] because the downward acceleration of the block is nearly equal to g. Taking the limit,

$$\lim_{M_{\mathbb{P}} \ll m} T = \lim_{M_{\mathbb{P}} \ll m} \frac{M_{\mathbb{P}}}{2\,m + M_{\mathbb{P}}} \times m\,g \to 0^{+} \times m\,g = 0^{+} \, .$$

confirms our prediction for the magnitude of the tension in the rope.

$M_{\mathbb{P}} \gg m$ Imagine that the disk-like pulley is much more massive than the block.

⋄   Our expectation is that the system will accelerate slowly. In fact,

$$\lim_{M_{\mathbb{P}} \gg m} a = \lim_{M_{\mathbb{P}} \gg m} \frac{1}{1 + \frac{1}{2}\frac{M_{\mathbb{P}}}{m}} \, g \to 0^{+} \, .$$

⋄   As the block is strongly restrained by the rope,

$$\lim_{M_{\mathbb{P}} \gg m} T = \lim_{M_{\mathbb{P}} \gg m} \frac{M_{\mathbb{P}}}{2\,m + M_{\mathbb{P}}} \, m\,g \to 1^{-} \times m\,g \to (m\,g)^{-} \, .$$

....

EXAMPLE [*A Two-Block and Pulley with Inertia System*]

Two blocks, labelled 1 and 2, with $M_1 > M_2$, are joined by a piece of ideal rope which passes over a pulley, as shown in Figure 41.3. The pulley has moment of inertia $I_{\mathbb{P}}$, total mass $M_{\mathbb{P}}$, and radius $R_{\mathbb{P}}$. No drag or frictional forces are operative, and the acceleration due to gravity is constant. To begin with, the blocks are held at rest, separated in height by a distance y_0. At time t_i the system is suddenly released. At a later time, t_f, Block 1 has fallen a distance h from its initial position, while Block 2 has risen by the same amount.

Q: What is the speed of the blocks at t_f?

A: Let's solve this by two different methods.

METHOD ONE: [*Following the dynamics recipe*]

1. Sketch—see Figure 41.3.

---

[1]Scale is a perennial issue; "small," here, is in comparison to the weight of the block.

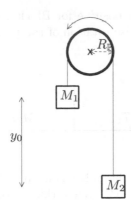

FIGURE 41.3 A Two-Block and Pulley System

2. There are three dynamical constituents: Block 1, Block 2, and the Pulley.

Block 1	**Block 2**
Weight 1 ○ $\vec{W_1} = M_1\, g\ [\downarrow]$.	Weight 2 ○ $\vec{W_2} = M_2\, g\ [\downarrow]$.
Tension 1 ○ $\vec{T_1} = T_1\ [\uparrow]$.	Tension 2 ○ $\vec{T_2} = T_2\ [\uparrow]$.

Pulley

Weight ○ $\vec{W_\mathbb{P}} = M_\mathbb{P}\, g\ [\downarrow]$.

Tension 1 ○ $\vec{T_1}' = T_1'\ [\downarrow]$.

Tension 2 ○ $\vec{T_2}' = T_2'\ [\downarrow]$.

Normal ○ $\vec{P}$ is the pulley support force.

Friction ○ The bearings are frictionless, and drag forces are negligible.

With the above enumeration of the forces on each dynamical constituent, we can draw the three separate FBDs illustrated in Figure 41.4.

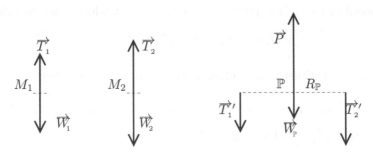

FIGURE 41.4 FBDs for Block 1, Block 2, and the Pulley

3. Coordinates: Let y be upward positive for Block 2 and downward positive for Block 1. Designate anti-clockwise rotation of the pulley to be positive.

4. $m\,\vec{a} = \vec{F}_{\text{NET}}$ and $I\,\vec{\alpha} = \vec{\tau}_{\text{net}}$

Block 1	Block 2
$M_1\,a_{1y} = F_{\text{NET}1y} = M_1\,g - T_1$	$M_2\,a_{2y} = F_{\text{NET}2y} = T_2 - M_2\,g$

Pulley

Vertical: No net force acts on the pulley since it is fixed in place: $M_{\mathbb{P}}\,a_{\mathbb{P}} = 0 = -M_{\mathbb{P}}\,g + T_1' + T_2' - P_y$. While this is true, it does not assist in answering the query.

Rotational: The pulley experiences angular acceleration due to the net torque, about the centre of its axle, produced by the four active forces.

$$I_{\mathbb{P}}\,\alpha = \tau_{\text{net}} = \tau_W + \tau_{T_1'} + \tau_{T_2'} + \tau_P \,.$$

Each of the torque contributions is examined below.

τ_W The weight of the pulley is deemed to act at its CofM which lies on the axis of rotation. Thus, the weight force has a vanishing moment arm. Hence, $\tau_W = 0$.

$\tau_{T_1'}$ The tension in the rope attached to Block 1 is applied at the leftmost edge of the pulley and is directed straight down, producing an anti-clockwise torque:

$$\tau_{T_1'} = T_1'\,R_{\mathbb{P}}\,\sin\left(\frac{\pi}{2}\right) = T_1'\,R_{\mathbb{P}}\,.$$

$\tau_{T_2'}$ The tension in the rope attached to Block 2 acts at the rightmost edge of the pulley and is also directed straight down, yielding a clockwise torque:

$$\tau_{T_2'} = -T_2'\,R_{\mathbb{P}}\,\sin\left(\frac{\pi}{2}\right) = -T_2'\,R_{\mathbb{P}}\,.$$

τ_P The support force acts on the axle, and hence $\tau_P = 0$.

The rotational motion of the pulley is governed by N2, which reads as follows:

$$I_{\mathbb{P}}\,\alpha = \tau_{\text{net}} = R_{\mathbb{P}}\left(T_1' - T_2'\right)\,.$$

5. The dynamical constituents are linked by the ideal rope and pulley.

Ideal Rope This rope transmits tension perfectly and does not stretch. Thus,

$$T_1 = T_1' \quad \text{AND} \quad T_2 = T_2'\,, \quad \text{WHILE} \quad a_{1y} = a_{2y} = a\,,$$

where a is the acceleration characteristic of the system as a whole.

ASIDE: Do note that $a_{1y} = a_{2y}$, rather than $a_{1y} = -a_{2y}$, because of our choices of y-axis orientation.

Pulley In order that the rope not slide, the tangential acceleration of points on the rim must conform precisely to a. Thus,

$$\alpha = \frac{a}{R_{\mathbb{P}}}$$

[as in the previous example]. With this realisation, N2 for the pulley becomes

$$I_{\mathbb{P}}\,\alpha = \tau_{\text{net}} \quad \Longrightarrow \quad I_{\mathbb{P}}\,\frac{a}{R_{\mathbb{P}}} = R_{\mathbb{P}}\,(T_1' - T_2') \quad \Longrightarrow \quad \frac{I_{\mathbb{P}}}{R_{\mathbb{P}}^2}\,a = T_1' - T_2' = T_1 - T_2\,.$$

In light of these linking relations, the dynamical equations may be written as

$$\left\{ \begin{array}{l} M_1\,a = M_1\,g - T_1 \\ M_2\,a = T_2 - M_2\,g \\ \dfrac{I_{\mathbb{P}}}{R_{\mathbb{P}}^2}\,a = T_1 - T_2 \end{array} \right\}\,, \text{ three equations for unknowns } \{a, T_1, T_2\}.$$

Solving these for the acceleration of the system, yields

$$a = \frac{M_1 - M_2}{M_1 + M_2 + \frac{I_{\mathbb{P}}}{R_{\mathbb{P}}^2}}\,g\,.$$

This acceleration is constant. Therefore, the constant acceleration kinematical formulae apply [serving to simplify the final steps needed to answer the question].

6. [Let's forego arguing the consistency of the result in favour of answering the question.] The "third" constant acceleration kinematical formula reads

$$v_f^2 - v_i^2 = 2\,a\,(\Delta x)\,,$$

With $v_i = 0$, $\Delta x = h$, and a as above, this becomes

$$v_f^2 = \frac{2\,(M_1 - M_2)\,g\,h}{M_1 + M_2 + \frac{I_{\mathbb{P}}}{R_{\mathbb{P}}^2}}\,.$$

The positive square root of the RHS directly above constitutes the sought-after answer to the question posed at the beginning of this example.

———————————————

METHOD TWO: [*Applying Energy Methods*]

The forces which act on the constituents of this system are conservative [gravity/weight], or internal [tensions], or do no net work on the system [normal/support]; and thus the total mechanical energy of the system is conserved.

The initial value of the total mechanical energy [at time t_i] is

$$E_i = K_i + U_i\,.$$

The initial kinetic energy is zero, because the blocks and the pulley are at rest when the system is released: $K_i = K_{1,i} + K_{2,i} + K_{\mathbb{P},i} = 0$. The initial potential energy consists of the gravitational potential energies of the three constituents,[2] $U_i = U_{1,i} + U_{2,i} + U_{\mathbb{P},i}$. WLOG, the reference height [assigned zero gravitational PE] may be chosen to coincide with the initial height of Block 1. Therefore,

$$U_{1,i} = 0, \qquad \text{AND} \qquad U_{2,i} = -M_2\,g\,y_0, \qquad \implies \qquad U_i = -M_2\,g\,y_0 + U_{\mathbb{P},i}.$$

Hence, the total mechanical energy at the initial time happens to be

$$E_i = -M_2\,g\,y_0 + U_{\mathbb{P},i}.$$

At time t_f, Block 1 has fallen a distance h, and Block 2 has risen by the same amount. The pulley's height has remained fixed. The blocks are each moving with speed v_f, and [to ensure that the rope does not slip on the pulley] the pulley is rotating with angular speed

$$\omega_f = \frac{v_f}{R_{\mathbb{P}}}.$$

The final value of the total mechanical energy is $E_f = K_f + U_f$. The final kinetic energy receives contributions from each of the blocks and the pulley:

$$K_f = K_{1,f} + K_{2,f} + K_{\mathbb{P},f} = \frac{1}{2}\,M_1\,v_f^2 + \frac{1}{2}\,M_2\,v_f^2 + \frac{1}{2}\,I_{\mathbb{P}}\,\omega_f^2 = \frac{1}{2}\left(M_1 + M_2 + \frac{I_{\mathbb{P}}}{R_{\mathbb{P}}^2}\right)v_f^2.$$

The final potential energy is $U_f = U_{1,f} + U_{2,f} + U_{\mathbb{P},f}$. According to the description of the final state of the system,

$$U_{1,f} = -M_1\,g\,h, \qquad U_{2,f} = -M_2\,g\,y_0 + M_2\,g\,h, \qquad U_{\mathbb{P},f} = U_{\mathbb{P},i},$$

and hence

$$U_f = -M_1\,g\,h + M_2\,g\,h - M_2\,g\,y_0 + U_{\mathbb{P},i} = -(M_1 - M_2)\,g\,h - M_2\,g\,y_0 + U_{\mathbb{P},i}.$$

The final value of the total mechanical energy, then, is

$$E_f = \frac{1}{2}\left(M_1 + M_2 + \frac{I_{\mathbb{P}}}{R_{\mathbb{P}}^2}\right)v_f^2 - (M_1 - M_2)\,g\,h - M_2\,g\,y_0 + U_{\mathbb{P},i}.$$

The total mechanical energy is a constant of the motion, *i.e.*, $E_f \equiv E_i$. Therefore, the difference between the final and initial values of the total energy must be exactly ZERO:

$$0 = E_f - E_i = \frac{1}{2}\left(M_1 + M_2 + \frac{I_{\mathbb{P}}}{R_{\mathbb{P}}^2}\right)v_f^2 - (M_1 - M_2)\,g\,h.$$

Hence, the final speed (squared) is

$$v_f^2 = \frac{2\,(M_1 - M_2)\,g\,h}{M_1 + M_2 + \frac{I_{\mathbb{P}}}{R_{\mathbb{P}}^2}},$$

in harmonious accord with the results of METHOD ONE!

---

[2]Since the pulley does not change height, it neither gains nor loses potential energy, and one is justified in discarding its contribution at the outset. We retain it so as to verify that it is safely neglectable!

Chapter 42

Angular Momentum

Angular momentum is the remaining foundational stone in [this wing of] the edifice that is CLASSICAL MECHANICS.[1]

ANGULAR MOMENTUM   The angular momentum, $\vec{L}$, of a point particle or system is determined with respect to a particular fixed point [or axis]. For a point particle possessing linear momentum $\vec{p}$,

$$\vec{L} = \vec{r} \times \vec{p},$$

where $\vec{r}$ is the vector from the fixed point to the [instantaneous] location of the particle.

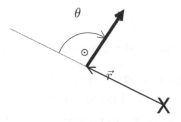

FIGURE 42.1   A Moving Particle and a Fixed Point

From the vantage point of the plane containing[2] both $\vec{r}$ and $\vec{p}$ [$\vec{p} = m\vec{v}$], illustrated in Figure 42.1, the angular momentum is expressable as

$$\vec{L} = \vec{r} \times \vec{p} = |\vec{r}|\,|m\vec{v}|\,\sin(\theta)\,[\,\text{RHR}\,] = m\,v\,r\,\sin(\theta)\,[\,\text{RHR}\,].$$

Three comments are warranted here.

- The SI units associated with angular momentum are $\mathbf{kg \cdot m^2/s}$. Foreshadowing, we observe that these units are equivalent to $\mathbf{N \cdot m \cdot s}$.

- The angular momentum depends crucially on the choice of the fixed point about which it is computed.

- It is not oxymoronic for a particle whose motion through space is along a perfectly straight line to possess non-zero angular momentum about some specified point. Consider the situation, shown in Figure 42.2, in which a particle is moving inertially [with constant velocity].

---

[1]While angular momentum is a useful classical construction, it is essential in quantum mechanics.
[2]We speak loosely of "containing," since $\vec{r}$ and $\vec{p}$ reside in different vector spaces.

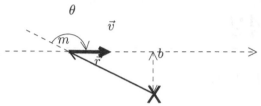

FIGURE 42.2   A Particle Moving Inertially May Possess Non-Zero Angular Momentum

The angular momentum of the particle, with respect to the fixed point, is

$$\vec{L} = \vec{r} \times \vec{p} = m\,v\,r\,\sin(\theta)\,[\otimes] = m\,v\,b\,[\otimes]\,,$$

where the **impact parameter**, b, is the distance of closest approach of the particle to the fixed point.

TORQUE AND ANGULAR MOMENTUM FOR A SINGLE POINT PARTICLE

The torque associated with a particular force is a measure of the ability of the force to effect rotation about a fixed point,

$$\vec{\tau} = \vec{r} \times \vec{F}\,,$$

where the vector $\vec{r}$ extends from the fixed point to the point of application of the force.

o   All forces acting on a point particle are exerted at exactly the same point.

o   Forces obey the PRINCIPLE OF LINEAR SUPERPOSITION.

o   The cross product is distributive over addition [*i.e.*, $(\vec{A}+\vec{B})\times\vec{C} = \vec{A}\times\vec{C}+\vec{B}\times\vec{C}$].

These three facts, taken together, enable the net torque to be simply expressed:

$$\vec{\tau}_{\text{net}} = \vec{r} \times \vec{F}_{\text{NET}}\,.$$

Furthermore, [the more general form of] N2 for translation reads

$$\frac{d\vec{p}}{dt} = \vec{F}_{\text{NET}}\,,$$

and thus, for the single point particle,

$$\vec{\tau}_{\text{net}} = \vec{r} \times \frac{d\vec{p}}{dt}\,.$$

The angular momentum of the particle [computed about the same fixed point as the torque] is

$$\vec{L} = \vec{r} \times \vec{p}\,.$$

Taking the time derivative[3] of the RHS of this expression requires the application of a **cross product rule**. Fortunately, the derivative distributes as expected:

$$\frac{d\vec{L}}{dt} = \frac{d\vec{r}}{dt} \times \vec{p} + \vec{r} \times \frac{d\vec{p}}{dt}\,.$$

---

[3]Physicists [almost] always want to know: "At what rate is ⟨whatever it is⟩ changing?"

[Care must taken to preserve the order of the factors, since $\vec{A} \times \vec{B} = -\vec{B} \times \vec{A}$.]

The first factor appearing in the first term of the expression for the time rate of change of the angular momentum is the time rate of change of $\vec{r}$. This is the velocity of the particle.

ASIDE: The only necessary condition for the veracity of this statement is that the fixed point be at rest. It need not be at the origin. Suppose that the "true" coordinates of the particle and the fixed point are $\vec{R}$ and $\vec{R}_{\text{fixed}}$ respectively. Then, by the definition of $\vec{r}$,

$$\vec{R} = \vec{R}_{\text{fixed}} + \vec{r}.$$

The velocity of the particle is the time rate of change of its position,

$$\vec{v} = \frac{d\vec{R}}{dt} = \frac{d}{dt}\left(\vec{R}_{\text{fixed}} + \vec{r}\right) = \frac{d\vec{R}_{\text{fixed}}}{dt} + \frac{d\vec{r}}{dt} = \frac{d\vec{r}}{dt},$$

by the presumed immobility of the fixed point.

Hence, the first term is equal to $\vec{v} \times \vec{p}$. However, the momentum of the point particle is parallel to its velocity, $\vec{p} = m\,\vec{v}$, and therefore this cross product vanishes.

Thus, the time rate of change of the point particle's angular momentum is [the second term],

$$\frac{d\vec{L}}{dt} = \vec{r} \times \frac{d\vec{p}}{dt}.$$

Substituting the translational N2 result yields

$$\frac{d\vec{L}}{dt} = \vec{\tau}_{\text{net}},$$

for a point particle. This constitutes a general formulation of [the rotational version of] N2.

The time rate of change of angular momentum of a point particle (*effect*) arises from the net torque acting on the particle (*cause*).

ASIDE: As in the presentation of linear momentum in Chapter 29, we shan't trumpet this result for point particles, since its extension to systems of particles and rigid bodies lies just ahead.

We now make the claim[4] that for a system comprised of $\mathcal{N}$ point(-like) particles, it is meaningful to construct the **total** angular momentum of the system by adding together the respective angular momenta of its constituents.

TOTAL ANGULAR MOMENTUM   Suppose that a system under consideration is comprised of $\mathcal{N}$ point(-like) particle constituents, each identified by a specific value of an index, $j \in 1...\mathcal{N}$. The mass, position, relative position with respect to a common fixed point, velocity, and momentum of the jth particle are given by $\{M_j\,,\ \vec{R}_j\,,\ \vec{r}_j\,,\ \vec{v}_j\,,\ \vec{p}_j\}$, respectively, where $\vec{p}_j = M_j\,\vec{v}_j$. The angular momentum of each constituent [all determined with respect to the same fixed point] is

$$\vec{L}_j = \vec{r}_j \times \vec{p}_j.$$

The aggregate angular momentum of the system is concisely expressed as

$$\vec{L}_{\text{Total}} = \sum_{j=1}^{\mathcal{N}} \vec{L}_j.$$

---

[4]The corresponding claim for linear momentum was made in Chapter 29.

Let us add together the torques due to all of the forces acting on all of the particles constituting the system[5] to determine the total net torque acting on the system.

$$\vec{\tau}_{\text{Total}} = \sum_{j=1}^{\mathcal{N}} \vec{\tau}_{\text{net},j} \,.$$

It is important to remember that all of the torques appearing in this sum are computed about the same fixed point or axis.

It is always possible to distinguish between the net torque which arises due to forces which are internal to the system and the net torque due to forces which are external. That is:

$$\vec{\tau}_{\text{net},i} = \vec{\tau}_{\text{net ext'l},i} + \vec{\tau}_{\text{net int'l},i} \,.$$

[Wow! It's *déjà vu* all over again!]

The external torque on the ith particle is a measure of the ability of the net external force acting on the ith particle to effect rotation about the fixed point. Similarly, the internal torque on the ith particle is a measure of the ability of the net internal force acting on the ith particle to effect rotation about the fixed point.

ASIDE: This separation into external and internal contributions amounts to merely rearranging terms entering into the vector sum determining the net torque acting on the ith particle. Assurance that we are freely able to do this stems from: the Principle of Linear Superposition applied to both forces and torques, the elementary/foundational properties of vectors and vector spaces expounded in Chapter 5, and the distributivity of vector addition through the cross product.

Thus, the total net torque acting upon the system of particles is

$$\vec{\tau}_{\text{Total}} = \sum_{j=1}^{\mathcal{N}} \vec{\tau}_{\text{net},j} = \sum_{j=1}^{\mathcal{N}} \left(\vec{\tau}_{\text{net ext'l},j} + \vec{\tau}_{\text{net int'l},j} \right) = \sum_{j=1}^{\mathcal{N}} \vec{\tau}_{\text{net ext'l},j} + \sum_{j=1}^{\mathcal{N}} \vec{\tau}_{\text{net int'l},j} \,.$$

Some thought reveals that in the sum over internal torques, we inevitably encounter action–reaction pairs of forces possessing a common line of action [with opposite orientation], and hence equal moment arms. These force pairs produce ZERO net torque when superposed. Thus,

$$\sum_{j=1}^{\mathcal{N}} \vec{\tau}_{\text{net int'l},j} = 0 \,,$$

and the total net torque acting on the system is contributed by the external forces only:

$$\vec{\tau}_{\text{Total}} = \sum_{j=1}^{\mathcal{N}} \vec{\tau}_{\text{net ext'l},j} = \vec{\tau}_{\text{net ext'l}} \,.$$

Finally, the generalised expression of N2 for rotation reads

$$\frac{d\vec{L}_{\text{Total}}}{dt} = \vec{\tau}_{\text{net ext'l}} \,.$$

This form of N2 is applicable to systems of point(-like) particles, and continuous distributions of matter, undergoing rotational motion.

---

[5]This is novel. Until now we have only added torques due to forces which act upon a single particle.

The LAW of CONSERVATION of ANGULAR MOMENTUM IF the net external torque acting on a system vanishes, THEN the total angular momentum of the system is conserved, *i.e.*, it is a constant of the motion.

    o   As was the case for the LAW OF MOMENTUM CONSERVATION promulgated in Chapter 31, this conservation law follows, tautologically, from the general expression of N2.

    o   This law applies to the system as a whole, and does not prevent exchanges of angular momentum among the constituents.

To complete the analogical association between rotational and translational dynamics, we need only establish the parallel with $\vec{p} = m\,\vec{v}$. Having established that $\vec{v} \leftrightarrow \vec{\omega}$, and $m \leftrightarrow I$, in Chapters 34 and 36 respectively, one might be inclined to propose: $\vec{L} = I\,\vec{\omega}$.

Q: Is $\vec{L} = I\,\vec{\omega}$ generally valid?

A: Not quite.

Consider a body rotating about a fixed axis, and choose a fixed point somewhere on the axis for use in determining the angular momentum. Let's compute the component of the total angular momentum of the body in the direction of the axis, which we shall identify as the z-axis, by means of the PARTITION, COMPUTE, and SUM strategy.

PARTITION   *Partition the body into $\mathcal{N}$ chunks*

The nth chunk in the $\mathcal{N}$-partition has mass Δm_n and it may be assigned a unique position vector, $\vec{r}_n$, extending from the fixed point to a representative point within [or on the surface of] the chunk. Let d_n denote the the perpendicular distance from the axis to the representative point. This chunk, the [z-]axis, and the fixed point are pictured together in Figure 42.3. The velocity of each chunk is directed tangent to its circular trajectory [lying in planes orthogonal to the axis], and its magnitude is equal to the product of the rigid rotation rate with d_n, *viz.*,

$$\vec{v}_n = v_n\,[\otimes] = \omega\,d_n\,[\otimes]\,.$$

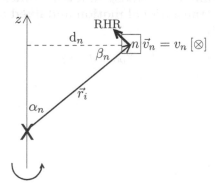

FIGURE 42.3   The Angular Momentum of a Chunk in a Partition of a Rigid Body

COMPUTE   *Compute the angular momentum of each chunk.*

The angular momentum of the nth chunk is

$$\vec{L}_n = \vec{r}_n \times \vec{p}_n = \vec{r}_n \times \left(\Delta m_n \, \vec{v}_n \right) = \Delta m_n \, v_n \, r_n \, [\mathrm{RHR}].$$

In deriving the final expression, we recognised that the velocity vector is orthogonal to the position vector [relative to the fixed point]. The direction specified by [RHR] is orthogonal to both $\vec{v}_n$ and $\vec{r}_n$ as indicated in Figure 42.3.

The magnitude of the z-component [along the axis of rotation] of $\vec{L}_n$ is obtained by projection onto the z-axis,

$$L_{n,z} = \left| \vec{L}_n \right| \sin(\alpha_n) = \Delta m_n \, v_n \, r_n \, \cos(\beta_n).$$

In deriving this last expression, it was recognised that the two angles $\{\alpha_n, \beta_n\}$ are complementary [sum to $\frac{\pi}{2}$] in order to rewrite the projection factor. Realising that $r_n \, \cos(\beta_n) = d_n$ and $v_n = \omega \, d_n$ for rigid-body rotation, leads to

$$L_{n,z} = \Delta m_n \, d_n^2 \, \omega.$$

SUM   *Sum the angular momentum contributions of the chunks.*

The z-component of the total angular momentum of the body about the fixed point is obtained by summing over the elements of the partition.

$$L_{\mathrm{Total}\, z} = \sum_{n=1}^{N} L_{n,z} = \sum_{n=1}^{N} \Delta m_n \, d_n^2 \, \omega = \left(\sum_{n=1}^{N} \Delta m_n \, d_n^2 \right) \omega = I \, \omega.$$

In the penultimate step, the [assumed] rigidity of the body ensures that all chunks move with the same angular speed. The last step merely invokes the definition of the moment of inertia [the mass-weighted squared distance from the axis].

The significance of this result is that the on-axis component of the total angular momentum is precisely equal to $I\omega$, but there is no fundamental reason[6] to presume that the off-axis components are zero.

The angular momentum of non-axisymmetric bodies is the sole aspect of the analogy between 1-d translational motion and fixed axis rigid rotation which is not generally valid.

Too bad, eh?

---

[6]Sufficient symmetry to ensure that the off-axis contributions evidently cancel would be reason enough.

Chapter 43

Rolling Motion

Our discussion of rolling shall be restricted to objects with circular cross-section [e.g., rings, cylinders, spheres, etc.] and sufficient symmetry to ensure that the CofM moves along a straight-line trajectory. Other objects [e.g., an egg or a gymnast] roll in more complicated manners.

ROLLING   A rolling object experiences two [correlated] types of motion simultaneously.

Translation	Translation of the CofM of the object
Rotation	Rotation about the CofM of the object

The correlation between the translational and rotational motions arises from a constraint:

IF an object rolls without sliding [a.k.a. pure rolling], THEN the distance travelled by the CofM is equal to the arc length unwound on the surface of the object.

When the rolling object has circular cross-section [radius = R] the constraint takes the form:

$$\left| \Delta \vec{R}_{\text{CofM}} \right| = \Delta s = R \, \Delta \theta \, .$$

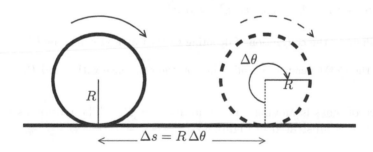

FIGURE 43.1   Rolling Combines Translation with Rotation

At each instant, the constraint relates the translational speed of the CofM to the angular speed of rotation:

$$V_{\text{CofM}} = \frac{ds}{dt} = R \frac{d\theta}{dt} = R\omega \, .$$

Furthermore, the magnitude of the acceleration of the centre of mass is proportional to the angular acceleration:

$$A_{\text{CofM}} = \frac{dV_{\text{CofM}}}{dt} = \frac{d^2 s}{dt^2} = R \frac{d\omega}{dt} = R\alpha \, .$$

ASIDE: The rolling object could slide as well, in which case the arc length along the surface would not be equal to the distance through which the CofM travels. There are two distinct possibilities.

$s > R\theta$ A sharply struck billiard ball, or a hard-thrown bowling ball, will often slide until it begins to roll at a rate which matches its forward speed.

$s < R\theta$ A car's tires may spin, while the car barely moves, under a variety of circumstances.

Two particular IRFs which provide complementary views of rolling motion are the CofM frame and that of an observer who is at rest with respect to the surface on which the rolling occurs, the lab frame.

CofM   In the CofM frame, the centre of mass is at rest, the BULK momentum vanishes, and the object experiences pure rotation [about the axis through the CofM]. All this is exhibited in Figure 43.2.

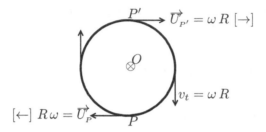

FIGURE 43.2   Rolling in the CofM Frame

The particular point on the object which is instantaneously in contact with the ground is labelled P. The velocity of P is $\vec{U}_P = R\omega\,[\leftarrow]$. Another point on the object, P', diametrically opposite to P, has velocity, $\vec{U}_{P'} = R\omega\,[\rightarrow]$.

[We adhere to the convention of denoting CofM frame velocities by $\vec{U}$.]

ASIDE: From the CofM perspective, all points on the rim move with speed $R\omega$.

LAB   The centre of mass is moving forward at $V_{\text{CofM}} = R\omega$. The motion as it appears in the laboratory IRF is illustrated in Figure 43.3.

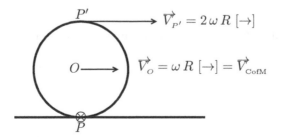

FIGURE 43.3   Rolling in the Lab Frame

The point of contact with the ground, P, is instantaneously at rest, while its diametrically opposite point, P', moves forward at twice the speed of the CofM.

The transformation[1] of velocities of particular points in the rotating body, Q, from the CofM frame to the lab frame, is accomplished[2] by

$$\vec{V_Q} = \vec{U_Q} + \vec{V}_{\text{CofM}} = \vec{U_Q} + \omega R \, [\rightarrow] \, .$$

A Digression on the Rotational Kinetic Energy of a Rolling Object

It was realised just above that the point of contact between the rolling object and the ground is instantaneously at rest. This suggests another way of thinking about rolling.

ROLLING   An object which is rolling [without sliding] undergoes pure rotation about its point of contact with the ground, *viz.*, about an axis which is moving forward at V_{CofM}.

The kinetic energy of rotation, about the axis through P, is

$$K = \frac{1}{2} I_P \, \omega^2 \, .$$

It may take a bit of thought to convince oneself that the angular speed of the body about the axis through P is the same as the angular speed about the axis through the CofM.

ASIDE: The forward speed of the CofM with respect to the point of contact matches exactly the backward speed of the point of contact WRT the CofM.

The Parallel Axis Theorem, stated and proven in Chapter 38, provides the means for determining the relevant moment of inertia,

$$I_P = I_{\text{CofM}} + M D^2 \, .$$

In this instance, the separation between the two axes is equal to the radius of the rolling object, $D = R$. Hence, the kinetic energy is

$$K = \frac{1}{2} \left(I_{\text{CofM}} + M R^2 \right) \omega^2 = \frac{1}{2} I_{\text{CofM}} \, \omega^2 + \frac{1}{2} M \left(R^2 \, \omega^2 \right) = \frac{1}{2} I_{\text{CofM}} \, \omega^2 + \frac{1}{2} M V_{\text{CofM}}^2 \, .$$

The first term in the final expression for the total kinetic energy is the internal, or relative, kinetic energy due to the rotational motion of the object about its CofM. The second term is the bulk kinetic energy of the object owing to the translational motion of its CofM.

The importance of rolling is recognised in the distinction made by ancient historians and anthropologists between societies whose technologies did, or did not, include the wheel.

Q: What is it about rolling that conveys a technological advantage?

A: For an object to roll, rather than slide, on a planar surface, static friction must be acting to prevent relative motion of those parts of the surfaces[3] which are in contact.

[Rolling matches the speed of translation of the CofM to the edge speed of the object.]

Therefore the following chain of reasoning applies.

> **Static friction does zero work.**
> **Thus it cannot change the mechanical energy of a system.**
> **Static friction effects the correlated motions in rolling.**
> **Mechanical Energy is conserved[4] for rolling motion!**

---

[1] To call this a "*translation* from one observer's language to another's" would be cute, but confusing.

[2] The quoted transformation works for the points P, P', and the CofM. Rigidity of the body ensures that it holds everywhere else, too.

[3] In our simplified analysis, this is represented by the point P. More generally, the contact occurs along a line or throughout an area.

[4] Assuming the absence of drag or other non-conservative forces, of course.

EXAMPLE [*A Cylinder Rolling down an Incline*]

A cylinder of mass M and radius R, with moment of inertia I_c about its axis of symmetry [passing through its CofM], is held at rest on an inclined plane. At time t_i the cylinder is suddenly released, and it rolls downward along the surface of the incline. At a later time, t_f, the cylinder has reached a point on the incline which is at distance x from the starting point. These details are illustrated in Figure 43.4. We make several assumptions: drag forces are negligible, the acceleration due to gravity is constant over the extent of the incline, the cylinder rolls without slipping, and it moves in a straight line.

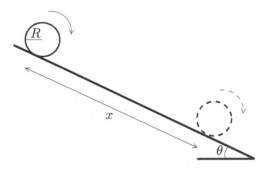

FIGURE 43.4 A Cylinder Rolls a Distance x on an Inclined Plane

Q1: What is the speed of the [CofM of the] cylinder when it has rolled a distance x?

Q2: What is the acceleration of the [CofM of the] cylinder when it has rolled a distance x?

A1: First things first, second thoughts afterward.

The initial mechanical energy of the cylinder, at time t_i, is the sum of its initial kinetic and potential energies. The cylinder is at rest at the instant it is released, so $K_i = 0$. The force of constant field gravity [a.k.a. weight of the particle] admits a potential energy of the form $U_g \sim m\,g\,y$. WLOG the reference height may be chosen to be the height of the CofM of the cylinder at t_i, in which case $U_i = U_{\mathrm{g},i} = 0$. Hence, $E_i = K_i + U_i = 0$.

> ASIDE: Recall that the value assigned to the mechanical energy depends upon the choices of IRF and reference position, and that only changes in this value are significant. ZERO is a convenient amount of initial mechanical energy.

The final mechanical energy of the cylinder, when it has travelled a distance x along the surface of the incline, is the sum of its kinetic and potential energies at time t_f. The kinetic energy may be decomposed into bulk [CofM] and internal [relative] contributions:

$$K_f = K_{\mathrm{CofM},f} + K_{\mathrm{relative},f} = \frac{1}{2}\,M\,V_{\mathrm{CofM}}^2 + \frac{1}{2}\,I_c\,\omega^2.$$

Rolling without slipping engenders the constraint $\omega = V_{\mathrm{CofM}}/R$, and thus

$$K_f = \frac{1}{2}\left(M + \frac{I_c}{R^2}\right)V_{\mathrm{CofM}}^2.$$

The final gravitational potential energy is

$$U_f = U_{\mathrm{g},f} = M\,g\,y_f,$$

where y_f is the final height of the cylinder with respect to its starting point, $y_i = 0$. Expressed in terms of the unique distance through which the CofM and the point of contact have both moved, $y_f = -x\,\sin(\theta)$.

ASIDE: Assumptions of circular cross-section and symmetry about the centre of mass were necessary for this last equivalence. The distribution of mass in the cylinder need not be uniform, so long as it is axially symmetric; it can depend on radius.

Hence, $U_f = -M g x \sin(\theta)$, and the total mechanical energy at t_f reads

$$E_f = K_f + U_f = \frac{1}{2} \left(M + \frac{I_c}{R^2} \right) V_{\text{CofM}}^2 - M g x \sin(\theta) .$$

In the model for this system, the only non-conservative forces acting on the cylinder are the normal force of contact between the cylinder and the incline and the frictional force which enables the cylinder to roll rather than slide. Neither of these forces does mechanical work on the cylinder. Thus, the total mechanical energy of the cylinder on the incline is conserved, *i.e.*, $E_f = E_i$. Hence,

$$\frac{1}{2} \left(M + \frac{I_c}{R^2} \right) V_{\text{CofM}}^2 - M g x \sin(\theta) = 0 \quad \Longrightarrow \quad V_{\text{CofM}}^2 = \frac{2 g \sin(\theta)}{1 + \frac{I_c}{M R^2}} x .$$

A2: Let's determine the acceleration at t_f two different ways.

METHOD ONE: [*The quick and dirty way*]

One might *guess* that the acceleration of the cylinder down the incline should be constant, and invoke the constant acceleration kinematical formula,

$$v_f^2 - v_i^2 = 2 \, a \, (\Delta x) .$$

Setting $v_i = 0$, one can match this formula with the above result for V_{CofM}^2, to obtain

$$a = \frac{g \sin(\theta)}{1 + \frac{I_c}{M R^2}} .$$

The quick and dirty way yields the correct result, but one may feel uneasy owing to the fact that we had to *know* or *assume* that the acceleration is constant.

METHOD TWO: [*A good and useful trick*]

Mechanical energy is conserved throughout the entire time interval $t_i \to t_f$ [and beyond]. As a direct consequence,

$$V_{\text{CofM}}^2 = \frac{2 g \sin(\theta)}{1 + \frac{I_c}{M R^2}} x$$

is generally valid.

ASIDE: Implicit time dependence enters through the explicit dependence on $x = x(t)$.

The trick stems from the realisation that the time derivative of the speed–squared equation yields a valid dynamical relation. Application of $\frac{d}{dt}$ to the LHS results in

$$\frac{d}{dt} \text{LHS} = \frac{d}{dt} V_{\text{CofM}}^2 = 2 V_{\text{CofM}} \frac{dV_{\text{CofM}}}{dt} = 2 V_{\text{CofM}} A_{\text{CofM}} .$$

The time derivative of the RHS reads

$$\frac{d}{dt} \text{RHS} = \frac{d}{dt} \left(\frac{2 g \sin(\theta)}{1 + \frac{I_c}{M R^2}} x \right) = 2 \frac{g \sin(\theta)}{1 + \frac{I_c}{M R^2}} \frac{dx}{dt} .$$

But $\frac{dx}{dt}$ is the time rate of change of the position of the point of contact of the cylinder with the plane, and this is precisely equal to the speed of the centre of mass. Hence,

$$\frac{d}{dt}\,\text{RHS} = 2\,V_{\text{CofM}}\,\frac{g\,\sin(\theta)}{1 + \frac{I_c}{M\,R^2}}\,.$$

Comparison of the time rates of change of the LHS and RHS reveals that the acceleration of the cylinder down the incline is

$$A_{\text{CofM}} = \frac{g\,\sin(\theta)}{1 + \frac{I_c}{M\,R^2}}\,,$$

which is in complete agreement with the result obtained the quick and dirty way.

There is an essential comment which must be made.

**The acceleration attained by the cylinder as it rolls down the incline
is less than it would experience sliding down a frictionless incline.**

Why this is so is evident from an energetics perspective. As the cylinder moves down the incline, an amount of potential energy is converted into kinetic energy.

Roll   The available kinetic energy is split between bulk and internal.

Slide   When the object slides without friction, all of the kinetic energy gained appears as bulk, while none is of the internal variety.

Because some kinetic energy is *soaked up* into the internal motion of the body, less is available for the motion of the CofM, and therefore the CofM speed of the rolling body is necessarily smaller than that of a sliding body which has absorbed the same amount of work.

EXAMPLE [*Rolling Race*]

A ring, a cylinder, and a sphere, all of the same total mass, M, and radius, R, roll down a plane inclined at angle θ. Determine the acceleration of each of the objects.

RING   For a uniform thin ring, $I_c = M\,R^2$, and $A_{\text{CofM}} = \frac{1}{2}\,g\,\sin(\theta)$.
  [This is one-half of the acceleration that the ring would experience on a frictionless incline.]

CYLINDER   For a uniform cylinder, $I_c = \frac{1}{2}\,M\,R^2$, and $A_{\text{CofM}} = \frac{2}{3}\,g\,\sin(\theta)$.

SPHERE   For a uniform sphere, $I_c = \frac{2}{5}\,M\,R^2$, and $A_{\text{CofM}} = \frac{5}{7}\,g\,\sin(\theta)$.

Chapter 44

Static Equilibrium

Let's first parse the title of this chapter.

$$\text{STATIC} \quad \Longleftrightarrow \quad \text{not changing in time}$$

$$\text{EQUILIBRIUM} \quad \Longleftrightarrow \quad \text{a state of "balance"}$$

STATIC EQUILIBRIUM (for a rigid extended body)   Two conditions must be met for an [extended] object to be in a state of static equilibrium. These are:

$\mathcal{F}$ORCE   The net external force must vanish, *i.e.*, $\vec{F}_{\text{NET ext'l}} = \vec{0}$.
    IF otherwise, **THEN**, by N2, there must exist a net rectilinear acceleration, which is incompatible with the system's being static.

$\mathcal{T}$ORQUE   The net external torque must vanish, *i.e.*, $\vec{\tau}_{\text{Net ext'l}} = \vec{0}$.
    IF otherwise, **THEN**, by N2, there must exist a net angular acceleration, which is incompatible with the system's being static.

EXAMPLE [*A Couple of Forces*]

There is no possibility whatsoever that a rigid body subject to two forces can be in static equilibrium unless **the forces are equal in magnitude and opposite in direction**. This is the only manner in which the $\mathcal{F}$ condition can be satisfied.

COUPLE   A couple consists of a pair of equal and opposite forces acting on the same rigid body.

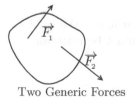
Two Generic Forces

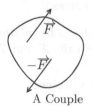
A Couple

FIGURE 44.1 Objects (at rest) Subjected to Pairs of Forces

Consider the situation, illustrated on the right in Figure 44.1, in which two equal and opposite forces act on a rigid body. In this instance, the net external torque is not zero, and we expect the object to rotate under the influence of the couple.

Q: What constraint on a couple is sufficient to ensure that the $\mathcal{T}$ condition is met?

A: The two forces must share a common line of action. The best way to see[1] this is with the aid of Figure 44.2.

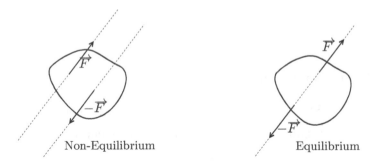

FIGURE 44.2   Force Couples Acting on Objects at Rest

EQUILIBRIUM   Consider the second object in Figure 44.2, *viz.*, the one claimed to be in static equilibrium. Let's choose the torque point to lie anywhere along the common line of action of the two forces. The moment arm corresponding to each of the forces is zero. Thus, each torque separately vanishes, and so must the net torque. Conditions $\mathcal{F}$ AND $\mathcal{T}$ are both met and we can be assured that the object is indeed in a state of static equilibrium.

NON-EQUILIBRIUM   Consider the first object in Figure 44.2, and choose the torque point to lie [somewhere] along the line of action of one of the forces. The moment arm corresponding to this force vanishes, while the moment arm of the other is equal to the minimum distance between the parallel lines of action of the two forces. Thus, one torque is necessarily zero, while the other is non-zero, and the net torque cannot possibly vanish. Although the force condition, $\mathcal{F}$, is satisfied, the torque condition, $\mathcal{T}$, is not, and we must accept that this object cannot possibly be in a state of static equilibrium.

EXAMPLE [*A Trio of Forces*]

In order for a rigid body to be in static equilibrium under the influence of three applied forces, $\{\vec{F_1}, \vec{F_2}, \vec{F_3}\}$, both the $\mathcal{F}$ and $\mathcal{T}$ conditions must be maintained.

$\mathcal{F}$ The net force vanishes,

$$\vec{F_1} + \vec{F_2} + \vec{F_3} = \vec{0}.$$

A direct consequence of this condition[2] is that the trio of forces must be CO-PLANAR [*i.e.*, spanning a 2-d plane rather than all of 3-d space].

----

[1] Saving approximately 10^4 words.
[2] The mathematical statement is that the set of forces is LINEARLY DEPENDENT.

$\mathcal{T}$ The net torque vanishes,

$$\vec{\tau_1} + \vec{\tau_2} + \vec{\tau_3} = \vec{0}.$$

It turns out that the necessary and sufficient condition for the net external torque to vanish is that the lines of action associated with each of the three forces intersect at a unique[3] point lying in the common plane.

Once again, the best evidence for this claim is pictorial and is found in Figure 44.3.

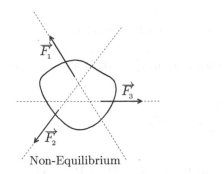

Non-Equilibrium

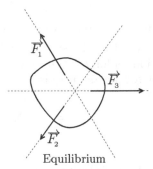

Equilibrium

FIGURE 44.3 Force Trios Acting on Objects

EQUILIBRIUM Consider the second object in Figure 44.3, and choose the torque point to lie at the intersection of all three lines of action of the forces. The moment arm corresponding to each of the forces is necessarily zero, and thus each torque separately vanishes. Hence, so too does the net torque. Conditions $\mathcal{F}$ AND $\mathcal{T}$ are both met, and thus the object is in a state of static equilibrium.

NON-EQUILIBRIUM Consider the first object in Figure 44.3, in which the lines of action of the three co-planar forces combine to form a triangle. Let's choose the fixed point to lie at one of the vertices of the triangle, *i.e.*, at the intersection of the lines of action of two of the forces. The moment arms corresponding to these two forces are necessarily zero, while the moment arm of the other is equal to the altitude [*a.k.a.* height] of the triangle. Therefore, two of the torques are zero while the third is non-zero. The net torque does not vanish, and hence this object cannot possibly be in a state of static equilibrium.

One might be less than completely satisfied with the arguments advanced in the preceding couple and trio examples. A source of concern is that particular torque points were chosen to determine whether or not the torque condition was met. A most natural query is:

Q: How does the choice of the torque point affect the result obtained for the torque?

A: Our answer to this question takes the form of a general theorem.

TORQUE UNIQUENESS THEOREM IF the net force acting on a body vanishes, THEN the value of the net torque is independent of the choice of the fixed point about which it is computed.

---

[3] There exist degenerate cases in which the three lines of action coincide and uniqueness is lost. Another curious case will be considered in Chapter 45.

PROOF of the TORQUE UNIQUENESS THEOREM

Suppose that a rigid body is acted upon by $\mathcal{N}$ external forces, $\{\vec{F_1}, \vec{F_2}, \ldots, \vec{F_{\mathcal{N}}}\}$, such that the net force vanishes, *viz.*, $\vec{F}_{\text{NET}} = \sum_{j=1}^{\mathcal{N}} \vec{F_j} = \vec{0}$. The net torque produced by these forces about some common point, P, is

$$\vec{\tau}_{\text{net}\,P} = \sum_{i=1}^{\mathcal{N}} \vec{\tau}_{P,i} = \sum_{i=1}^{\mathcal{N}} \vec{r}_{P,i} \times \vec{F_i},$$

where each vector $\vec{r}_{P,i}$ extends from P to the point at which $\vec{F_i}$ is applied.

The torque produced by the same forces about a different point, Q, is determined by the same construction as for P. With each $\vec{r}_{Q,i}$ extending from Q to the point at which $\vec{F_i}$ is applied, the net torque about Q is

$$\vec{\tau}_{\text{net}\,Q} = \sum_{i=1}^{\mathcal{N}} \vec{\tau}_{Q,i} = \sum_{i=1}^{\mathcal{N}} \vec{r}_{Q,i} \times \vec{F_i}.$$

The relative position of Q with respect to P is uniquely specified, $\vec{R}_Q = \vec{R}_P + \overrightarrow{PQ}$, and, with a nod to Figure 44.4, one may re-express each of the vectors $\vec{r}_{P,i} = \overrightarrow{PQ} + \vec{r}_{Q,i}$.

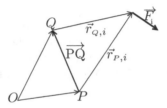

FIGURE 44.4   The Origin O, Two Torque Points $\{P, Q\}$, and a Force

Let's now re-compute the net torque about P.

$$\vec{\tau}_{\text{net}\,P} = \sum_{i=1}^{\mathcal{N}} \vec{r}_{P,i} \times \vec{F_i} = \sum_{i=1}^{\mathcal{N}} \left[\left(\overrightarrow{PQ} + \vec{r}_{Q,i} \right) \times \vec{F_i} \right] = \left[\sum_{i=1}^{\mathcal{N}} \overrightarrow{PQ} \times \vec{F_i} \right] + \left[\sum_{i=1}^{\mathcal{N}} \vec{r}_{Q,i} \times \vec{F_i} \right].$$

The above analysis has merely distributed the cross product through each term in the sum and regrouped. In the first term on the right, $\overrightarrow{PQ}$ is constant, and the linearity of the cross product enables writing

$$\text{First term} = \overrightarrow{PQ} \times \left(\sum_{i=1}^{\mathcal{N}} \vec{F_i} \right) = \overrightarrow{PQ} \times \vec{F}_{\text{NET}}.$$

The lone restriction that we have placed upon the collection of forces will cause this term to vanish when it is enforced. The second term is the torque computed about the point Q.

The practical consequence of this theorem is that any point may be employed to determine whether the conditions of static equilibrirum are satisfied.

Chapter 45

Statics: Levers and Ladders

Let's analyse a playground scenario involving a seesaw [*a.k.a.* teeter–totter, and more aptly, **massless lever**].

EXAMPLE [*Massless Lever*]

Two kids, PK and NK, with masses M_P and M_N, respectively, are seated on a seesaw. It shall be assumed that the bar and seats are massless, and that the bar rests on a pointlike fulcrum, labelled O in Figure 45.1.

FIGURE 45.1 PK and NK Seeing and Sawing whilst Teetering and Tottering

Q: What circumstances are necessary for the system, comprised of PK, NK, and the bar, to be in static equilibrium with the bar lying horizontally?

A: Let's work through the dynamics recipe, paying particular attention to the $\mathcal{F}$ and $\mathcal{T}$ conditions for the system.

(1) The physical situation is sketched in Figure 45.1.

(2) In addition to enumerating the forces acting on the three constituents, we shall distinguish between internal and external forces.

PK

Weight ○ $\vec{W_P} = M_P\, g\ [\downarrow]$.   EXT'L

Normal ○   The force of contact of the bar on PK, $\vec{N_P} = N_P\ [\uparrow]$.   INT'L

NK

Weight ○ $\vec{W_N} = M_N\, g\ [\downarrow]$.   EXT'L

Normal ○   The force of contact of the bar on NK, $\vec{N_N} = N_N\ [\uparrow]$.   INT'L

Bar

Weight ○ $\vec{W_{Bar}} = M_B\, g\ [\downarrow] = \vec{0}$, by the [presumed] masslessness of the bar.   EXT'L

PK ○   The force of contact of PK on the bar, $\vec{C_P} = N_P\ [\downarrow]$.   INT'L

NK ○   The force of contact of NK on the bar, $\vec{C_N} = N_N\ [\downarrow]$.   INT'L

Support ○   The force of contact of the fulcrum on the bar, $\vec{N_O} = N_O\ [\uparrow]$.   EXT'L

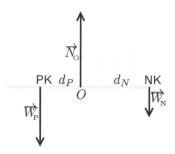

FIGURE 45.2   FBD for the PK, NK, and Bar System

The FBD displayed in Figure 45.2 exhibits only external forces, $\{\vec{W_P}, \vec{W_N}, \vec{N_O}\}$, and these are drawn originating at the points at which they act.

(3-4-5) For the system to be in static equilibrium, both the $\mathcal{F}$ and $\mathcal{T}$ conditions must be satisfied. That the net force must vanish finds its particular expression as

$$N_O - M_P\, g - M_N\, g = 0\,.$$

Consideration of $\mathcal{T}$ requires that a torque point be chosen.[1]

> ASIDE: The Torque Uniqueness Theorem [*cf.* Chapter 44], assures that the net torque is independent of the choice of torque point.

That the bar and riders rotate about the fulcrum when the system is in motion, and that the *a priori* unknown support force acts at the fulcrum, militate for choosing the fulcrum as the torque point. Condition $\mathcal{T}$ then reads

$$\vec{0} = \vec{\tau}_{\text{net}} = \vec{\tau_P} + \vec{\tau_O} + \vec{\tau_N}\,.$$

Each of the three external forces must be analysed to determine their respective torques about the fulcrum:

$$\vec{\tau_P} = \vec{r}_P \times \vec{W_P} = d_P\, M_P\, g\,[\odot] = d_P\, M_P\, g\,[\,\text{anti-clockwise}\,]\,,$$
$$\vec{\tau_O} = \vec{r}_o \times \vec{N_O} = \vec{0}\,,$$
$$\vec{\tau_N} = \vec{r}_N \times \vec{W_N} = d_N\, M_P\, g\,[\otimes] = d_N\, M_P\, g\,[\,\text{clockwise}\,]\,.$$

Therefore, the torque condition simplifies:

$$0 = \tau_{\text{net}} = d_P\, M_P\, g - d_N\, M_P\, g \qquad \Longrightarrow \qquad M_P\, d_P = M_P\, d_N\,.$$

This final relation is familiar to all who have studied levers or ridden on seesaws!

(6) That this result is reasonable is self-evident from the fact that the heavier of the two kids must sit nearer to the fulcrum for the system to be in equilibrium.

ASIDE: **Q:** Does this result change if the seesaw is not horizontal?
A: It does not matter unless the seesaw is tipped up vertically.

---

[1] In the abstract consideration of trios of forces in the previous chapter, it was argued that the net torque could not vanish unless the lines of action of all three forces intersected at a single point. Here the lines of action (of the weights and the fulcrum force) are all parallel and thus non-intersecting! This potential conundrum is avoided by stating that the parallel lines DO intersect: at THE POINT AT INFINITY.

EXAMPLE [*Hoisting a Half-Pint*]

With a mountain of papers and problem sets due and exams to prepare for, Albert and Bertal Steinein set off for a nearby pub. While there, they developed a model for the biomechanical system comprised of an arm holding a half-pint of Guinness. After some debate[2] they settled upon the following details and assumptions.

⊢   The system is in static equilibrium.

⊢   The forearm and hand together act like a rigid bar with mass $m_A = 4\,\text{kg}$ and CofM located a distance $d_A = 20\,\text{cm}$ from the elbow joint. The humerus and the elbow are both rigidly fixed. The forearm is extended horizontally, the upper arm is vertical.

⊢   The Guinness and its container have mass $m_G = 0.5\,\text{kg}$, and are held at an effective distance $d_G = 40\,\text{cm}$ from the elbow.

⊢   The bicipital tendon connects the biceps muscle to the forearm. It is attached a distance $d_B = 5\,\text{cm}$ from the elbow. When the forearm is horizontal, the bicipital tendon is vertical.

⊢   The bicipital tendon conveys a force, $\vec{F_B}$, upon the forearm. The humerus exerts a supporting force, $\vec{P}$, on the forearm at the elbow joint.

Many of these model details are illustrated in Figure 45.3.

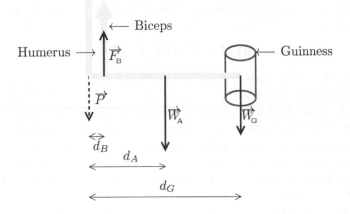

FIGURE 45.3   FBD for the Model of the Elbow

Q: What are the magnitudes of the bicipital and support forces?
A: The four forces, $\{\vec{P}, \vec{F_B}, \vec{W_A}, \vec{W_G}\}$, must cancel to satisfy $\mathcal{F}$. The two weights are known, while the remaining two forces are as yet undetermined. The $\mathcal{T}$ condition,

$$\vec{0} = \vec{\tau}_{\text{net}} = \vec{\tau}_P + \vec{\tau}_B + \vec{\tau}_A + \vec{\tau}_G\,,$$

must also be met. Choosing the elbow as the torque point for the usual reasons [that the torque produced by one of the unknown forces automatically vanishes, and that this is the point about which the system actually rotates] yields

---

[2] Arm-wrestling, as it were.

$$\vec{\tau}_P = \vec{0}$$
$$\vec{\tau}_A = d_A \, m_A \, g \, [\text{clockwise}] = -d_A \, m_A \, g$$

$$\vec{\tau}_B = d_B \, F_B \, [\text{anti-clockwise}] = +d_B \, F_B$$
$$\vec{\tau}_G = d_G \, m_G \, g \, [\text{clockwise}] = -d_G \, m_G \, g$$

Collecting these torques and inserting them into the $\mathcal{T}$ condition gives

$$0 = d_B \, F_B - d_A \, m_A \, g - d_G \, m_G \, g, \qquad \Longrightarrow \qquad F_B = \frac{d_A \, m_A + d_G \, m_G}{d_B} \, g \, .$$

With the particular values assumed for the masses and moment arms, the necessary biceps force is determined to be

$$F_B = \frac{20 \, [\, \text{cm} \,] \, 4 \, [\, \text{kg} \,] + 40 \, [\, \text{cm} \,] \, 0.5 \, [\, \text{kg} \,]}{5 \, [\, \text{cm} \,]} \times 10 \, [\, \text{m/s}^2 \,] = 200 \, \text{N} \, .$$

Thus, the biceps must exert a force of $200\,\text{N}$ on the forearm merely to hold it and the Guinness in place. This force is considerably more than the combined weights of the arm and Guinness. The support force is ascertained through satisfaction of $\mathcal{F}$:

$$\vec{P} = -\vec{F_B} - \vec{W_A} - \vec{W_G} = -200 \, [\uparrow] - 40 \, [\downarrow] - 5 \, [\downarrow] \qquad \Longrightarrow \qquad \vec{P} = 155 \, \text{N} \, [\downarrow] \, .$$

Whew! Who would have thought that imbibing was such a strenuous activity?!

EXAMPLE [*Hey, What's Your Sign?*]

Bertal Steinein elects to go into business for himself. One of his first tasks is to fabricate and hang his shingle.

~PHYSICIST FOR HIRE~

Including the squiggles at the ends and the spaces between words, the sign must allow for 20 characters.[3] Making each character [nominally] 10 centimetres wide,[4] the sign must be two metres long, $L = 2\,\text{m}$. Bertal paints the letters on a thin uniform piece of steel which acts like a rigid bar and has a mass of $M = 6\,\text{kg}$. [The paint adds negligibly to the weight.]

Bertal intends to hang his sign, protruding horizontally from a vertical wall, as illustrated in Figure 45.4, with the aid of an anchor at the wall end and a guide wire at the far end. The guide wire is also to be affixed to the wall, $2\,\text{m}$ above the bar, making the two corner angles both equal to $\pi/4\,\text{rad}$.

Q1: What is the tension in the guide wire? [Should B. use nylon fishing line or steel cable?]

Q2: What force is exerted at the wall? [Should B. employ glue or cement-in anchor bolts?]

A: Let's examine the $\mathcal{T}$ and $\mathcal{F}$ conditions to determine the answers.

(1) Figure 45.4 contains a sketch.

(2) Three forces act on the bar: its weight, the tension in the guide wire, and the wall support force. The weight is deemed to act at the CofM of the bar, which by assumption, is taken to be its geometric centre, while the tension and the support force act at their respective ends.

---

[3]The proprietor is quite a character too, but this is not accounted for here.

[4]Uniformly-sized type looks like that produced by an old-fashioned typewriter, giving the sign a *retro* look. In typeset text, the space allotted to each letter is varied.

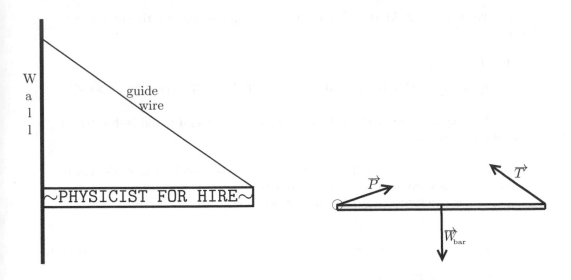

FIGURE 45.4   Bertal Steinein's Sign and Its FBD

Bar

Weight ∘ $\vec{W}_{\text{bar}} = 60\,\text{N}\ [\downarrow]$.

Tension ∘ $\vec{T} = T\ [\nwarrow]$.

Support ∘ $\vec{P} = P\ [\hat{P}]$.

The FBD in Figure 45.4 shows the points at which these forces are applied.

(3-4-5)   Static equilibrium entails that the two conditions $\mathcal{F}$ and $\mathcal{T}$ must be met.

Consider first $\mathcal{T}$:

$$\vec{0} = \vec{\tau}_{\text{net}} = \vec{\tau}_W + \vec{\tau}_T + \vec{\tau}_P \,.$$

With the torque point chosen to be at the wall end of the bar, $\vec{\tau}_P = \vec{0}$, irrespective of the magnitude and direction of the support force, leaving

$$\vec{0} = \vec{\tau}_{\text{net}} = \vec{\tau}_W + \vec{\tau}_T \,.$$

The torques produced by the [known] weight and the [unknown] tension are

$$\vec{\tau}_W = \vec{r}_W \times \vec{W} = (1)\,(60)\ [\otimes] = 60\ [\,\text{clockwise}\,]\,,$$
$$\vec{\tau}_T = \vec{r}_T \times \vec{T} = (2)\,T\,\sin(3\,\pi/4)\ [\odot] = \sqrt{2}\,T\ [\,\text{anti-clockwise}\,]\,,$$

where the implicit units are $[\,\text{m}\,]$ for distances, $[\,\text{N}\,]$ for forces, and $[\,\text{N}\cdot\text{m}\,]$ for torques. Substituting, the torque condition for the sign reads

$$\vec{0} = 60\ [\otimes] + \sqrt{2}\,T\ [\odot] \qquad \Longrightarrow \qquad T = 30\,\sqrt{2}\,.$$

A1: The tension in the guide wire needed to hold the sign in place is $T \simeq 42.4\,\text{N}$.

With the tension now known, $\mathcal{F}$ determines the support force acting at the wall, via

$$\vec{0} = \vec{F}_{\text{NET}} = \vec{F}_{\text{W}} + \vec{F}_{\text{T}} + \vec{F}_{\text{P}}$$

$$\implies \vec{F}_{\text{P}} = -\vec{F}_{\text{W}} - \vec{F}_{\text{T}} = 60 \ [\uparrow] + 30\sqrt{2} \ [\searrow] = \left(30 \ [\rightarrow] + 30 \ [\uparrow]\right) = 30\sqrt{2} \ \text{N} \ [\nearrow].$$

A2: The support force at the wall has magnitude $P \simeq 42.4\,\text{N}$, and is directed at $45°$ above horizontal.

(6) Estimates for the tension and support forces necessary to hold Bertal's sign in place have thus been obtained. These estimates, along with product information from the manufacturers, will assist him in deciding which materials to use.

....

EXAMPLE [*The Ladder Problem*]

A ladder of length L and mass M is at rest, standing on a horizontal floor and leaning against a vertical wall. The coefficient of static friction acting between the foot of the ladder and the floor is $\mu_{\text{s}} > 0$. The frictional force between the head of the ladder and the wall is assumed to vanish.[5]

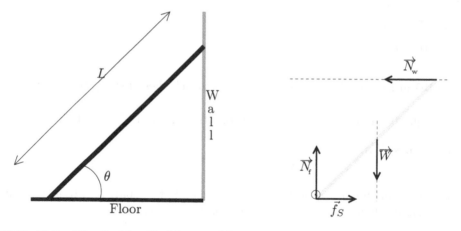

FIGURE 45.5 The Ladder Problem and Its FBD

Q: What is the minimum angle, $\theta_{\min}$, at which the ladder remains at rest [does not slip]?

A: As long as the ladder is not slipping, the system is in a state of static equilibrium. Furthermore, the force of static friction has a maximum upper bound, and it stands to reason that this bound is saturated when the angle is minimised. Let's apply the static equilibrium conditions, $\mathcal{F}$ and $\mathcal{T}$, to the ladder, and enforce the constraint that the static frictional force attains its maximal value.

Five forces act on the ladder.

---

[5] Perhaps the wall is very smooth and/or the contact force acting between the surfaces is relatively small. Either way, it is not good practice to rely on friction with the wall to support a ladder.

$\vec{W}$ Weight $\vec{W} = M g \,[\downarrow]$

The weight force acts downward at the location of the CofM of the ladder, which is assumed to coincide with its midpoint.

$\vec{N_f}$ Floor Normal $\vec{N_f} = N_f \,[\uparrow]$

The floor exerts a normal force acting upward on the ladder at its foot.

$\vec{f_S}$ Floor Friction $\vec{f_S} = f_S \,[\rightarrow]$

The floor exerts a static frictional force acting on the foot of the ladder, directed horizontally toward the wall. The maximum value of this force is

$$f_{S,\text{max}} = \mu_{\text{s}} \, N_f \,,$$

in keeping with the model developed to describe static friction.

$\vec{N_w}$ Wall Normal $\vec{N_w} = N_w \,[\leftarrow]$

The wall exerts a normal force on the head of the ladder acting to the left.

$\vec{f_w}$ Wall Friction $\vec{f_w} = f'_S \,[\uparrow] = \vec{0}$

This force is assumed to vanish.

All of the force information is incorporated into the generalised FBD in Figure 45.5. Even at the critical [minimum] angle, the ladder is at rest and remains at rest. Hence, it is in static equilibrium. That the net external force vanishes [*i.e.*, the condition $\mathcal{F}$] necessitates that the floor support the entire weight of the ladder, and that the normal force at the wall exactly cancel the force of static friction acting at the foot. Analysis of components yields:

$$\vec{0} = \vec{F}_{\text{NET}} = \vec{W} + \vec{N_f} + \vec{f_S} + \vec{N_w} \qquad \Longrightarrow \qquad \left\{ \begin{array}{l} 0 = N_f - M g \\ 0 = f_S - N_w \end{array} \right\}.$$

These results yield an explicit formula for the maximum magnitude of the normal force exerted by the wall,

$$N_w = f_S \leq f_{S,\text{max}} = \mu_{\text{s}} \, N_f = \mu_{\text{s}} \, M g\,.$$

At the minimum angle, the frictional force is saturated,[6] and hence $N_w = \mu_{\text{s}} \, M g$.

The optimal choice for the torque point is the foot of the ladder, because, with this choice, two of the four forces extant have vanishing moment arms. Thus,

$$\vec{0} = \vec{\tau}_{\text{net}} = \vec{\tau}_W + \vec{\tau}_{Nw} + \vec{\tau}_{Nf} + \vec{\tau}_f = \vec{\tau}_W + \vec{\tau}_{Nw}\,.$$

The torque about the foot generated by the weight of the ladder, $\vec{\tau}_W$, is computable by either of two different methods.

METHOD ONE: [*Cross-Product Definition*]

$$\begin{aligned} \vec{\tau}_W &= \vec{r}_W \times \vec{W} \\ &= \left(\frac{L}{2}\right) (M g) \sin\left(\pi - (\pi/2 - \theta_{\text{min}})\right) \,[\otimes] \\ &= \frac{1}{2} M g L \sin\left(\frac{\pi}{2} + \theta_{\text{min}}\right) \,[\otimes] \\ &= \frac{1}{2} M g L \cos(\theta_{\text{min}}) \,[\otimes]\,. \end{aligned}$$

---

[6]If this were this not the case, then the angle could be reduced without the ladder slipping.

METHOD TWO: [*Moment Arm Definition*]

$$\vec{\tau}_W = \left(\text{moment arm}\right) W\ [\otimes] = \left(\frac{L}{2}\cos(\theta_{\min})\right) M\,g\ [\otimes] = \frac{1}{2}\,M\,g\,L\,\cos(\theta_{\min})\ [\otimes]\,.$$

Unsurprisingly, the two methods yield identical results.

The torque about the foot of the ladder produced by the force of contact with the wall, $\vec{\tau}_{Nw}$, is obtained in a similar manner.

METHOD ONE: [*Cross-Product Definition*]

$$\begin{aligned}
\vec{\tau}_{Nw} = \vec{r}_{Nw} \times \vec{N}_{w} &= (L)\,(N_{w})\,\sin\left(\pi - \theta_{\min}\right)\ [\odot]\\
&= L\,(\mu_{s}\,M\,g)\,\sin\left(\theta_{\min}\right)\ [\odot]\\
&= \mu_{s}\,M\,g\,L\,\sin(\theta_{\min})\ [\odot]\,.
\end{aligned}$$

METHOD TWO: [*Moment Arm Definition*]

$$\vec{\tau}_{Nw} = \left(\text{moment arm}\right) N_{w}\ [\odot] = \left(L\,\sin(\theta_{\min})\right)(\mu_{s}\,M\,g)\ [\odot] = \mu_{s}\,M\,g\,L\,\sin(\theta_{\min})\ [\odot]\,.$$

The two methods yield identical results in this case, too.

There are two non-zero oppositely-directed contributions to the net torque. That they cancel requires that their magnitudes be equal. That is,

$$\frac{1}{2}\,M\,g\,L\,\cos(\theta_{\min}) = \mu_{s}\,M\,g\,L\,\sin(\theta_{\min}) \qquad \Longrightarrow \qquad \tan(\theta_{\min}) = \frac{1}{2\,\mu_{s}}\,.$$

Three pertinent observations follow below.

  ⚠   The minimum angle depends on **NEITHER** the mass **NOR** the length of the ladder.

  ⚠   The rougher the floor, the smaller the minimum angle at which the ladder rests.

  ⚠   On a perfectly smooth floor, $\mu_{s} = 0$, the ladder must stand vertically.

Chapter 46

Step It Up

EXAMPLE [*The Cylinder and the Step Problem*]

In joyous celebration of their smashing successes on a physics midterm exam, Albert and Bertal Steinein threw a party. To foster the festivities they purchased a keg of (root)beer. [A full keg has total mass M, and is approximately cylindrical with radius R and length L. It is reasonable to assume that its centre of mass coincides with its geometric centre.] Unhappily, the promised curbside delivery was to the road side rather than the sidewalk. Thus, they were faced with the daunting prospect of getting the keg up [a height h] onto the sidewalk.[1]

Their impeccable physical intuition led them to realise that they would have more success with a ROLL up the step-like curb, rather than a dead lift, so they carefully tipped the keg onto its side. Due to the presence of obstacles, the lads were only able to push and pull horizontally on the topmost surface of the recumbent keg, as illustrated in Figure 46.2.

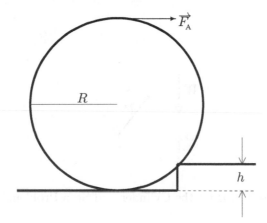

FIGURE 46.1   The Cylinder and Step Problem

Q: What minimum applied force [exerted horizontally at its top] is needed to roll the keg up the curb?

A: Until the keg begins to roll, the system is in a state of static equilibrium. Furthermore, the normal force of contact between the ground and the keg will vanish just as the keg begins to move. These are the constraints which enable one to ascertain the minimum necessary horizontally applied force.

First, we enumerate the forces acting on the cylinder.

---

[1] An engineering solution to their predicament is to tap the keg where it stands.

$\vec{W}$ Weight $\circ$ $\vec{W} = M g$ [↓].

The weight of the cylinder acts at its midpoint.

$\overset{\rightarrow}{F_A}$ Applied Force $\circ$ $\overset{\rightarrow}{F_A} = F_A$ [→].

In this example, the applied force is restricted to act horizontally at the topmost part of the cylinder.

> ASIDE: The lads realise that the keg can be rolled with less effort by pushing and pulling tangentially to the point on the rim directly opposite to the point of contact with the curb.

$\overset{\rightarrow}{N_O}$ Normal Force at the point of contact with the curb $\circ$ $\overset{\rightarrow}{N_O} = N_O$ [↖].

The edge of the curb exerts a normal force on the cylinder which is directed somewhat upward and to the left. [The direction is not precisely determined *a priori*, because the curb is modelled as though it were a step. Hence, the line of contact is the intersection of two planes.]

$\overset{\rightarrow}{N_B}$ Normal Force Beneath $\circ$ $\overset{\rightarrow}{N_B} = N_B$ [↑] = $\vec{0}$.

This force vanishes just as the cylinder is about to move.

Second, we combine all of the force information into the generalised FBD found in Figure 46.2. Third, we apply the $\mathcal{F}$ and $\mathcal{T}$ conditions.

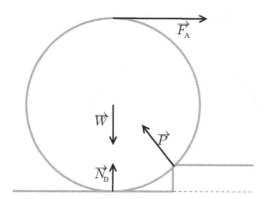

FIGURE 46.2   Generalised FBD for the Cylinder and Step Problem

At the critical [minimum necessary] force, the static equilibrium conditions still apply. The vanishing of both $\overset{\rightarrow}{N_B}$ and the net external force determines $\overset{\rightarrow}{N_O}$:

$$\vec{0} = \vec{F}_{\text{NET}} = \vec{W} + \overset{\rightarrow}{F_A} + \overset{\rightarrow}{N_O} + \overset{\rightarrow}{N_B}$$

$$\implies \quad \left\{ \begin{array}{l} 0 = F_A - N_{O_x} \\ 0 = N_{O_y} - M g \end{array} \right\} \implies \overset{\rightarrow}{N_O} = F_A \; [\leftarrow] + M g \; [\uparrow].$$

Were the cylinder to move, it would begin to rotate about the point of contact with the step. Choosing this as the torque point, the torque condition, $\mathcal{T}$, reads:

$$\vec{0} = \vec{\tau}_{\text{net}} = \vec{\tau}_W + \vec{\tau}_A + \vec{\tau}_{N_O} + \vec{\tau}_{N_B} \,,$$

where the last two terms are automatically zero.

The torque about the pivot point produced by the weight of the keg is most easily computed by means of the moment arm analysis:

$$\vec{\tau}_W = (\text{moment arm}) \, W \, [\odot] = (d) \, M g \, [\odot] .$$

The magnitude of the moment arm, d, is ascertained by examining Figure 46.3.

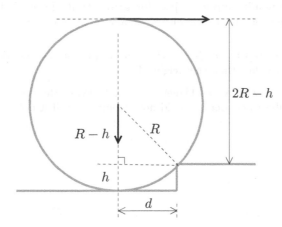

FIGURE 46.3 The Moment Arm Associated with the Weight of the Cylinder

The line of action of the weight is directed straight down from the CofM located at the centre of the cylinder. The three sides of a right triangle are:

VERTICAL   a segment of this vertical line

HYPOTENUSE   the line segment from the centre of the cylinder to the point of contact with the step

HORIZONTAL   a horizontal segment extending from the contact point to the line of action of the weight force

By construction, the length of the vertical side is equal to the radius less the height of the step, $R - h$. The radial line [the hypotenuse] has length R. The length of the horizontal side is equal to the sought-after d. Hence,

$$R^2 = (R - h)^2 + d^2 \qquad \Longrightarrow \qquad d^2 = R^2 - (R - h)^2 = (2R - h)\,h .$$

Therefore the torque produced by the weight about the point of contact with the step is

$$\vec{\tau}_W = (d) \, M g \, [\odot] = \sqrt{(2R - h)\,h} \; M g \, [\odot] .$$

The torque produced by the applied force [about the same point] is computed in a similar manner.

$$\vec{\tau}_A = (\text{moment arm}) \, F_A \, [\otimes] = (2R - h) \, F_A \, [\otimes] .$$

The line of action of the applied force passes horizontally at distance $2R$ above the roadway. Thus, the moment arm is $2R$, less the height of the curb, h.

Substituting these expressions into the $\mathcal{T}$ condition enforces equality of the magnitudes of the two remaining [oppositely directed, non-zero] torques. This yields an expression for the minimum force in terms of the properties of the keg and the height of the curb.

$$\vec{0} = \vec{\tau}_{\text{net}} \quad \Longrightarrow \quad \sqrt{(2R - h)h}\, Mg = (2R - h)\, F_A$$

$$\Longrightarrow \quad F_A = \sqrt{\frac{h}{2R - h}}\, Mg\,.$$

While this is the general result, suppose, just for grins, that the curb height is one-fifth the radius [as would be the case for a keg with diameter 60 cm, and a 6 cm step]. With these dimensions, $F_A = \frac{1}{3} M g$.

> ASIDE: Provided that the step height is less than the radius of the cylinder, the minimum horizontal force needed is less than the weight of the keg!

Faced with a step higher than the CofM of the keg,
Albert and Bertal would not attempt to roll it up.

Chapter 47

Universal Gravitation

It is rather *awkward* [at this late stage] to recollect that the very first line of Chapter 1 reads:

PHYSICS concerns itself with the <u>fundamental processes</u> which .....

(<u>emphasis</u> added). Thus far, we have considered phenomenological forces, *viz.*, normal, tension, contact, spring, friction, and drag, and "applied." Weight [*a.k.a.* constant field gravity] entered too, but it is in this chapter that we at last encounter a fundamental force.

Amongst Newton's many discoveries in his *Annus Mirabilis*[1] of 1666 was the NEWTONIAN LAW OF UNIVERSAL GRAVITATION [NUG].

LAW of UNIVERSAL GRAVITATION (for Two Point Particles)

The gravitational force exerted **by** a point[2] particle of mass M_1 at position $\vec{r}_1$, **on** another point particle of mass M_2 located at $\vec{r}_2$, is

$$\vec{F}_{\text{G},12} = -\frac{G\,M_1\,M_2}{\left|\vec{r}_{12}\right|^2}\,\hat{r}_{12}.$$

The relative position vector, $\vec{r}_{12}$, extends **from** particle 1 **to** 2 and is equal to the difference in the respective position vectors, $\vec{r}_{12} = \vec{r}_2 - \vec{r}_1 = r_{12}\,\hat{r}_{12}$, as illustrated in Figure 47.1.

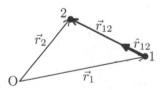

FIGURE 47.1   The Law of Universal Gravitation for Two Point Particles

---

[1] *Annus Mirabilis* translates as "wonderful year."
[2] In the next chapter, this law will be extended so as to apply to **point-like** extended bodies.

A slightly simplified expression for the Newtonian gravitational force, $\vec{F}_{G,12}$, is

$$\vec{F}_{G,12} = -\frac{G\,M_1\,M_2}{r_{12}^2}\,\hat{r}_{12}\,.$$

The various factors appearing on the RHS are discussed below.

G Newton's Gravitational Constant

The gravitational constant is, to the best of our current knowledge, a fundamental constant of nature. An approximate value of G, sufficient for our purposes, is

$$G \simeq 6.67 \times 10^{-11}\ \frac{\text{N}\cdot\text{m}^2}{\text{kg}^2} = 6.67 \times 10^{-11}\ \frac{\text{m}^3}{\text{kg}\cdot\text{s}^2}\,.$$

There is much [very much, actually] that could be said about G. However, we'll confine ourselves to only a few remarks.

- The value of G is quite small when expressed in SI. Gravity is the weakest of the known fundamental forces. The effects of gravitation are strongly felt when the sources are tremendously massive [*cf.* the skydiving adventure recounted in Chapter 10].

- The first successful attempt to measure G is attributed to Cavendish[3] (*circa* 1798) more than a full century after NUG was proposed. Prior to this measurement, the masses of the Sun and the Earth were not known, except through the combinations $G\,M_\odot$ and $G\,M_\oplus$ inferred by studying the orbits of the planets and the Moon.

- Areas of ongoing research include study of models in which G varies with time, or with the length scale under consideration. Experimental bounds on possible variations of G continue to be tightened.

- There are only a few fundamental physical constants. Two others are:

The speed of light *in vacuo*	$c \simeq 2.998 \times 10^8\ \text{m/s}\,.$
$\dfrac{\text{Planck's constant}}{2\,\pi}$	$\hbar \simeq 1.055 \times 10^{-34}\ \text{J}\cdot\text{s}\,.$

From these three quantities, it is possible to form unique combinations with dimensions of length and time only,[4] *viz.*,

$$\text{Planck length}\qquad l_P = \sqrt{\frac{\hbar\,G}{c^3}} = 1.616 \times 10^{-35}\ \text{m}\,.$$

$$\text{Planck time}\qquad t_P = \sqrt{\frac{\hbar\,G}{c^5}} = \frac{l_P}{c} = 5.391 \times 10^{-44}\ \text{s}\,.$$

There is much speculation that the Planck length and time represent the scales upon which spacetime[5] might cease to appear smooth and continuous and instead exhibit granular or foamy properties.

> ASIDE: The mere fact space and time seem continuous does not enervate these models any more than swimming in a lake nixes the notion of water molecules.

---

[3]Henry Cavendish (1731–1810) was a British scientist whose experiments led to advances in chemistry as well as physics.

[4]There is a combination with dimensions of mass, too.

[5]Einstein argued for uniting space and time in a more general structure dubbed "spacetime."

M_1, M_2 The Inertial Masses of the Particles

Q: How can it be that the inertia of the particles acts as gravitational charge?

A: We haven't the foggiest idea. Thus, we are reduced to stating this fact as a principle, the EQUIVALENCE PRINCIPLE, discussed below.

r_{12}^{-2} Reciprocal of the Squared-Distance between the Particles

For this reason, NUG is called an **inverse-square law**. Several beautiful consequences of this particular form of spatial dependence shall be encountered in Chapters 48 through 50.

$-\hat{r}_{12}$ Unit Vector

The gravitational force acts along the line which joins the two particles, and it is directed toward the one which exerts the force. Colloquially: *The force of gravitation is attractive.*[6] The appellation "**central**" is given to interaction forces directed along the line joining the particles.

EQUIVALENCE PRINCIPLE   The Equivalence Principle [EP] states that the gravitational charge of a particle is equal to its mass.

A colloquial expression of the EP is:

Objects fall at a rate which is independent of their composition.

There is much [very much, actually] that could be said about the EP. However, we'll confine ourselves to only a few remarks.

- The colloquial version is attributed to Galileo, who is reputed to have dropped a solid iron ball and a hollow iron ball having the same size and shape from the Leaning Tower of Pisa simultaneously, and observed that they struck the ground at the same instant.

  [The story is apocryphal.[7] He really rolled various balls along shallow ramps.]

- The Equivalence Principle is not yet explained in terms of a more fundamental aspect of nature. This state of affairs is not due to a lack of effort on the part of physicists throughout the past four hundred years (or so).

- The EP is the source of the universality of Newton's Law of Gravitation, and it is an essential facet of Einstein's theory of GENERAL RELATIVITY [GR].[8]

---

[6]We cannot help but concur with both senses of this statement!

[7]Why let the facts get in the way of such a good story, eh?

[8]Careful consideration reveals that there are a number of differing versions of the EP possible in GR, including Strong (SEP), Weak (WEP), and Einstein (EEP) forms.

Q: The force that 1 exerts on 2 is given by $\vec{F}_{G,12}$. What force, if any, does 2 exert on 1?

A: The designations "1" and "2" are labels, so it stands to reason that the gravitational force exerted **by** particle 2 **on** particle 1 is

$$\vec{F}_{G,21} = -\frac{G\,M_2\,M_1}{\left|\vec{r}_{21}\right|^2}\,\hat{r}_{21}\,,$$

where $\hat{r}_{21}$ is the unit vector pointing from 2 toward 1. Clearly then,

$$\vec{F}_{G,21} = -\vec{F}_{G,12}\,,$$

in exact accord with N3.

EXAMPLE [*"When a body meets a body ..."*]

With apologies to Robert Burns

When PK met NK they were seated across a [massless] table. Completely smitten and overcoming his natural reserve,[9] PK asked, "Do you know how strongly I am attracted to you right now?"

NK fluttered her eyelashes and softly breathed, "No. Please do tell me."

"Alright then," chirped PK, grabbing a nearby napkin and uncapping his pen with a flourish, "approximating each of us as a point mass, $M_P = 85\,\text{kg}$ and $M_N = 60\,\text{kg}$ respectively, and estimating that we are about 1 metre apart,

$$F_{\text{NP}} = \frac{G\,M_N\,M_P}{r_{\text{NP}}^2} \approx \frac{6.67 \times 10^{-11} \times 60 \times 85}{1^2} \simeq \frac{20}{3} \times 5100 \times 10^{-11} = 3.4 \times 10^{-7}\,\text{N}\,,$$

or, roughly speaking, about one-third of a micronewton."

There are at least four important points made in this example.

- The gravitational force of interaction between macroscopic [human–sized] objects is quite miniscule. Gravity is a very weak force.

- The version of NUG employed in the computation applies to point particles, while PK and NK are both extended objects. For now, we'll put off worrying about this by recognising that the computation is only intended to be approximate. [In Chapter 48, spherically symmetric extended gravitational sources are explicated in detail. In Chapter 1 of VOLUME II, the effects of gravity on extended objects are considered.]

  ASIDE: Some additional solace is derived from the observation that any bounded distribution of matter can be made to look like a point particle by viewing it from a sufficiently large distance away.

- PK was *moved* by NK, even though they were separated and not in contact. This constitutes the **action at a distance** conundrum discussed next in this chapter.

- Furthermore, NUG presupposes instantaneous propagation of influences. This has the potential to give rise to acausal physical behaviours. A more complete theory must take into account the speed with which the gravitational influence propagates from the exerter to the exertee, and the concomitant time lags.

- On occasion, the nerd does get the girl.

---

[9]The snarky among us might inquire, "Reserve of what?"

ACTION AT A DISTANCE   Roughly speaking, the expression "action at a distance" is applied to any interaction between bodies which are not in contact.

ASIDE: More precise definitions of action at a distance often insist that the bodies be bounded [finite in extent] and non-interpenetrating [disjoint] as well.

One can opt to live with action at a distance as just another crude approximation, not unlike the frictionless plane, the ideal rope, the ideal spring, *etc.* However, philosophical difficulties arise when one attempts to claim that such a force is fundamental in nature. Two representative problems are:

♯   Cognitive dissonance with the Aristotelian idea that, in order for an influence to be conveyed, there must be a material connection.

♯   **Q:** What if NK moves to another table[10] while PK is busy gesticulating at an unresponsive waiter? How will PK know that the force acting on him has changed?

To obviate these concerns, physics has adopted the principle that

> ## All physics is local!

Adherence to this new principle[11] necessitates introduction of gravitational fields.

GRAVITATIONAL FIELD   The gravitational field of a particular body [acting as the **source** of the field], is the local gravitational force per unit mass [experienced by a putative responder] at any point in space, produced by the source body.

The formal procedure for defining the field produced by body 1 commences with the introduction of a point(-like) test particle with mass, m_0, at the point in space, with relative position $\vec{r}_{10}$, where one wishes to know the value of the field. After ascertaining the gravitational force produced by body 1 on the test particle, the mass of the test particle is divided out. Furthermore, the limit in which the magnitude of the test mass approaches zero is taken, *viz.*,

$$\vec{g}_1(\vec{r}_{10}) = \lim_{m_0 \to 0} \frac{\vec{F}_{G,10}}{m_0}.$$

Notable properties of the gravitational field include the following.

- The units of the gravitational field are newtons per kilogram, $[\text{N}/\text{kg}] = \text{m}/\text{s}^2$.

- It is a vector-valued function of position [and other things perhaps].

- As a function over space, the gravitational field exists everywhere.[12]

- The reason for taking the limit as the test mass approaches zero is to eliminate possible effects arising from the gravitational influence of the test mass on the source of the field.

---

[10]It's perfectly understandable, eh?

[11]An unrelated social maxim is: *"All politics is local."*

[12]CAVEAT: For point masses, the field does not exist at the single point occupied by the particle, so as to avoid self-interaction.

RE-IMAGINING THE GRAVITATIONAL FORCE ACTING BETWEEN TWO POINT PARTICLES
IN TERMS OF FIELDS

Two point particles, $\star$ and $\diamond$, interact gravitationally. The $\star$ particle has mass $M_\star$ and position $\vec{r}_\star$, while the corresponding quantities for $\diamond$ are $M_\diamond$ and $\vec{r}_\diamond$. The relative position of $\star$ with respect to $\diamond$ is $\vec{r}_{\diamond\star}$, and conversely, that of $\diamond$ with respect to $\star$ is $\vec{r}_{\star\diamond} = -\vec{r}_{\diamond\star}$.

The gravitational force experienced by the $\star$ particle is equal to the product of its mass with the gravitational field of $\diamond$ evaluated at the location of $\star$, *viz.*,

$$M_\star \, \vec{g}_\diamond(\vec{r}_{\diamond\star}) \, .$$

There is no action at a distance; $\star$ feels the field [established by $\diamond$] at the point where $\star$ resides, without concern as to the source of the field.

The field produced by $\diamond$ at the location of $\star$ is determined via the limiting procedure described just above. Since $\diamond$ is a point particle, the gravitational force that it exerts on a test mass m_0 at relative position $\vec{r}$ is prescribed by NUG,

$$\vec{F}_{\text{G},\diamond 0} = -\frac{G\,M_\diamond\,m_0}{|\vec{r}|^2}\,\hat{r} \, .$$

Substituting this force into the defining expression for the gravitational field yields

$$\vec{g}_\diamond(\vec{r}) = -\frac{G\,M_\diamond}{|\vec{r}|^2}\,\hat{r} \, .$$

The arrows in the first panel in Figure 47.2 indicate the relative magnitude and direction of the [vectorial] gravitational field at the location occupied by the base of the arrow. The field permeates all of space. The force on $\star$ arising from the field produced by $\diamond$ is

$$\vec{F}_{\text{G},\diamond\star} = M_\star\,\vec{g}_\diamond(\vec{r}_{\diamond\star}) = -\frac{G\,M_\diamond\,M_\star}{|\vec{r}_{\diamond\star}|^2}\,\hat{r}_{\diamond\star} \, .$$

This is exactly the same result as was obtained by direct application of NUG.

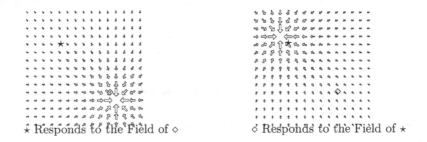

$\star$ Responds to the Field of $\diamond$ $\diamond$ Responds to the Field of $\star$

FIGURE 47.2   Two Point Particles Interacting via Gravitational Fields

By a precisely parallel line of reasoning, the gravitational force exerted by $\star$ on $\diamond$ can be interpreted as $\diamond$ responding to the local value of the gravitational field established by $\star$, as illustrated in the second panel of Figure 47.2. The argument goes as follows,

$$\vec{F}_{\text{G},\star\diamond} = M_\diamond\,\vec{g}_\star(\vec{r}_{\star\diamond})\,, \qquad \text{AND} \qquad \vec{g}_\star(\vec{r}) = -\frac{G\,M_\star}{|\vec{r}|^2}\,\hat{r} \qquad \Longrightarrow \qquad \vec{F}_{\text{G},\star\diamond} = -\frac{G\,M_\star\,M_\diamond}{|\vec{r}_{\star\diamond}|^2}\,\hat{r}_{\star\diamond} \, .$$

Cool, eh?

Chapter 48

Extended Sources and Energetics

Upon re-re-imagining the gravitational interaction between two point particles, subsequent to the last chapter, one may be tempted to dismiss the field as an unnecessarily circumloquacious means of deriving results which could be more straightforwardly gleaned by direct application of NUG. This temptation should be resisted for at least three reasons.

✓   All physics should be local!

✓   There are physical situations in which NUG is impossibly awkward to apply, whereas field methods remain robust.

✓   Much later we'll see, in the context of electricity and magnetism,[1] that fields are themselves dynamical objects capable of bearing energy and momentum.

Here, we shall first consider the gravitational field produced at a single point by a distributed source [a spherical shell], and then examine the energetics associated with NUG.

Q: How does one compute the gravitational field produced by an extended body?

A: One applies the PARTITION, COMPUTE and SUM procedure. That is, partition the extended body into $\mathcal{N}$ chunks, treat each as a point particle, compute the gravitational field produced by each of the chunks at the common field point, and sum over the chunks in the partition.

With only slight loss of generality, we shall consider a thin spherical shell of matter, with mass M, radius R, and uniform areal mass density σ_0. These are related via

$$M = 4\pi R^2 \sigma_0 \,,$$

as expected for a uniform spherical shell. Furthermore, we assume here that the field point is external to the shell. Later, we'll repeat the analysis for a field point interior to the shell.

PARTITION   Chop the shell into $\mathcal{N}$ tiny chunks [labelled by $i = 1, \ldots, \mathcal{N}$]. The qualifier "tiny" ensures that a representative position, $\vec{r}_i$, and well-defined vector area element, $\Delta\vec{A}_i$, are associated with each of the chunks. The mass of the ith chunk is

$$\Delta M_i = \sigma_0 \left| \Delta\vec{A}_i \right| .$$

[In anticipation of summing, the adjacent chunks should aggregate into circular ribbons.]

COMPUTE   For the computations, we shall employ spherical polar coordinates centred on the shell, with polar axis running through the field point, as illustrated in Figure 48.1.

---

[1]Coming soon in VOLUME III.

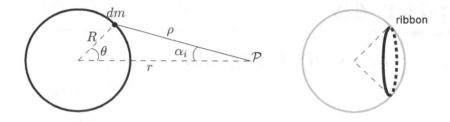

FIGURE 48.1   The Spherical Shell, the Mass Element dm, and the Field Point $\mathcal{P}$

The distance from the centre of the shell to $\mathcal{P}$ is denoted by r. The direction from the centre of the shell to the field point is along the polar axis. Hence, the spherical polar coordinates, $(r\,,\theta\,,\phi)$, of the field point are

$$\vec{r}_{\mathcal{P}} = (r\,,0\,,0)\,.$$

[The azimuthal coordinate is undefined along the polar axis; its value is set to zero.]

The coordinates of the ith chunk are $(R\,,\theta_i\,,\phi_i)$. The radial component, R, has the same value for all of the chunks comprising the shell. Referring to Figure 48.1, the distance from the source chunk to the field point is denoted by ρ, while the direction is [partially] parameterised by the angle α_i. The contribution of each source chunk to the net gravitational field at the field point is

$$\Delta\vec{g}_i = \frac{G\,\Delta M_i}{\rho_i^2}\,[\,-\hat{\rho}_i\,]\,.$$

SUM   In conformity with the notion of linear superposition, the net field is obtained by summing over the chunks. In the $\mathcal{N}\to\infty$ limit, this becomes a three-dimensional integral of a complicated vector-valued function of position. Fortunately, in the present situation, organising the chunks into ribbons assists in simplifying the analysis.

Chunks with a prescribed polar angle, θ, and any azimuthal angle, $0\le\phi_i<2\,\pi$, form a circular ribbon which is symmetrically draped from the vantage of the field point.

[The quantities ρ_i and α_i are constant for each ribbon, while differing between ribbons.]

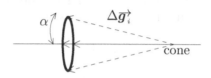

FIGURE 48.2   The Cone of Field Contributions from the Ribbon with Open Angle α

The directed magnitudes of differential contributions to the net field from the chunks in each ribbon form a symmetric cone about the polar axis, as shown in Figure 48.2. Thus, the off-axis contributions must all cancel to zero, while the on-axis bits add with no cancellation whatsoever. Hence, each ribbon contributes

$$\Delta \vec{g_\theta} = \left(\frac{G \, \Delta M_\theta}{\rho^2} \cos(\alpha) \, , \, -\pi \, , \, 0 \right) = \left(\Delta g_\theta \, , \, -\pi \, , \, 0 \right) ,$$

to the net gravitational field at the location of the field point. The mass associated with each ribbon, denoted by ΔM_θ in the above expression, is

$$\Delta M_\theta = \sigma_0 \times \left(2 \, \pi \, R \, \sin(\theta) \right) \times \left(R \, \Delta\theta \right) = 2 \, \pi \, \sigma_0 \, R^2 \, \sin(\theta) \, \Delta\theta .$$

Therefore, the magnitude of the on-axis contribution to the field due to the ribbon is

$$dg = \frac{G \, 2 \, \pi \, \sigma_0 \, R \left(R \, \sin(\theta) \, d\theta \right) \cos(\alpha)}{\rho^2} ,$$

where we have anticipated taking the limit $\mathcal{N} \to \infty$ by converting Δ's to d's. Furthermore, we've eliminated the θ subscript on dg, and have grouped terms in preparation for the next phase of the analysis.

The impediment to further progress is that the quantities $\{\alpha, \theta, \rho\}$ are all related. For instance,

$$2 \, r \, R \, \cos(\theta) = r^2 + R^2 - \rho^2 , \qquad \text{by the law of cosines, while}$$

$$\cos(\alpha) = \frac{r - R \, \cos(\theta)}{\rho} = \frac{r - \left(\frac{r^2 + R^2 - \rho^2}{2r} \right)}{\rho} = \frac{1}{2 \, r \, \rho} \left(r^2 + \rho^2 - R^2 \right) ,$$

by definition (the first equality), and substitution for $\cos(\theta)$ (subsequently). Furthermore,

$$\frac{d}{d\theta} \left(2 \, r \, R \, \cos(\theta) \right) = \frac{d}{d\theta} \left(r^2 + R^2 - \rho^2 \right)$$

$$-2 \, r \, R \, \sin(\theta) \, d\theta = -2 \, \rho \, d\rho$$

$$\implies \qquad R \, \sin(\theta) \, d\theta = \frac{\rho}{r} \, d\rho .$$

Combining these results, in order to convert the original expression into one involving constants and factors of ρ only, yields

$$dg = \frac{G \, 2 \, \pi \, \sigma_0 \, R}{\rho^2} \left\{ \frac{\rho}{r} \, d\rho \right\} \left[\frac{1}{2 \, r \, \rho} \left(r^2 + \rho^2 - R^2 \right) \right] = G \, \pi \, \sigma_0 \, \frac{R}{r^2} \left[1 + \frac{r^2 - R^2}{\rho^2} \right] d\rho .$$

This is amenable to integration, yielding the total on-axis field,

$$g = \int_{\rho_-}^{\rho_+} dg = G \, \pi \, \sigma_0 \, \frac{R}{r^2} \left[\rho - \frac{r^2 - R^2}{\rho} \right] \Bigg|_{\rho_-}^{\rho_+} ,$$

where $\rho_\pm$ denote the upper/lower limits of integration, respectively.

For field points which are external to the shell, the lower limit is $\rho_- = r - R$, while the upper limit is $\rho_+ = r + R$. Substitution of these into the expression for g gives,

$$g = G \, \pi \, \sigma_0 \, \frac{R}{r^2} \Big[(r+R) - (r-R) - (r-R) + (r+R) \Big] = G \, \pi \, \sigma_0 \, \frac{R}{r^2} \Big[4 \, R \Big] = \frac{G \, 4 \, \pi \, R^2 \, \sigma_0}{r^2} = \frac{G \, M}{r^2} .$$

Clearly then,

$$\vec{g}_{\text{shell}} = -\frac{G \, M}{r^2} \, \hat{r} .$$

The gravitational field external to the spherical shell is *indistinguishable* from that of a point mass located at its centre.

The INTERIOR CASE

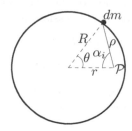

FIGURE 48.3   The Spherical Shell, the Mass Element, and the Interior Field Point

Consider the gravitational field produced by the same thin spherical shell of matter as above, at a point located in the interior space of the shell. The situation is shown in Figure 48.3. The most important thing to realise is that all of the particulars of the partition of the shell are completely independent of the location of the field point, and so the PARTITION and COMPUTE analyses carry over exactly, as does the SUM, **except for the limits of integration**. The upper limit is still $\rho_+ = r + R$, but the lower limit is $\rho_- = R - r$ in the interior case. This small change in the limit has a profound effect on the integral, causing it to vanish!

$$
g = G\,\pi\,\sigma_0\,\frac{R}{r^2}\left[\rho - \frac{r^2 - R^2}{\rho}\right]\Bigg|_{R-r}^{r+R}
$$

$$
= G\,\pi\,\sigma_0\,\frac{R}{r^2}\Big[(r + R) - (r - R) - (R - r) - (R + r)\Big] = 0\,,
$$

The gravitational field inside the spherical shell vanishes!

GAUSS'S LAW FOR GRAVITY   Gauss's Law is an important mathematical result with profound implications for physical models involving fields. The general form of Gauss's Law will be studied in depth in VOLUME III in the context of electric and magnetic fields. It is sufficient for our present needs to quote a restricted version of the law:

○·   Outside a spherically symmetric shell of matter, with total mass M, the gravitational field due to the shell is indistinguishable from that of a point source, also of mass M, located at the centre of the shell.

⊙   Within a spherically symmetric shell of matter, with total mass M, the gravitational field of the shell vanishes.

PROOF of (the restricted version of) GAUSS'S LAW

The outside and inside computations directly above constitute proof of the quoted form of Gauss's Law.

Gauss's Law provides the operational definition of a **point-like** gravitational source.

POINT-LIKE GRAVITATING BODY   A gravitating body is point-like insofar as it is [to a sufficient degree] spherically symmetric.

⊙·   IF the field point is external to the body, THEN the entire mass of the body contributes to the point-like field.

⊙   IF the field point is located within the body, THEN the effective mass is that of the substance located within the radius of the field point.

....

EXAMPLE [*The Gravitational Field of the Earth at Its Surface*]

The Earth is [approximately] spherically symmetric, and thus its surface gravitational field is [nearly] identical to that of a point source, $M_\oplus$, located at the geometric centre of the Earth. The distance from the centre of the Earth to the surface is its radius, $R_\oplus$, and hence the surface field is

$$\vec{g_\oplus} = -\frac{G\,M_\oplus}{R_\oplus^2}\,\hat{r}\,.$$

Plugging in the various parameter values [with implicit consideration of units], fixes

$$\left|\vec{g_\oplus}(R_\oplus)\right| = \frac{G\,M_\oplus}{R_\oplus^2} \simeq \frac{6.67 \times 10^{-11} \times 6 \times 10^{24} \left[\frac{\text{m}^3}{\text{kg}\cdot\text{s}^2}\,\text{kg}\right]}{(6.38 \times 10^6)^2\,[\text{m}^2]} \simeq 9.8 \text{ m/s}^2 \sim 10 \text{ m/s}^2\,.$$

Wow! **The acceleration due to gravity employed in the analysis of projectile motion in Chapter 7, and incorporated into the weight force in Chapter 11 *et seq.*, is just the surface value of the Newtonian gravitational field of the Earth!**

<hr/>

GRAVITATIONAL ENERGETICS

The most important consequence of the inverse-square law property is that NUG is conservative, and therefore a system comprised of two or more gravitating bodies has an associated gravitational potential energy.

ASIDE: **Q:** Didn't we uncover the gravitational potential energy function

$$U_g(x,y) = M\,g\,y\,,$$

where y is the coordinate aligned with the vertical direction, in Chapter 25?
A: Upon recollection, U_g is the potential energy function associated with locally constant field gravity, rather than NUG.

Consider two point-like bodies, M_1 and M_2, where the initial relative position of 2 with respect to 1 is $\vec{r}_{12,i} = r_i\,\hat{r}_i$. Suppose that body 2 is moved [without impinging upon 1] to a final relative position $\vec{r}_{12,f} = r_f\,\hat{r}_f$. The goal of the next section is to compute the work done by the NUG force produced by 1 acting on 2 during its displacement:

$$W_{if}\left[\vec{F_\text{G}}\right] = \int_{\vec{r}_i}^{\vec{r}_f} -\frac{G\,M_1\,M_2}{r^2}\,\hat{r}\cdot d\vec{s} = -G\,M_1\,M_2 \int_{\vec{r}_i}^{\vec{r}_f} \frac{1}{r^2}\,\hat{r}\cdot d\vec{s}\,.$$

En route to the generic result, let's consider two particular instances: radial displacements which occur along the line joining the masses, and displacements confined to spherical shells centred on 1.

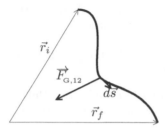

FIGURE 48.4   Gravitational Work Associated with Displacement $i \to f$

Purely radial displacements have fixed values of $\{\theta, \phi\}$ while r varies[2] from r_i to r_f. In Figure 48.5, $\vec{r}_f$ is shown below $\vec{r}_i$ for clarity, whereas in fact they are overlaid.

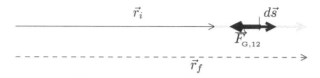

FIGURE 48.5   Gravitational Work Associated with Radial Displacement $i \to f$

Along radial displacements, each $d\vec{s}$ is parallel to $\hat{r}$ everywhere, and hence

$$\hat{r} \cdot d\vec{s} = dr \, ,$$

throughout the path. Substituting this result into the integral for the work, one obtains

$$W_{if}\left[\vec{F_G}\right]_{(\mathrm{radial})} = -G\,M_1\,M_2 \int_{r_i}^{r_f} \frac{1}{r^2}\,dr = G\,M_1\,M_2\left[\frac{1}{r}\right]\Bigg|_{r_i}^{r_f} = G\,M_1\,M_2\left[\frac{1}{r_f} - \frac{1}{r_i}\right] .$$

Displacements constrained to lie within the spherical shell have constant radial component while the angular parts $\{\theta, \phi\}$ vary freely from their initial to final values. Figure 48.6 illustrates one such path.

Each $d\vec{s}$ is perpendicular to $\hat{r}$ everywhere along the displacement. Thus the integrand,

$$\hat{r} \cdot d\vec{s} = 0 \, ,$$

vanishes throughout the domain of integration. Therefore, so too does the work,

$$W_{if}\left[\vec{F_G}\right]_{(\mathrm{shell})} \equiv 0 \, .$$

---

[2]The radial variation must be continuous, but it need not be monotonic or differentiable.

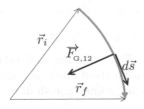

FIGURE 48.6 Gravitational Work Associated with Tangential Displacement $i \to f$

The explicit analyses of the purely radial and spherical paths, carried out just above, enable us to claim that the work done by NUG in the case of an arbitrary displacement is determined solely by the net change in the radial component of the spatial position. [The generalisation hinges on the fact that any single-step displacement may be decomposed into its radial (1-d) and tangential (2-d) projections.] Furthermore, with a few *Gedanken* experiments [not reproduced here] one can easily convince oneself that the mechanical work done by the NUG force is independent of the path taken from the initial position to the final position. Therefore, NUG is a conservative force, and thereby possesses an associated potential energy function.

The change in the NUG potential energy is merely "minus the work done by the force acting through the displacement from initial to final," making it possible to infer a potential energy function.

$$\Delta U_\mathrm{G} = -W_{if}\left[\vec{F}_\mathrm{G}\right]$$

$$U_\mathrm{G}(\vec{r}_f) - U_\mathrm{G}(\vec{r}_i) \quad = \quad -G\,M_1\,M_2\left[\frac{1}{r_f} - \frac{1}{r_i}\right].$$

The simplest form of the Newtonian gravitational potential energy function consistent with the expression for the gravitational work is

$$U_\mathrm{G}(\vec{r}) = U_\mathrm{G}(r,\theta,\phi) = -\frac{G\,M_1\,M_2}{r}.$$

Comments:

- For two widely separated particles [*a.k.a.* asymptotic behaviour],

$$\lim_{r\to\infty} U_\mathrm{G}(\vec{r}) = 0\,, \quad \text{irrespective of the direction } \{\theta,\phi\}.$$

IF the two interacting bodies are infinitely far apart, THEN the potential energy of the system vanishes.

- The short-distance behaviour is divergent,

$$\lim_{r\to 0} U_\mathrm{G}(\vec{r}) \to -\infty\,, \quad \text{irrespective of the direction } \{\theta,\phi\}.$$

Gravitational forces grow unboundedly as point particles approach each other.

[One anticipates that finite-size effects will cut off this divergent energy in actual systems.]

EXAMPLE [*"When a body meets a body ..."* (Part 2)]

Recovering quickly from the shock of PK's startling declaration, NK replied tongue-in-cheek, "Isaac Newton assures me that this attraction must be mutual. However, I cannot help but think that you, or rather we, are being somewhat shallow, as I reckon that we are bound by

$$U_{\text{NP}} = -\frac{G\,M_N\,M_P}{r_{\text{NP}}} = -\frac{6.67 \times 10^{-11} \times 60 \times 85}{1} \simeq -\frac{20}{3} \times 5100 \times 10^{-11} = -3.4 \times 10^{-7} \text{ J},$$

or, roughly speaking, about one-third of a microjoule."

EXAMPLE [*Gravitational Potential Energy near the Surface of the Earth*]

The change in the gravitational potential energy of a particle of mass m, undergoing a small vertical displacement, h, near the surface of the Earth, is expressed as

$$\Delta U_{\text{G}} = G\,M_\oplus\,m \left[\frac{1}{r_i} - \frac{1}{r_f} \right] \simeq G\,M_\oplus\,m \left[\frac{1}{R_\oplus} - \frac{1}{R_\oplus + h} \right].$$

While this result is exact as it stands, it is worthwhile to take the Taylor [or binomial, if one prefers] approximation, to obtain

$$\Delta U_{\text{G}} \simeq G\,M_\oplus\,m\,\frac{h}{R_\oplus^2} = m\,\frac{G\,M_\oplus}{R_\oplus^2}\,h = m\,g_\oplus\,h.$$

This is precisely the form of the gravitational potential energy function previously obtained in the context of locally constant field gravity.

[Any continuous field is locally constant on sufficiently small scales.]

ADDENDUM: The Gravitational Potential

GRAVITATIONAL POTENTIAL   The gravitational potential function associated with a source is the gravitational potential energy per unit responder mass at a point in space. The potential is a scalar-valued function of position with SI units: $\text{J/kg} = \text{m}^2/\text{s}^2$. The fundamental definition of the potential function associated with a gravitating body, "1," say, is via the line integral of the field produced by 1 along a path from some reference position, $\vec{R}_{\text{ref}}$, at which the potential is deemed to vanish, to the point of interest, $\vec{r}$, *viz.*,

$$V_{\text{G},1}(\vec{r}) = -\int_{\vec{R}_{\text{ref}}}^{\vec{r}} \vec{g}_1 \cdot d\vec{s}.$$

Potential is to Potential Energy   as   Field is to Force.

Chapter 49

Gravitational Effects and Dynamics

In these explorations of the dynamics and energetic elements of NUG we have, thus far, neglected the kinematical aspect.

Q: What do the trajectories of particles subject to NUG look like?

A: Good question! Let's examine a particular and popular case.

Consider the scenario in which a planet of mass m and a star of mass $M \gg m$ form a system. It is not unreasonable to assume that the star is at rest and to restrict attention to the motion of the planet under the influence of the star's gravitational field.

> ASIDE: A more thorough and general treatment, [leading to an appreciation of the experimental evidence favouring the existence of planets in orbit about distant stars], is the sort of thing one encounters in an intermediate mechanics class.

Since the force of gravity is central, the trajectory of the planet is confined to the unique plane[1] established by the centres of the bodies and and the planet's velocity vector. Hence, the *a priori* 3-d dynamics of the system may always be reduced to 2-d. WLOG, a polar coordinate system, (r, θ), with its origin coincident with the centre of the star, shall be employed to describe the motion of the planet.

In Chapter 8 the kinematics were self-consistently[2] worked out for [2-d] polar coordinates:

$$\vec{v} = \dot{r}\,\hat{r} + \omega r\,\hat{\theta}, \qquad \text{and} \qquad \vec{a} = \left[\ddot{r} - \omega^2 r\right]\hat{r} + \left[2\,\omega\,\dot{r} + \dot{\omega}\,r\right]\hat{\theta},$$

where the "dot" accent denotes a time derivative, $r = r(t)$, $\theta = \theta(t)$, and $\omega = \dot{\theta}$.

While a complete solution consists of knowing $r(t)$ and $\theta(t)$ simultaneously, the path of the planet is not so obvious in this parameterised form.

> ASIDE: A similar situation arose in the study of projectile motion in Chapter 7. Our recourse then was to recast the trajectory entirely in terms of spatial variables:
>
> $$y = y(x) = \tan(\theta_0)\,x - \frac{g}{2\,v_0^2\,\cos^2(\theta_0)}\,x^2,$$
>
> which is clearly a parabola. Here, we shall seek to determine $r = r(\theta)$, an as yet unknown function.

Realisation that the gravitational force is central, and thus there is no cause for acceleration in the $\hat{\theta}$ direction, leads to

$$2\,\omega\,\dot{r} + \dot{\omega}\,r = 0, \qquad \Longrightarrow \qquad \omega\,r^2 = l,$$

where l is a constant.[3] From this result, one is able to determine $\frac{d\omega}{d\theta}$, via

$$0 = \frac{d}{d\theta}\left(\omega\,r^2\right) = \frac{d\omega}{d\theta}\,r^2 + 2\,\omega\,r\,\frac{dr}{d\theta}, \qquad \Longrightarrow \qquad \frac{d\omega}{d\theta} = -2\,\frac{\omega}{r}\,\frac{dr}{d\theta}.$$

---

[1] A degenerate case arises when the velocity of the planet is directed parallel to the force.

[2] Recall that the polar basis vectors depend on position, and therefore time also.

[3] The units of l, $[\mathrm{m^2/s}]$, are those of **angular momentum per unit mass**. The relation, $\omega\,r^2 = l$, is equivalent to the statement of Kepler's Second Law, to come in Chapter 50.

The time derivatives of the radial component may be rewritten in terms of angle. First,

$$\dot{r} = \frac{dr}{d\theta}\frac{d\theta}{dt} = \frac{dr}{d\theta}\,\omega \,,$$

and second,

$$\ddot{r} = \frac{d}{dt}\,\dot{r} = \frac{d}{d\theta}\left(\frac{dr}{d\theta}\,\omega\right)\frac{d\theta}{dt} = \left(\frac{d^2r}{d\theta^2}\,\omega^2 + \frac{dr}{d\theta}\frac{d\omega}{d\theta}\,\omega\right) = \left[\frac{d^2r}{d\theta^2} - \frac{2}{r}\left(\frac{dr}{d\theta}\right)^2\right]\omega^2 \,.$$

Plugging these into the expression for the radial component of the acceleration yields

$$a_r \equiv \left[\frac{d^2r}{d\theta^2} - \frac{2}{r}\left(\frac{dr}{d\theta}\right)^2 - r\right]\omega^2 \,.$$

Knowing that the cause of this acceleration is the gravitational field of the star allows us to write:

$$\left[\frac{d^2r}{d\theta^2} - \frac{2}{r}\left(\frac{dr}{d\theta}\right)^2 - r\right]\omega^2 = -\frac{GM}{r^2} \,.$$

Dividing by ω^2, and recognising that $\omega = \frac{l}{r^2}$, the denominator on the RHS simplifies to

$$\frac{1}{(\omega r)^2} = \frac{r^2}{l^2} \,,$$

and the dynamical equation becomes a **second-order non-linear ordinary inhomogeneous differential equation**:

$$\left[\frac{d^2r}{d\theta^2} - \frac{2}{r}\left(\frac{dr}{d\theta}\right)^2 - r\right] = -\frac{GM}{l^2}\,r^2 \,.$$

Solutions of this differential equation, $r(\theta)$, if they exist, may be amenable to interpretation as trajectories of planets in a planet–star system.

Non-linear ODE's are notoriously difficult to solve. To proceed, we resort to an *Ansatz*:[4]

$$r(\theta) = \frac{R}{1 - \epsilon\,\cos(\theta)}$$

where R is a constant with dimension of length, and ϵ is a dimensionless constant. There are two independent parameters in the *Ansatz* (corresponding to the two constants of integration which one expects to encounter in solutions to a second order ODE).

For this particular $r(\theta)$,

$$\frac{dr}{d\theta} = -\frac{R\epsilon\,\sin(\theta)}{(1 - \epsilon\,\cos(\theta))^2} = -\frac{\epsilon\,\sin(\theta)}{1 - \epsilon\,\cos(\theta)}\,r \,,$$

$$\frac{d^2r}{d\theta^2} = \frac{2\,R\,\epsilon^2\,\sin^2(\theta)}{(1 - \epsilon\,\cos(\theta))^3} - \frac{R\epsilon\,\cos(\theta)}{(1 - \epsilon\,\cos(\theta))^2} = \frac{\epsilon^2\left(2 - \cos^2(\theta)\right) - \epsilon\,\cos(\theta)}{R\left(1 - \epsilon\,\cos(\theta)\right)}\,r^2 \,,$$

---

[4]One makes an "inspired guess" as to the mathematical form of the solution in terms of some arbitrary parameters. By substitution of the *Ansatz* into the differential equation, one obtains algebraic equations for the parameters. One hopes that [some of] the solutions for the parameters will yield trajectories exhibiting physically interesting behaviours. This is our first recourse to this time-tested method in these notes and will certainly not be the last!

and hence the dynamical equation reduces to

$$
\left[\frac{\epsilon^2 \left(2 - \cos^2(\theta) \right) - \epsilon \cos(\theta)}{R \left(1 - \epsilon \cos(\theta) \right)} - \frac{2}{r} \left(-\frac{\epsilon \sin(\theta)}{1 - \epsilon \cos(\theta)} \right)^2 - \frac{1}{r} \right] r^2 = -\frac{GM}{l^2} r^2 .
$$

Amazingly, the LHS simplifies to $-\frac{r^2}{R}$, as is seen in the following sequence of equalities.

$$
\begin{aligned}
\frac{\text{LHS}}{r^2} &= \frac{1}{r} \frac{1}{\left(1 - \epsilon \cos(\theta) \right)^2} \left[\epsilon^2 \left(2 - \cos^2(\theta) - 2 \sin^2(\theta) \right) - \epsilon \cos(\theta) \right] - \frac{1}{r} \\
&= \frac{1}{r} \frac{1}{\left(1 - \epsilon \cos(\theta) \right)^2} \left[\epsilon^2 \cos^2(\theta) - \epsilon \cos(\theta) \right] - \frac{1}{r} \\
&= \frac{1}{r} \left[\frac{-\epsilon \cos(\theta) \left(1 - \epsilon \cos(\theta) \right)}{\left(1 - \epsilon \cos(\theta) \right)^2} - 1 \right] \\
&= \frac{1 - \epsilon \cos(\theta)}{R} \left[\frac{-\epsilon \cos(\theta)}{1 - \epsilon \cos(\theta)} - 1 \right] \\
&= \frac{1 - \epsilon \cos(\theta)}{R} \left[\frac{-1}{1 - \epsilon \cos(\theta)} \right] \\
&= \frac{-1}{R}
\end{aligned}
$$

What this means is that, upon imposing the *Ansatz* with the two unspecified parameters, the radial component of N2 reduces to an algebraic relation,

$$
-\frac{1}{R} = -\frac{GM}{l^2} \qquad \Longrightarrow \qquad R = \frac{l^2}{GM} ,
$$

and thus the *Ansatz* is completely and incontrovertibly consistent with the inverse-squared force law prescribed by NUG. The actual values of the [*a priori* undetermined] solution parameters $\{l, \epsilon, R\}$ are to be fixed by the appropriate data pertinent to the planet–star system.

ASIDE: **Q:** But what of the parameterised trajectories, $\vec{r} = (r(t), \theta(t))$?

A: The relation, $\frac{d\theta}{dt} = \omega = l/r^2$, exploited above, can be combined with the *Ansatz* and integrated to yield an expression for $t(\theta)$. This cumbersome relation—not quoted here—provides an implicit solution $(r(\theta(t)), \theta(t))$ to the dynamical equations.

Rather than attempt to determine the $\{R, \epsilon\}$ parameters for a particular system, let's investigate the variegated possibilities afforded by choosing particular values of these constants. As those of us who remember our algebraic geometry might recall,

the loci of $r(\theta)$ are conic sections!

Note that the radial motion is precisely periodic.[5] The parameter R sets the scale for the motion of the planet. The dependence on ϵ is rather more complicated.

$\epsilon = 0$ IF $\epsilon = 0$, THEN $r(\theta) = \frac{R}{1 - \epsilon \cos(\theta)} \big|_{\epsilon = 0} = R$, and hence the trajectory is circular. The angular speed, $\omega = l/r^2 = l/R^2$, is constant, too. Thus, IF the motion is circular, THEN it is also necessarily uniform.

---

[5]That is to say that the orbits do not **precess**. However, this periodicity in angle need not give rise to periodicity in the temporal behaviour of the planet, as we'll see momentarily.

$0 < \epsilon < 1$ IF $0 < \epsilon < 1$, THEN the trajectory is elliptical. The maximal and minimal values of r, called the **apsides**, are attained at $\theta = 0$ and $\theta = \pi$,

$$r(0) = \frac{R}{1 - \epsilon} \quad \text{and} \quad r(\pi) = \frac{R}{1 + \epsilon},$$

respectively. As $\epsilon \to 0$, the orbits become more circular, whereas as $\epsilon \to 1$, they become more and more elongated.

$\epsilon = 1$ IF $\epsilon = 1$, THEN $r(\theta) = \frac{R}{1 - \cos(\theta)}$, and the trajectory is parabolic. The apsides are $r \to R/2$, as $\theta \to \pi$, and $r \to \infty$, as $\theta \to 0$, respectively. One may think of a parabola as the limiting case of an ellipse which closes at infinite distance. Such a trajectory is not periodic in time, since the orbit is of infinite length.

$1 < \epsilon$ IF $1 < \epsilon$, THEN the trajectory is hyperbolic. The domain of angles is reduced from $(0, 2\pi)$ to that subset, centred on $\theta = \pi$ for which the radial distance is positive [$r \to +\infty$ before changing sign at $\pm \cos^{-1}(1/\epsilon)$].

Newton developed clever geometric proofs that the solutions to equations of motion involving central forces with inverse-squared radial coordinate dependence were conic sections.

[Newton did this to convince those who were skeptical of his newfangled calculus techniques.]

EXAMPLE [*Falling from Rest in a Gravitational Field*]

A test body of mass m_0 is released from rest at $t = 0$ in the gravitational field of a point-like object of [large] mass M. When released, the radial separation between the centres of the bodies is r_0. At a later time, t, the radial separation is reduced to r.

Q: What is the [relative] speed of the test mass at time t?

A: To model the system, we shall assume that M remains at rest, even though the bodies are interacting.[6] There are no non-conservative forces extant, so the total mechanical energy is conserved. The initial energy is

$$E_i = K_i + U_{G,i} = 0 - \frac{G M m}{r_0} = -\frac{G M m}{r_0}.$$

The final energy is

$$E_f = K_f + U_{G,f} = \frac{1}{2} m v^2 - \frac{G M m}{r}.$$

Setting $E_f = E_i$ yields

$$\frac{1}{2} m v^2 = G M m \left[\frac{1}{r} - \frac{1}{r_0} \right] \quad \Longrightarrow \quad v = \sqrt{2 G M \left(\frac{1}{r} - \frac{1}{r_0} \right)}.$$

Thus the speed of the infalling test body depends on its radial distance from the point-like source.

---

[6]This tactic was discussed and justified in the skydiving example found in Chapter 10.

TOTAL MECHANICAL ENERGY in a PLANETARY SYSTEM

The total mechanical energy of a planet orbiting a star [assumed to be fixed in space] receives a positive contribution from the kinetic energy of the planet, and a negative[7] contribution from the potential energy of the system:

$$E = K + U_{\mathrm{G}} = \frac{1}{2} m v^2 - \frac{GMm}{r}.$$

The total mechanical energy is strongly correlated to the orbital kinematics.

Total Energy	Conic Section
$E < 0$	Ellipse
$E = 0$	Parabola
$E > 0$	Hyperbola

Example [*Virial Theorem*]

Suppose that a planet of mass m is in circular orbit, of radius R, about an immobile star of mass $M \gg m$.

Q: What is the total mechanical energy of the planet?

A: The centripetal $m\,\vec{a}$ for the circular orbit of the planet is effected by NUG:

$$m \frac{v^2}{R} = \frac{GMm}{R^2}.$$

The *trick* is to recognise that the LHS can be converted into the expression for the kinetic energy of the planet via multiplication by $R/2$. Thus

$$K_{\mathrm{planet}} = \frac{1}{2} m v^2 = \frac{1}{2} \left(m \frac{v^2}{R} \right) R = \frac{1}{2} \left(\frac{GMm}{R^2} \right) R = \frac{1}{2} \frac{GMm}{R}.$$

Meanwhile the gravitational potential energy of the system is

$$U_{\mathrm{G}} = -\frac{GMm}{R}.$$

Thus the kinetic energy of the planet cancels precisely one-half of its gravitational binding energy, leaving one-half of the binding energy as the total mechanical energy of the system.

$$E = K_{\mathrm{planet}} + U_{\mathrm{G}} = \frac{1}{2} U_{\mathrm{G}} = -\frac{GMm}{2R}.$$

[This result is the first instance of what are known as **Virial Theorems**.]

----

[7]Adhering to the convention that $U_{\mathrm{G}} \to 0$ at infinite separation of the constituents.

EXAMPLE [*Message in a Bottle*]

Declan Steinein seeks an interplanetary penpal. He composed a note and stuffed it into a sturdy bottle. He now wishes to launch the letter into space.

Q: With what minimum initial speed must he fire off his missive?

A: Consider a crude model with the following features:

★ The bottle [plus contents] acts as a point-like body of mass m

★ The bottle is initially at rest[8]

★ The bottle is immersed in the gravitational field of the Earth only

★ Drag and other dissipative effects are ignored

Declan knows that for the bottle to go arbitrarily far into interplanetary space it must follow a trajectory with total mechanical energy equal to, at minimum, zero. The initial energy of the bottle [at rest on the Earth's surface] is

$$E_i = -\frac{G\,M_\oplus\,m}{R_\oplus}\,.$$

Declan must provide enough kinetic energy at the launch to cancel this gravitational potential [*a.k.a.* binding] energy. Therefore,

$$K_{\min} = \frac{1}{2}\,m\,v_l^2 = +\frac{G\,M_\oplus\,m}{R_\oplus}\,,$$

and hence the necessary launch speed is

$$v_l = \sqrt{\frac{2\,G\,M_\oplus}{R_\oplus}} = \sqrt{\frac{2 \times 6.67 \times 10^{-11} \times 6 \times 10^{24}}{6.38 \times 10^6}} \simeq \sqrt{\frac{8}{6.38}} \times 10^4 \simeq 1.12 \times 10^4 \ \text{m/s}\,.$$

Thus, to send his message in a bottle into interplanetary space, Declan must project it at approximately 11.2 km/s.

There are a number of comments which must be made.

• The launch speed does not depend on the mass of the particle/payload.

  [This is a manifestation of the EQUIVALENCE PRINCIPLE and the universality of NUG.]

• The launch speed is awfully large. Back in the day, people thought that this result precluded the possibility of ever sending objects into space. They were wrong, because they failed to realise that as one climbs, the effective escape speed diminishes.

  [CAVEAT: There is no escape whatsoever from the constant field gravity of Chapter 7.]

• Although the bottle escapes the Earth, it will still be gravitationally bound to the Solar System, and thus not likely to reach interstellar space.

  ASIDE: To escape the gravitational field of the Sun as well, Declan will have to impel it with even greater speed, unless he is very astute and employs a **gravitational sling-shot effect** involving interaction with other planets.

---

[8]The rotation of the Earth about its axis can be exploited to give the letter a bit of a boost, saving Declan a wee bit of effort.

Chapter 50

Kepler's Laws

Tycho Brahe (1546–1601) was an avid astronomer and telescope builder. Over the course of many years he made very precise observations of objects in the night skies. His notebooks contained comprehensive records of

date	time	object	right ascension	declination
⋮	⋮	⋮	⋮	⋮

where **right ascension** and **declination** are angular coordinates [rather like latitude and longitude] specifying positions on the "celestial sphere."

Johannes Kepler (1571–1630) was an itinerant scholar[1] to whom Brahe entrusted his data. Over the course of a decade, he subjected Brahe's planetary observations to an exhaustive numerical analysis, sifting the data for evidence of regularities. By this arduous method, Kepler uncovered the three empirical laws which bear his name.

Kepler's First Law [K1]

K1   Planets move in elliptical orbits, with the Sun at one of the focal points.

To appreciate K1, let's first recall a bit of information about ellipses and their properties. In the first panel of Figure 50.1, an ellipse appears in standard form, while in the second it is drawn so as to be evocative of a planetary orbit.

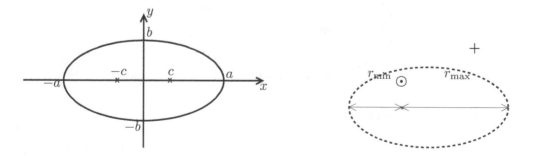

FIGURE 50.1   An Ellipse in Standard Form and an Elliptical Orbit

---

[1]This was not at all an unusual career arc for academics or musician/composers in that era. The nearest modern equivalent might be a "peripatetic postdoc," *i.e.,* a person who, upon completing his Ph.D., works as a research scientist for a number of different groups or institutions.

The **semi-major axis** has length a, the **semi-minor axis** has length b, and the focal points are at $\pm c = \pm\sqrt{a^2 - b^2}$ along the semi-major axis. The mathematical description of this ellipse is the locus of points, (x, y), in the Cartesian plane for which

$$\frac{x^2}{a^2} + \frac{y^2}{b^2} = 1.$$

In polar form, (r, θ), centred on one focus, the [locus of points comprising the] ellipse is defined by the relation[2]

$$r(\theta) = \frac{a\left(1 - \epsilon^2\right)}{1 - \epsilon \cos(\theta)},$$

where ϵ is the **eccentricity** of the ellipse. The polar form better corresponds to the image of the orbit in the second panel of Figure 50.1.

Two comments bear mentioning.

o   A circle is a degenerate[3] ellipse in which $a = b$, or equivalently, $\epsilon = 0$.

♯   Kepler's claim was contentious because of the prevailing neo-Platonic notions about the supposed perfection of circular orbits in the old-fashioned Ptolemaic, and then-novel Copernican, models of the heavens.

PROOF of KEPLER'S FIRST LAW

As was shown in Chapter 49, the general solutions of N2 for a central force with an inverse-squared radial dependence are the conic sections. The subset of trajectories which evince periodic behaviour are those for which the eccentricity, $0 \le \epsilon < 1$, *i.e.*, ellipses.

Kepler's Second Law [K2]

K2   During equal time intervals the vector from the Sun to the planet sweeps out equal areas irrespective of where the planet is in the course of its orbit.

An illustration of K2 is provided in Figure 50.2, in which the areas swept out in three equal time intervals are indicated.

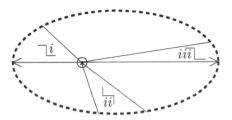

FIGURE 50.2   Illustration of K2

The idea is that when the satellite is closer to the focus [*i.e.*, the Sun], it moves more quickly and thus its path subtends a larger angle in a time interval of particular duration. When the satellite is farther away, it moves more slowly, and hence its trajectory subtends a smaller angle. Amazingly, these two effects precisely cancel!

---

[2] Writing $R = a\left(1 - \epsilon^2\right)$ converts this expression to the *Ansatz* employed in Chapter 49.
[3] A snarky person might crack wise that a circle is "an ellipse from the wrong side of the tracks."

PROOF of KEPLER'S SECOND LAW

The only force acting on the satellite is presumed to be NUG. Consider the torque produced by the gravitational force acting on the satellite, computed about the focal point at the centre of the Sun.

$$\vec{\tau}_{\text{net}} = \vec{\tau}_G = \vec{r} \times \vec{F}_G = \vec{0}.$$

The torque must vanish because the force is central, that is, along the line joining the centres of the Sun and the planet. Recall that the generalised expression for the rotational version of N2 equates the net torque with the time rate of change of the angular momentum. Thus, the absence of net torque ensures constancy of the angular momentum of the planet [computed about the centre of the Sun]. The angular momentum of the satellite is

$$\vec{L} = \vec{r} \times \vec{p} = \vec{r} \times (m\,\vec{v}) = m\,\vec{r} \times \vec{v} = m\,|\vec{r}|\,|\vec{v}|\,\sin(\theta)\,[\odot],$$

where the details are found in Figure 50.3.

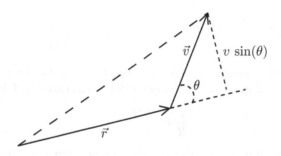

FIGURE 50.3 The Proof of K2

The differential area, dA, swept out in time dt is given by the area of a triangle with base equal to $|\vec{r}|$, and height equal to $|\vec{v}|\,dt\,\sin(\theta)$. See Figure 50.3 for details. Thus,

$$\frac{dA}{dt} = \frac{1}{2}|\vec{r}|\,|\vec{v}|\,\sin(\theta) = \frac{1}{2m}|\vec{L}| = \text{a constant!}$$

Therefore, through any finite time interval, Δt, the area swept out by the radius vector amounts to

$$\Delta A = \frac{|\vec{L}|}{2m}\,\Delta t,$$

and is independent of the position of the satellite in its orbit.

Kepler's Third Law [K3]

K3 The square of the orbital period of a planet is proportional to the cube of its semi-major axis,

$$T^2 \propto a^3.$$

PROOF of KEPLER'S THIRD LAW (**in the case of circular orbits**)

METHOD ONE: [*Centripetal* $m\vec{a} = \vec{F}_G$]
The isotropy of the [assumed circular] orbit, along with K2, constrains the satellite to move uniformly. The force of gravity is the cause of the centripetal $m\,\vec{a}$ experienced by the satellite. Recalling the expression for centripetal acceleration developed in Chapter 8, and ascribing its cause to NUG, one obtains

$$m\,\frac{v^2}{r} = \frac{G\,M\,m}{r^2}\,,$$

[*cf.* the VIRIAL THEOREM example in Chapter 49]. Uniformity of the motion ensures that the orbital [tangential] speed is equal to the distance travelled in one revolution (circumference), divided by the time interval required to complete one revolution (period). Thus, the speed is

$$v = \frac{2\,\pi\,r}{\mathcal{T}}\,.$$

Hence,

$$\frac{\left(\frac{2\,\pi\,r}{\mathcal{T}}\right)^2}{r} = \frac{G\,M}{r^2} \qquad \Longrightarrow \qquad \mathcal{T}^2 = \frac{4\,\pi^2}{G\,M}\,r^3\,.$$

METHOD TWO: [*Exploiting the analysis of trajectories in Chapter 49*]
According to the analysis of gravitational trajectories undertaken in Chapter 49,

$$\frac{1}{R} = \frac{G\,M}{l^2}\,,$$

while, in this case of circular orbits, the angular momentum per unit mass is evidently constant, $l = \omega\,R^2$. Combining these two results yields

$$1 = \frac{G\,M}{\omega^2\,R^3}\,.$$

In addition, the angular speed is constant and equal to $2\,\pi/\mathcal{T}$. Therefore,

$$R^3 = \frac{G\,M}{4\,\pi^2}\,\mathcal{T}^2$$

is incontrovertibly true for circular trajectories subject to NUG.

Whether one's preferences lie with METHOD ONE or TWO, each offers a proof of the veracity of K3 in this restricted case of circular orbits.

PROOF of KEPLER'S THIRD LAW (**general case**)

The general proof relies on application of Kepler's First and Second Laws, along with the realisation that the area swept out by the radial line to the satellite in a time interval of one orbital period is equal to the entire area bounded by the elliptical orbit.

Elementary geometry informs us that the area of an ellipse with semi-major axis, a, and semi-minor axis, b, is $\mathcal{A} = \pi\,a\,b$. The axes are related via the eccentricity, $b = a\,\sqrt{1-\epsilon^2}$, and so the area can be written as

$$\mathcal{A} = \pi\,a^2\,\sqrt{1-\epsilon^2}\,.$$

Careful inspection of Figure 50.1 reveals that twice the semi-major axis is equal to the sum of the maximum[4] and minimum[5] distances of the satellite from a focus. Hence,

$$a = \frac{1}{2}\left(r_{\max} + r_{\min}\right) = \frac{1}{2}\left(r(0) + r(\pi)\right).$$

These apsidal distances were expressed in terms of the radius, R, and eccentricity, ϵ, in Chapter 49:

$$r(0) = \frac{R}{1 - \epsilon} \qquad \text{and} \qquad r(\pi) = \frac{R}{1 + \epsilon}.$$

Thus, $\{a, R, \epsilon\}$ are related via

$$a = \frac{1}{2}\left(\frac{R}{1-\epsilon} + \frac{R}{1+\epsilon}\right) = \frac{R}{1-\epsilon^2} \qquad \Longleftrightarrow \qquad \frac{1-\epsilon^2}{R} = \frac{1}{a}.$$

We shall need this to complete the proof.

Kepler's Second Law states that the rate at which area is swept out by the radius vector is constant. The value assumed by this constant, although not directly relevant for the discussion of K2, is needed here. Consider,

$$\frac{dA}{dt} = \frac{|\vec{L}|}{2\,m} = \frac{|\vec{r} \times \vec{v}|}{2} = \frac{1}{2}\,r\,v_\perp = \frac{1}{2}\,r\,(\omega\,r) = \frac{1}{2}\,\omega\,r^2.$$

The moment arm formulation of the angular momentum cross-product led to the introduction of $v_\perp$, the component of the satellite velocity in the direction orthogonal to the radial vector. This component then finds expression in terms of the angular speed of the satellite [viewed from the perspective of the gravitating body at the focus]. The absence of torque about the focus, as discussed in Chapter 49, ensures that there be a constant of the motion. This constant, denoted "l," precisely equals $\omega\,r^2$. Also,

$$l^2 = G\,M\,R$$

follows from the radial component of N2.

Starting with the constancy of the rate at which areas are swept out,

$$\frac{dA}{dt} = \frac{l}{2},$$

the total area enclosed by the elliptical orbit is obtained by integrating through the time span corresponding to one complete orbital period. *I.e.*,

$$A = \int_0^T \frac{dA}{dt}\,dt = \frac{l}{2}\int_0^T dt = \frac{l\,T}{2}.$$

Taking the two separate determinations of the area bounded by the ellipse, squaring them (to clear square root factors), and equating the results yields:

$$\left.\begin{cases} A^2 = \pi^2\,a^4\,(1-\epsilon^2) \\[2mm] A^2 = \dfrac{l^2\,T^2}{4} = \dfrac{G\,M\,R\,T^2}{4} \end{cases}\right\} \qquad \Longrightarrow \qquad T^2 = \frac{4\,\pi^2}{G\,M}\,\frac{a^4\,(1-\epsilon^2)}{R} = \frac{4\,\pi^2}{G\,M}\,a^3.$$

Thus, $T^2 \propto a^3$ holds for generic elliptical orbits, and the constant of proportionality scales inversely with the mass of the gravitating body at the focus.

---

[4]**Apogee** and **aphelion** for satellites orbiting the Earth and Sun, respectively. The generic term is **apoapsis**.

[5]For terrestrial and solar orbits the minimal distances are called **perigee** and **perihelion**, respectively. Generically, the term used is **periapsis**.

Epilogue

In the very first chapter of this VOLUME, a dimensional argument determined salient aspects of the mathematical form of K3 [predicated solely on the notion that $G M_\odot$ is a significant quantity for solar system dynamics]. In the very last chapter, we derived K3, including all of its various factors, from Newton's Laws of Motion and the Law of Universal Gravitation.

> *Tell Dexeter, we've come full circle!*
> *Full Circle*, Doctor Who, Season 18, BBC Television

Let's indulge in one additional final word.

A budding biologist, an up-and-coming chemist, and a proto-physicist were walking back to their lodgings after an arduous study session [in preparation for finals] when they espied a cow in the university quad. They stopped and, owing to surprise and fatigue, each was unable to recollect the word "COW." After a moment's pause, all three reached into their respective knapsacks.

The biologist hauled a massive tome, COMPENDIUM ANIMALIA, *126th ed.*, from the depths of his bag and began grazing through the indices and charts and tables therein.

"Phylum Chordata, Subphylum Vertebrata ... Class Mammalia, ... hmmmm ... hmmmm ... It looks to be some sort of bovine."

CHOC FULL: THE COMPLETE HANDBOOK OF ORGANIC CHEMISTRY emerged from the chemist's bookbag and was opened to well-worn pages. Using several fingers as well as slips of paper to mark pages, she began flitting from one section to another, whilst ruminating:

"A complex series of biochemical processes .... yielding a large amount of methane ... It is clearly a living organism, and perhaps a ruminant."

The physicist withdrew a pad of paper and a mechanical pencil. Finding a blank page and surveying the scene, he pronounced:

"From here, I can model it as a sphere."

Part II

Mechanics Problems

K

Kinematics Problems

Useful Data

Unless otherwise specified, here, and throughout the other collections of problems:

SI   All times are measured in seconds, **s**; distances in metres, **m**; masses in kilograms, **kg**; forces in newtons, **N**; and angles in radians, **rad**.

$\vec{g}$ Take the magnitude of the acceleration due to gravity at the Earth's surface to be $10\,\mathrm{m/s^2}$, and its direction to be vertically downward.

c Take the speeds of sound and light through air to be $340\,\mathrm{m/s}$ and $3.0 \times 10^8\,\mathrm{m/s}$, respectively.

K.1   A **light-year** is defined to be that distance travelled by the leading edge of a light-beam [passing through vacuum spacetime] in one year.[1]

(a) Given that there are 60 seconds in a minute, 60 minutes in an hour, 24 hours in a day, and roughly 365.23 days in a year, compute the approximate duration of a year in seconds.

(b) The speed of light in vacuum is measured to be $c \simeq 2.998 \times 10^8$ metres per second. Compute the length of a light-year in metres. Comment.

K.2   The acre–foot is a unit often used in the measurement of large volumes of water. Determine the conversion factor from acre–feet to cubic metres, given that there are 2.54 centimetres in an inch, 12 inches in a foot, 5280 feet in a mile, and 640 acres in a square mile.

K.3   A furlong is a distance equal in length to one-eighth of a mile. A fortnight is a two-week time period. There are 2.54 cm in an inch, 36 inches in a yard, and 1760 yards in a mile.

(a) Determine the conversion factors   (i) from **m/s** to furlongs/fortnight, and
(ii) from furlongs/fortnight to **m/s**.

(b) Comment on the relation between these conversion factors.

K.4   The wavelength of yellow–orange light is about 500 nm. An object which is 5 μm across can be discerned under a light microscope. How many yellow–orange wavelengths would it take to span the extent of the object?

K.5   The density of water is commonly quoted as one gram per cubic centimetre. Convert this to kilograms per cubic metre.

---

[1] Snarky scientists opine that a light-year has fewer Calories than a regular year.

K.6 The density of seawater was measured to be $1.07\,\text{g/cm}^3$. Express this density in kilograms per cubic metre.

K.7 How far does a light beam advance in one picosecond?

K.8 Ascertain the number of cubic millimetres in one cubic metre.

K.9 Convert $90\,\text{km/h}$ to its equivalent in m/s.

K.10 Momentum, p, and kinetic energy, K, are physical quantities with dimensions

$$[\,p\,] = \left[\frac{\text{Mass} \times \text{Length}}{\text{Time}}\right] \quad \text{and} \quad [\,K\,] = \left[\frac{\text{Mass} \times \text{Length}^2}{\text{Time}^2}\right],$$

respectively. On the basis of dimensional analysis alone, propose a candidate formula expressing kinetic energy in terms of momentum and mass (m).

K.11 Angular momentum, L, moment of inertia, I, and kinetic energy, K, have dimensions

$$[\,L\,] = \left[\frac{\text{Mass} \times \text{Length}^2}{\text{Time}}\right], \quad [\,I\,] = \left[\text{Mass} \times \text{Length}^2\right], \quad [\,K\,] = \left[\frac{\text{Mass} \times \text{Length}^2}{\text{Time}^2}\right],$$

respectively. On the basis of dimensional analysis alone, suggest a candidate formula expressing kinetic energy in terms of angular momentum and moment of inertia.

K.12 Bernoulli's Equation (*circa* 1738), governing the flow of ideal fluids, states that the local fluid pressure plus two additional terms is everywhere equal to a phenomenological constant. The additional terms depend on factors of fluid density, ρ, speed, v, height above some reference point, h, and the magnitude of the local acceleration due to gravity, g. The units associated with each of these quantities are

$$[\,\rho\,] = \frac{\text{kg}}{\text{m}^3}, \quad [\,v\,] = \frac{\text{m}}{\text{s}}, \quad [\,h\,] = \text{m}, \quad [\,g\,] = \frac{\text{m}}{\text{s}^2}.$$

Employ dimensional analysis to propose the forms of the two distinct Bernoulli terms, given that the units of pressure are

$$[\,P\,] = \frac{\text{kg}}{\text{m} \cdot \text{s}^2}.$$

K.13 Quantities α, ω, and r have units s^{-2}, s^{-1}, and m, respectively. Another quantity, Q, with units m^2/s^2, is believed to be related to α, ω, and r, via

$$Q = \sqrt{r^a\,\alpha^b + r^c\,\omega^d},$$

for a set of exponents: $\{a, b, c, d\}$. Ascertain the values of the exponents.

K.14 Planck's constant, $\hbar \simeq 1.055 \times 10^{-34}\,\text{J} \cdot \text{s}$, sets the scale for the quantum aspects of nature. According to Einstein, the speed of light [*in vacuo*] has a constant value, $c \simeq 2.998 \times 10^8\,\text{m/s}$, in all inertial reference frames. Newton's Law of Universal Gravitation has a coupling constant, $G \simeq 6.67 \times 10^{-11}\,\frac{\text{N} \cdot \text{m}^2}{\text{kg}^2}$.

(a) Given that $1\,\text{N} = 1\,\text{kg} \cdot \text{m/s}^2$ and $1\,\text{J} = 1\,\text{N} \cdot \text{m}$, determine multiplicative combinations involving $\hbar$, c, and G, with dimensions of   (i) time,   (ii) length, and   (iii) mass.

(b) Evaluate these expressions to obtain estimates, in SI units, for the so-called Planck time, Planck length, and Planck mass. Comment.

K.15 In 2009, Usain Bolt sprinted 100.0 metres in 9.58 seconds.

(a) Determine his average speed in the race (in metres per second).

(b) Convert the speed obtained in (a) to kilometres per hour.

K.16 Newton thought of linearity in terms of "equal proportion increments" [geometrically, similar triangles]. Consider three functions: (a) $F_{(a)}(t) = at + f_0$, (b) $F_{(b)}(t) = bt^2 + f_0$, and (c) $F_{(c)}(t) = \frac{c}{t} + f_0$. Here, $\{a, b, c\}$ and f_0 are non-zero constants, and the values of the parameter t lie within the interval $T_i < t < T_f$. [We explicitly exclude from consideration the parameter value $t = 0$ in part (c).] Show how, for any allowed t_i and $t_f = t_i + \Delta t$, where $T_i < t_i < t_f < T_F$, Newton's notion of linearity applies, or fails to apply, to the functions (a), (b), and (c).

K.17 The locations of a particle [confined to move in one dimension] at selected times are given in the table below.

TIME [s]	0	5	10	20	25
POSITION [m]	10	17.5	7.5	−2.5	7.5

(a) Determine the displacement of the particle throughout each of the following time intervals: (i) $t = 0 \to 5$, (ii) $t = 5 \to 10$, (iii) $t = 10 \to 20$, (iv) $t = 20 \to 25$, (v) $t = 0 \to 25$, and (vi) $t = 0 \to 10$.

(b) Determine the average velocity of the particle through each of the time intervals considered in (a).

K.18 PK would like—someday—to walk the *Camino de Santiago* [Way of Saint James] pilgrimage route from Paris to Santiago de Compostela, Spain. To get into shape for this grand hike across a substantial part of Western Europe, PK plans to walk a distance of 1000 km in 25 km daily increments. He mapped out a straightline 12.5 km path and follows it, back and forth, six days a week (with every seventh day off to recover).

(a) How many days, including rest days, will it take PK to go 1000 km?

(b) Determine PK's estimated average speed (i) including only walking days and (ii) including rest days too for the thousand-kilometre training exercise.

(c) On the last day, PK's boots fell apart on the return leg of his walk when he was 1.5 km from the farthest extent of his path and he could walk no more. Determine PK's (i) actual average speed [in km/day], (ii) displacement, and (iii) average velocity throughout the entire duration of his training.

K.19
The adjacent figure shows the trajectory [*i.e.,* position *vs.* time] of a particle moving in one dimension. The position, x, is expressed in metres from a reference point, while the times are in minutes elapsed from $t = 0$.
(a) From the graph, determine the position of the particle at the instants (in minutes):
(i) $t = 4$, (ii) $t = 6$, (iii) $t = 8$,
(iv) $t = 10$, (v) $t = 12$, (vi) $t = 14$.

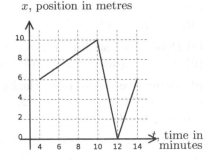

(b) Determine the particle's average velocity during the following time intervals:
(i) $4 \le t \le 10$, (ii) $4 \le t \le 12$, (iii) $4 \le t \le 14$, (iv) $10 \le t \le 12$,
(v) $10 \le t \le 14$, (vi) $12 \le t \le 14$.

(c) Estimate the average acceleration of the particle during the following time intervals:
(i) $6 \le t \le 11$, (ii) $6 \le t \le 13$, (iii) $11 \le t \le 13$, (iv) $9 \le t \le 11$.

K.20
The nearby figure shows the trajectory [po-
sition *vs.* time] of a particle moving in one
dimension. The position, x, is expressed
in metres from a reference point, while the
times are in seconds elapsed from $t = 0$.
(a) From the graph, determine the position
of the particle at the instants: (i) $t = 4$,
(ii) $t = 6$,   (iii) $t = 8$,   (iv) $t = 10$,
(v) $t = 12$,   (vi) $t = 14$.

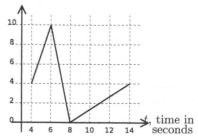

x, position in metres

time in seconds

(b) Determine the particle's average velocity during: (i) $4 \le t \le 6$ (ii) $4 \le t \le 8$,
(iii) $4 \le t \le 14$,   (iv) $6 \le t \le 8$,   (v) $6 \le t \le 14$,   (vi) $8 \le t \le 14$.

(c) Estimate the average acceleration of the particle during the following time intervals:
(i) $5 \le t \le 7$,   (ii) $7 \le t \le 9$,   (iii) $7 \le t \le 11$,   (iv) $5 \le t \le 11$.

K.21   A 5-hour train trip on a long straight stretch of track unfolds in the following manner.
For the first hour, the train travels with constant speed 60 km/h. During the next 30 minutes,
the speed is constant at 120 km/h. The following 90 minutes are spent at a 40 km/h pace,
while the speed during the final two hours is 20 km/h. [It is safe to assume that a negligible
amount of time is spent accelerating and decelerating as the speed of the train changes.]

(a) Sketch a graph of speed *vs.* time for the train.

(b) Determine the total distance travelled in the course of the 5-hour trip.

(c) Determine the average speed of the train in the time intervals (specified in hours
elapsed from the start):   (i) $0 \le t \le 5$,   (ii) $0 \le t \le 2$,   (iii) $0 \le t \le 4$,
(iv) $1 \le t \le 3$,   (v) $2 \le t \le 5$,   (vi) $4 \le t \le 5$.

(d) Sketch a graph of position *vs.* time during the trip.

K.22   A particle moves in one dimension along a trajectory given by

$$x(t) = \frac{1}{4}t^4 - \frac{1}{2}t^3 + 1, \quad \text{where } -1 \le t \le 4.$$

(a) Determine the position of the particle at: (i) $t = 0$,   (ii) $t = 1$,   (iii) $t = 2$,
(iv) $t = 3$.

(b) Determine the instantaneous velocity of the particle at: (i) $t = 0$,   (ii) $t = 1$,
(iii) $t = 2$,   (iv) $t = 3$.

(c) Determine the instantaneous acceleration of the particle at the times: (i) $t = 0$,
(ii) $t = 1$,   (iii) $t = 2$,   (iv) $t = 3$.

(d) Determine the average velocity of the particle in the following time intervals:
(i) $0 \le t \le 1$,   (ii) $0 \le t \le 2$,   (iii) $0 \le t \le 3$,   (iv) $1 \le t \le 2$,   (v) $1 \le t \le 3$,
(vi) $2 \le t \le 3$.

(e) Determine the average acceleration of the particle in the following time intervals:
(i) $0 \le t \le 1$,   (ii) $0 \le t \le 2$,   (iii) $0 \le t \le 3$,   (iv) $1 \le t \le 2$,   (v) $1 \le t \le 3$,
(vi) $2 \le t \le 3$.

(f)   (i) Infer an approximate value of the average acceleration throughout the time
interval $0.5 \le t \le 2.5$, employing the average velocity results obtained in part (d).
(ii) Justify, on the basis of common sense and a mathematical theorem, your attempt
in part (i).

K.23 A particle moves in one dimension along a trajectory described by the function

$$x(t) = 2\,t^3 + 3\,t^2 + 5\,t + 7.$$

(a) Determine the position of the particle at: (i) $t = -2$,   (ii) $t = 0$,   (iii) $t = 2$, (iv) $t = 4$.

(b) Determine the particle's average velocity during the intervals: (i) $-2 \le t \le 0$, (ii) $0 \le t \le 2$,   (iii) $2 \le t \le 4$,   (iv) $-2 \le t \le 4$.

K.24 A particle moves in one dimension along a trajectory described by the function

$$x(t) = 2\,t^3 + 3\,t^2 + 5\,t + 7.$$

Derive formulae for the particle's instantaneous   (a) velocity and   (b) acceleration.

K.25 A particle at the point $x = 2$ at time $t = 2$ is moving in the forward [positive x] direction.

(a) Sketch representative trajectories for this particle, were it to have: (i) constant velocity,   (ii) constant positive acceleration,   (iii) diminishing positive acceleration, (iv) constant negative acceleration,   (v) negative acceleration with diminishing magnitude.

(b) Repeat (a) for the case in which the particle is initially moving backward.

K.26 The trajectory of a particle moving in one dimension is illustrated below.

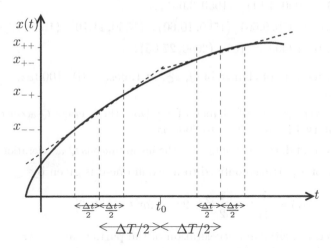

Formally, the positions of the particle at the times specified in the figure may be labelled (from left to right):

$$x_{--} = x\left(t_0 - \frac{\Delta T}{2} - \frac{\Delta t}{2}\right), \qquad x_{-+} = x\left(t_0 - \frac{\Delta T}{2} + \frac{\Delta t}{2}\right),$$

$$x_{+-} = x\left(t_0 + \frac{\Delta T}{2} - \frac{\Delta t}{2}\right), \qquad x_{++} = x\left(t_0 + \frac{\Delta T}{2} + \frac{\Delta t}{2}\right).$$

In words, "x_{-+} is the position of the particle at the instant $t = t_0 - \frac{\Delta T}{2} + \frac{\Delta t}{2}$."

(a) Determine the average velocity of the particle in the time interval of duration Δt centred on   (i) $t_0 - \frac{\Delta T}{2}$, and   (ii) $t_0 + \frac{\Delta T}{2}$.

(b) From the results for (a), estimate the average acceleration, a_{av}, throughout the time interval of duration ΔT centred on t_0.

(c) Specialise the formula in (b) to the case in which $\Delta T = \Delta t$, using

$$x_- = x(t_0 - \Delta t) \qquad x_0 = x(t_0) \qquad x_+ = x(t_0 + \Delta t),$$

and thereby obtain an estimate of the acceleration at time t_0.

K.27 Employ the formalism developed in the problem just above to estimate the average acceleration experienced by a particle whose trajectory is $x(t) = b t^3$, in an open time interval including $1 \le t \le 2$.   [CAVEAT: The results obtained in this question simplify in an unrealistically spectacular manner. The intention is to show consistency (internally, with K.26, and with the example found at the end of Chapter 2) without complexity and/or approximation. For a somewhat less trivial analysis, try $x(t) = b t^4$ on the interval $t \in [1, 2]$.]

(a) Set $t_0 = 1.5$, $\Delta T = 1$, and either take the limit $\Delta t \to 0$, or fix it at some small value, *i.e.*, $\Delta t = 0.002$.

(b) (i) A crude estimate of the instantaneous acceleration at $t = 1.5$ s is obtained by setting $\Delta T = 1 = \Delta t$.   (ii) Refine the estimate by taking $\Delta T = 0.02 = \Delta t$.

(c) Compute the instantaneous acceleration at $t = 1.5$ s (by taking the second derivative of the trajectory at this instant). Comment on your result.

K.28 An experimental device provides digital output of its measurements of a quantity, Q, for computerised collection and analysis. The output stream consists of pairs of numbers (time in milliseconds , Q). Quoted below are three snippets from the avalanche of data.

$$\ldots,\ (950, 2.70),\ (1000, 3.14),\ (1050, 3.65),\ldots$$
$$\ldots,\ (1400, 8.60),\ (1450, 9.60),\ (1500, 10.60),\ (1550, 11.70),\ (1600, 12.85),\ldots$$
$$\ldots,\ (1950, 23.30),\ (2000, 25.15),\ (2050, 27.05),\ldots$$

(a) Estimate the time rate of change of Q, v_Q, at times: (i) 1000 ms,   (ii) 1500 ms, and   (iii) 2000 ms.

(b) Estimate the average rate of change of v_Q (*i.e.*, the average Q-acceleration, $a_{av,Q}$), throughout the interval from 1000 ms to 2000 ms.

(c) Estimate the second derivative of Q (*i.e.*, the instantaneous Q-acceleration), at 1500 ms.

K.29 The trajectory of a particle confined to move in one-d is given by

$$x(t) = \frac{1}{24} t^4 + \frac{1}{2} t^2 + 2t, \quad \text{for } t \in [-2, 3].$$

Determine the position, velocity, and acceleration of the particle at $t = 2$ s.

K.30 [See also K.50.] A particle confined to one-d moves according to

$$x(t) = 3\, \frac{\sin(\pi t)}{t}, \quad t > 0.$$

Compute the   (a) position,   (b) velocity, and   (c) acceleration   of the particle at the instants: (i) $t = 1$,   (ii) $t = 2$,   (iii) $t = 5$, and also   (iv) in the limit as $t \to \infty$.

K.31 A particle moving in one-d has trajectory $x(t) = A e^{-\gamma t} \cos(\omega t + \phi)$, where A, γ, ω, and ϕ are all positive, real-valued constants.

(a) Ascertain the dimensions of A, γ, ω, and ϕ.

(b) Determine the velocity and the acceleration functions for the particle.

K.32 A particle, confined to 1-d, has trajectory $x(t) = V_0\, t \exp\left(-a\left(t^2 - t_0^2\right)^2\right)$, for positive real constants $\{V_0, a, t_0\}$.

(a) Ascertain the dimensions of each of the constants. Comment.

(b) Determine the particle's velocity and the acceleration functions.

(c) Determine the limiting behaviour of the trajectory in the three cases: (i) $t \to -\infty$, (ii) $t \to 0$, and (iii) $t \to +\infty$.

(d) Determine the velocity as (i) $t \to -\infty$, (ii) $t \to 0$, and (iii) $t \to +\infty$.

(e) Determine the acceleration as (i) $t \to -\infty$, (ii) $t \to 0$, and (iii) $t \to +\infty$.

K.33

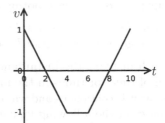

The adjacent sketch shows a graph of velocity *vs.* time for a particle confined to move in one-d. At $t = 0\,$s the particle is located at the origin, $x = 0\,$m.

(a) Infer the velocity of the particle at the times: (i) $t = 0$, (ii) $t = 4$, (iii) $t = 5$, (iv) $t = 8$, (v) $t = 10$.

(b) Estimate the particle's average acceleration during the intervals: (i) $0 \le t \le 4$, (ii) $0 \le t \le 5$, (iii) $0 \le t \le 8$, (iv) $0 \le t \le 10$.

(c) Estimate the instantaneous acceleration of the particle at the instants: (i) $t = 3$, (ii) $t = 5$, (iii) $t = 8$.

(d) At what time(s) is the particle at rest?

K.34 Average velocities, during successive one-second intervals, of a particle [moving in 1-d] are exhibited in the histogram below. Data for the $3 \to 4$ second interval is missing.

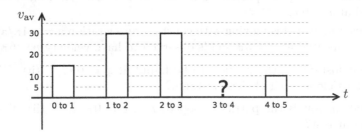

(a) Determine the displacement of the particle in each of the intervals for which data is available: (i) $0 < t < 1$, (ii) $1 < t < 2$, (iii) $2 < t < 3$, (iv) $4 < t < 5$.

(b) At time $t = 5$ it was noted that the particle was exactly 100 m from its starting point (at $t = 0\,$s). Determine the particle's average velocity throughout the time interval $0 < t < 5\,$s.

(c) Infer the average velocity of the particle in the time interval $3 < t < 4\,$s.

K.35 During a thunderstorm, you happen to be standing alongside a long straight segment of railroad track. Off in the distance is a water tower, also beside the track. A bolt of lightning hits the tower just as a train passes by it. At the sight of the flash, you begin to count seconds. At $t = 5\,$s, you hear the rumble of thunder, and at $t = 68\,$s, the train passes by you. [Assume at first that the light reaches you instantaneously.]

(a) Through what distance did the train travel from the water tower to alongside you?

(b) What was the average speed of the train as it moved from the tower to you?

(c) Roughly how long did the light take to reach your position? Comment.

K.36 You are watching Joe Athlete and Akira Ateretesaki compete in a biathlon from a vantage point alongside a straight section of the ski trail approaching a shooting area. Just as Joe passes by you, you see the muzzle flash from Akira's gun. One and one-half seconds later you hear the sound of the shot. Fifty-one seconds after the flash, Joe reaches the rifle range. [Hint: First, assume that the light from the muzzle flash reaches you instantaneously.]

(a) How far are you from the rifle range?

(b) What was Joe's average speed (in metres per second) during the time that he skied from your position to the shooting area?

(c) Roughly how long did it take the light from the muzzle flash to travel from the shooting range to your position? Comment.

K.37 A graph displays the instantaneous velocity of a particle moving in one dimension as a function of time for times in the interval $0 \le t \le t_f$. [This particular curve appears to be differentiable, as it is continuous and without cusps or kinks.] Describe in words how one might obtain estimates of the following kinematic quantities from the graph:

(a) The displacement throughout the time interval $t = 0 \to t_f$

(b) The velocities at $t = 0$, $t = t_f$, and any intermediate time

(c) The average velocity throughout the time interval $t = 0 \to t_f$

(d) The average acceleration throughout the time interval $t = 0 \to t_f$

(e) The instantaneous acceleration at any intermediate time $0 < t < t_f$

K.38 A particle residing in one dimension is observed to be moving with constant velocity $v_0 = -3\,\mathrm{m/s}$. At time $t = 5\,\mathrm{s}$, the particle is located at $x = 6\,\mathrm{m}$. At what instant is the particle to be found at $x = 0\,\mathrm{m}$?

K.39 A particle residing in 1-d is moving with constant acceleration $a = -4\,\mathrm{m/s^2}$. At time $t = 2\,\mathrm{s}$, the particle is located at $x = 3\,\mathrm{m}$ and is observed to have velocity $v = 6\,\mathrm{m/s}$.

(a) (i) At what instants is the particle to be found at $x = 0\,\mathrm{m}$?   (ii) What is the velocity of the particle at each of these instants?

(b) (i) At what instant is the particle stopped, *i.e.*, $v = 0\,\mathrm{m/s}$?   (ii) Where is the particle (briefly) at rest?

K.40 A particle moves in 1-d with constant acceleration $a_0 = 6\,\mathrm{m/s^2}$. At time $t = 0$ s the particle is located at $x = 6$ m, and it is moving with velocity $v_0 = -12$ m/s.

(a) Express the particle's   (i) velocity and   (ii) position   as functions of time.

(b) Ascertain the position of the particle at times
      (i) $t = 1\,\mathrm{s}$,   (ii) $t = 2\,\mathrm{s}$,   (iii) $t = 3\,\mathrm{s}$, and   (iv) $t = 4\,\mathrm{s}$.

(c) (i) Does the particle's trajectory pass through the point $x = 0$?   (ii) If the answer to (i) is 'yes', then determine the times at which crossings through $x = 0$ occur. If 'no', then determine the distance of closest approach to the origin.

K.41 A particle confined to move in one dimension has constant acceleration $a = +4\,\mathrm{m/s^2}$. At time $t = 0$ s, the particle is located at $+5\,\mathrm{m}$ with respect to the origin and is moving with velocity $-1\,\mathrm{m/s}$.

(a) Determine the position and velocity of the particle as functions of time, t.

(b) Sketch the acceleration, velocity, and position functions.

K.42 A particle moves in one dimension with constant acceleration $a = 4\,\mathrm{cm/s^2}$.

(a) Suppose that at $t = 0$ s the particle is at $x = 2$ cm and is moving with velocity $v = 3$ cm/s. Determine the position and velocity of the particle at: (i) $t = -2$, (ii) $t = 0$,   (iii) $t = 2$,   (iv) $t = 4$.

(b) Suppose instead that at $t = 1$ s the particle is at $x = 5$ cm and is moving with velocity $v = -3$ cm/s. Determine the position and velocity of the particle at times:
(i) $t = -2$,   (ii) $t = 0$,   (iii) $t = 2$,   (iv) $t = 4$.

K.43 A particle confined to one dimension has a time-varying acceleration described by

$$a(t) = 12\,t^2 + 18\,t + 2, \quad \text{for } t \in [0\,, 10]\,.$$

At $t = 0$, the particle happens to be at rest at the origin.

(a) Derive formulae for the velocity and the position of the particle [valid for $t \in [0\,, 10]$].

(b) Ascertain the position, velocity, and acceleration of the particle at $t = 2$ s.

K.44 A particle confined to one dimension has trajectory:

$$x(t) = \begin{cases} t^3 - \frac{1}{8}t^4\,, & -1 < t < 2 \\ -26 + 24\,t - 4t^2\,, & 2 < t < 4 \end{cases}.$$

(a) Determine the position of the particle at   (i) $t = 0$,   (ii) $t \to 2^-$,   (iii) $t \to 2^+$, and   (iv) $t = 3$.

(b) Compute the velocity of the particle for all times in the interval $-1 < t < 4$, except at $t = 2$.

(c) Determine the velocity of the particle at: (i) $t = 0$,   (ii) $t \to 2^-$,   (iii) $t \to 2^+$, and   (iv) $t = 3$.

(d) Compute the acceleration of the particle for all times in the interval $-1 < t < 4$, except at $t = 2$.

K.45 A particle moving in one-d experiences the time-varying acceleration
$$a(t) = \{\ 0, \quad t < 0; \quad 2 - \tfrac{2}{3}t, \quad 0 \le t \le 6; \quad 0, \quad 6 < t\ \}.$$
[It is assumed that the units are consistent in this expression for the acceleration.]

(a) Suppose that the particle is at rest at the origin at $t = 0$. Determine the particle's (i) velocity and   (ii) trajectory   in each of the three regimes, *i.e.*, for $t < 0, 0 \le t \le 6$, and $6 < t$.

(b) Suppose that the particle is moving past $x = 3$ m with velocity 2 m/s [forward] at $t = 0$. Determine the particle's   (i) velocity and   (ii) trajectory   in each regime.

K.46 A Triumph Spitfire and a sensible sedan are side by side and at rest at a stoplight. At $t = 0$ s, the light changes and the cars both accelerate at 5 m/s^2 [forward]. The sedan and the Spitfire accelerate for four and five seconds, respectively, after which each maintains a constant velocity.

(a) Determine the speed of the sedan at times   (i) $t = 4$ s,   (ii) $t = 5$ s,   and (iii) $t = 10$ s.

(b) Ascertain the speed of the Spitfire at times   (i) $t = 4$ s,   (ii) $t = 5$ s,   and (iii) $t = 10$ s.

(c) Determine the distance through which the sedan has travelled at time   (i) $t = 4$ s, (ii) $t = 5$ s, and   (iii) $t = 10$ s.

(d) Ascertain the distance through which the Spitfire has travelled at time   (i) $t = 4$ s, (ii) $t = 5$ s, and   (iii) $t = 10$ s.

(e) For each of the two cars, sketch graphs of   (i) velocity *vs.* time and   (ii) position *vs.* time.

K.47   A Triumph Spitfire and a goods-laden lorry [transport truck] are side by side and at rest at a stoplight. At $t = 0$ s, the light changes. The Spitfire nimbly accelerates forward at $5\,\mathrm{m/s^2}$ for 6 s, while the lorry lumbers forward at $2\,\mathrm{m/s^2}$ for a total of 15 s. After each vehicle's acceleration phase, it maintains a constant velocity.

(a) Ascertain the velocity of the Spitfire in the intervals   (i) $0 \leq t \leq 6$,   (ii) $t > 6$.

(b) Ascertain the displacement of the Spitfire for   (i) $0 \leq t \leq 6$,   (ii) $t > 6$.

(c) Sketch graphs illustrating the acceleration, velocity, and displacement of the Spitfire as functions of time for $0 \leq t \leq 20$.

(d) Determine the lorry's velocity in the time intervals   (i) $0 \leq t \leq 15$,   (ii) $t > 15$.

(e) Determine the lorry's displacement in the intervals   (i) $0 \leq t \leq 15$,   (ii) $t > 15$.

(f) Sketch graphs illustrating the acceleration, velocity, and displacement of the lorry as functions of time for $0 \leq t \leq 20$.

K.48   PK thought it a lark to hang a banner reading *"Poor Spellers of the World, Untie!"* high on a radio transmission tower. Pausing for breath on a landing 125 m above ground, PK accidently dropped the knapsack containing the tightly rolled-up banner over the side. [Assume that the knapsack falls straight downward under the influence of gravity only (air resistance is insignificant), and that the knapsack was at rest on the platform just before it fell.] Determine the knapsack's velocity and its displacement from the platform at the following times (after it began to fall).

(a) $t = 1$ s   (b) $t = 2$ s   (c) $t = 3$ s   (d) $t = 4$ s   (e) $t = 5$ s

(f) Sketch a histogram illustrating the distance fallen by the knapsack in successive one-second time intervals, *i.e.,* 0–1, 1–2, 2–3, 3–4, and 4–5. Comment.

K.49   Three particles confined to move in 1-d have the velocity functions listed below:

$$v_s = 2 + \tfrac{1}{3}t^2 \qquad\qquad v_l = 2 + t \qquad\qquad v_r = 2 + \sqrt{3\,t}$$

(a) Show that $v_s = v_l = v_r$,   (i) as $t \to 0$ and   (ii) as $t \to 3$.

(b) Sketch these curves and verify that, everywhere throughout $0 < t < 3$, $v_s < v_l < v_r$.

(c) The average velocity throughout the time interval is obtained by integrating $v(t)$ from $t = 0$ to $t = 3$ and dividing the result by the duration, *viz.,* 3 s. Do this for each of the three particles.

(d) The average velocity is, kinematically, $\Delta x/\Delta t$. What inference can be made about the relative displacement of each of the particles during the time interval under consideration? [*Cf.* the Addendum to Chapter 3.]

K.50   [See also K.30.] A particle moving in one dimension has the trajectory

$$x(t) = X_0 \,\frac{\sin(\pi\, t)}{t}, \quad \text{for } t > 0 \text{ and constant } X_0 < 0.$$

(a) Determine the particle's position, velocity, and acceleration at:
    (i) $t = 1/2$,   (ii) $t = 1$,   (iii) $t = 3/2$,   (iv) $t = 2$.

(b) Sketch the position as a function of time for $0 < t < 5$.

K.51 A particle residing in one dimension moves in a manner described by

$$x(t) = X_0 \, e^{-\gamma t}, \quad \text{where } X_0 = 5 \, \text{m} \text{ and } \gamma = \frac{1}{20} \, \text{s}^{-1}.$$

(a) Sketch $x(t)$ *vs.* t for $t \in [0, 60]$.

(b) Determine (i) $v(t)$ and (ii) $a(t)$ for the particle.

(c) Determine the instant at which the particle passes through the point $x = 1 \, \text{m}$.

K.52 A particle residing in one dimension moves in a manner described by

$$x(t) = X_0 \, e^{-\gamma t} \sin(\omega t), \quad \text{where } X_0 = 5 \, \text{m}, \ \gamma = \frac{1}{20} \, \text{s}^{-1}, \text{ and } \omega = \frac{\pi}{6} \, \text{rad/s}.$$

(a) Sketch $x(t)$ *vs.* t for $t \in [0, 30]$.

(b) Determine (i) $v(t)$ and (ii) $a(t)$ for the particle.

(c) Estimate the time after which the particle is confined within a band of width $4 \, \text{m}$, centred on the origin [*viz.*, the earliest time, t_2, such that $|x(t)| < 2$ for all $t > t_2$].

K.53 A particle has a velocity described by the function $v(t) = X_0/\sqrt{a^2 + t^2}$, where X_0 and a are both positive constants, for times $0 < t < 10 \, a$.

(a) Work out the dimensions of the constants.

(b) Describe the behaviour of the velocity function (for $t > 0$).

(c) Compute the particle's displacement in intervals: (i) $[3 \, a, \, 4 \, a]$ and (ii) $[6 \, a, \, 7 \, a]$.

(d) Are your results for (b) and (c) consistent?

K.54 A particle has a velocity described by the function $v(t) = X_0/\sqrt{a^2 - t^2}$, where X_0 and a are both positive constants, for times $0 < t < a$.

(a) Work out the dimensions of the constants.

(b) Describe the behaviour of the velocity function (for $t > 0$).

(c) Compute the particle's displacement for intervals: (i) $[\frac{a}{4}, \frac{a}{3}]$ and (ii) $[\frac{2a}{3}, \frac{3a}{4}]$.

(d) Are your results for (b) and (c) consistent?

K.55 A particle found at rest at the origin at $t = 0 \, \text{s}$ experiences an acceleration,

$$a(t) = 2 \, e^{-t/3}, \quad \text{subsequent to } t = 0.$$

(a) Determine the (i) velocity and (ii) position of the particle for $t > 0$.

(b) Characterise the behaviour of the (i) velocity and (ii) position as $t \to \infty$.

K.56 A particle, initially at rest at the origin, subsequently experiences an acceleration $a(t) = A_0 \, e^{-t/\tau}$, where both A_0, the initial acceleration, and τ, establishing the time scale, are positive constants.

(a) Determine the (i) velocity and (ii) position of the particle for $t > 0$.

(b) Characterise the behaviour of the (i) velocity and (ii) position as $t \to \infty$.

K.57 A particle confined to move in one dimension is at rest at the origin at $t = 0 \, \text{s}$. From $t = 0$ to $t = 2 \, \text{s}$ the particle experiences an acceleration of $2 \, \text{m/s}^2$ in the positive x-direction. From $t = 2 \, \text{s}$ onward, the magnitude of the acceleration is doubled and its direction reversed, *i.e.*, $a_{II} = -4 \, \text{m/s}^2$.

(a) Sketch the acceleration as a function of time for the domain $0 \le t \le 5$.

(b) Determine both the velocity and the position of the particle at times: (i) $t = 0$,
(ii) $t = 1$,   (iii) $t = 2$,   (iv) $t = 3$,   (v) $t = 4$,   (vi) $t = 5$.

K.58
A particle which is at rest at the
origin at time $t = 0$ s is subjected
to the piecewise constant acceleration displayed in the figure and
expressed in the formula.

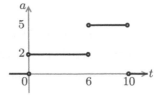

$$a(t) = \begin{cases} 0, & t < 0 \\ 2, & 0 < t < 6 \\ 5, & 6 < t < 10 \\ 0, & 10 < t \end{cases}$$

Determine the trajectory of the
particle, *i.e.*, $x(t)$.

K.59
A particle which is at rest at the
origin at time $t = 0$ s is subjected to the piecewise acceleration displayed in the figure and
expressed in the formula.
Determine the trajectory of the
particle, *i.e.*, $x(t)$.

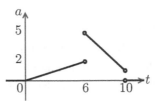

$$a(t) = \begin{cases} 0, & t < 0 \\ \frac{1}{3}t, & 0 < t < 6 \\ 11 - t, & 6 < t < 10 \\ 0, & 10 < t \end{cases}$$

K.60   A particle constrained to move in one dimension is at rest at the origin until $t = 0$ s,
at which time it is subjected to a constant acceleration, $a_I = 2\,\text{m/s}^2$. At $t = 3$ s, the
acceleration suddenly changes to $a_{II} = -3\,\text{m/s}^2$, and remains so until $t = 5$ s, after which
the acceleration vanishes. Determine the position and velocity of the particle at $t = 5$ s.

K.61   A particle constrained to move in one dimension is at rest at the origin until $t =
0$ s, at which time it is subjected to an increasing acceleration, $a_I = t$. At $t = 2$ s, the
acceleration suddenly begins to diminish, $a_{II} = -6 + 2\,t$. At $t = 3$ s, the acceleration has
dropped to zero and remains zero at later times, $a_{III} = 0$. Determine the position and
velocity of the particle at $t = 5$ s.

K.62   A particle is at rest at the origin at time $t = 0$. In the time interval from $t = 0$ to
$t = 2$ seconds, the particle accelerates at $2 - t$ metres per second-squared, while from $t = 2$
seconds to $t = 3$ seconds, it decelerates at $2\,(t - 2)$ metres per second-squared. Determine
the   (a) velocity   and   (b) position   of the particle at: (i) $t = 0$,   (ii) $t = 1$,
(iii) $t = 2$, and   (iv) $t = 3$.

K.63
A particle constrained to move in one dimension experiences variable acceleration $a(t)$, defined nearby. At $t = 0$ s, the particle is found
to be at rest at the point $x = 2$ m. Determine the
position of the particle at $t = 4$ s.

$$a(t) = \begin{cases} -3t^2 & 0 \le t \le 2 \\ 0 & 2 < t \le 3 \\ 1 & 3 < t \le 4 \\ 0 & 4 < t, \text{ and } t < 0 \end{cases}$$

K.64   A unit vector in the xy-plane lies at an angle $\pi/6$ with respect to the positive x-axis.
Determine the x- and y-components of this vector.

K.65   A velocity vector has xy-components $\vec{v} = (5, -12)\,\text{m/s}$.   (a) Determine the magnitude of this vector.   (b) Express its direction in terms of the angle that it makes with
respect to the positive x-axis.

K.66   Two vectors lying in a 2-d Euclidean space have Cartesian components $\vec{R} = (12, 5)$
and $\vec{S} = (4, -3)$. Determine the Cartesian components of the following vectors.

(a) $\vec{A} = \vec{R} + \vec{S}$ (b) $\vec{B} = \vec{R} - \vec{S}$ (c) $\vec{C} = 3\vec{R} + 5\vec{S}$
(d) $\vec{D} = (\vec{R} \cdot \vec{S})\vec{R}$ (e) $\vec{E} = (\vec{R} \cdot \hat{\imath})\,\hat{\imath} + (\vec{S} \cdot \hat{\jmath})\,\hat{\jmath}$ (f) $\vec{F} = (\vec{R} \cdot \hat{\jmath})\,\hat{\imath} + (\vec{S} \cdot \hat{\imath})\,\hat{\jmath}$

K.67   Two velocity vectors [residing in 2-d Euclidean space] have magnitudes $|\vec{R}| = 5\,\text{m/s}$

and $\left|\vec{S}\right| = 13\,\text{m/s}$, respectively. The acute angle lying between the vectors, when they are placed tail to tail, is $\theta = \pi/6\,\text{rad}$.

(a) Compute $\vec{R} \cdot \vec{S}$.

(b) Determine the component of $\vec{R}$ in the direction of $\vec{S}$.

K.68 Compute the (a) cross and (b) dot products of the following pairs of vectors.
(i) 13 [in xy-plane, at 22.62° WRT x-axis] and 5 [in xy-plane, at 53.13° WRT y-axis],
(ii) $(12, 5, 0)$ and $(-4, 3, 0)$, (iii) 6 [↓] and 5 [→].

K.69 Consider two vectors in three dimensions, $\{\vec{a}, \vec{b}\}$, where $\vec{a}$ has Cartesian components $(1, -1, 0)$, while $\vec{b}$ has length $\sqrt{2}$ and is directed in the xy-plane at 45° from the x-axis. Determine the following quantities.

(a) $\vec{a} + \vec{b}$, (b) $\vec{a} - \vec{b}$, (c) $\vec{a} \cdot \vec{b}$, (d) $\vec{a} \times \vec{b}$.

K.70 Eight 2-d vectors, expressed WRT a Cartesian basis, are enumerated below.

$$\vec{A} = \left(\sqrt{2}, 0\right) \qquad \vec{B} = (1, 1) \qquad \vec{C} = \left(0, \sqrt{2}\right) \qquad \vec{D} = (-1, 1)$$
$$\vec{E} = \left(-\sqrt{2}, 0\right) \qquad \vec{F} = (-1, -1) \qquad \vec{G} = \left(0, -\sqrt{2}\right) \qquad \vec{H} = (1, -1)$$

Without doing any explicit computations, determine the sum of all of these vectors:

$$\vec{S} = \vec{A} + \vec{B} + \vec{C} + \vec{D} + \vec{E} + \vec{F} + \vec{G} + \vec{H} = \ ?$$

[Be sure to explain your reasoning.]

K.71 Prove the TRIANGLE INEQUALITY: $\left|\vec{A} + \vec{B}\right| \le \left|\vec{A}\right| + \left|\vec{B}\right|$, ∀ vectors $\{\vec{A}, \vec{B}\}$.

K.72 Two vector quantities, $\{\vec{A}, \vec{B}\}$, whose directions can be altered at will, have a sum, $\vec{A} + \vec{B}$, whose magnitude falls between 0 and 200 (in appropriate units). Determine the magnitudes of $\vec{A}$ and $\vec{B}$.

K.73 Five vectors, expressed in terms of Cartesian components, are as follows:

$$\vec{A} = (5, 1) \qquad \vec{B} = (4, -2) \qquad \vec{C} = (3, 3) \qquad \vec{D} = (2, -4) \qquad \vec{E} = (1, 5).$$

(a) Plot (and label) each of these vectors (in a suitable xy-plane).

(b) Compute the length of each of the vectors, and its angle relative to the positive x-axis. Express the angles in both radians and degrees.

(c) Express the sum $\vec{F} = \vec{A} + \vec{B} + \vec{C} + \vec{D} + \vec{E}$, in terms of its (i) Cartesian coordinates and (ii) magnitude and direction.

(d) Express the sum $\vec{G} = \vec{A} - 2\vec{B} + 3\vec{C} - 4\vec{D} + 5\vec{E}$, in terms of its (i) Cartesian coordinates and (ii) magnitude and direction.

K.74 Six vectors, $\vec{V}_{1\text{-}6}$, each of length 6 cm, reside in a plane. The angles that these vectors make with the x-direction are:

$$\theta_1 = 30°, \qquad \theta_2 = 90°, \qquad \theta_3 = 150°, \qquad \theta_4 = 210°, \qquad \theta_5 = 300°, \qquad \theta_6 = -30°.$$

(a) Sketch (and label) each of these vectors in an xy-plane.

(b) Determine the Cartesian coordinates of each of the vectors.

(c) The vectors are arranged in such a way that the second begins at the tip of the first, the third begins at the tip of the second, and so on. The first vector originates at the origin. Determine the position of the tip of the sixth vector in terms of its
(i) Cartesian coordinates and (ii) magnitude and direction.

(d) How might one, by changing the angle of one vector only, make it so that, when the vectors are combined as in part (c), the tip of vector six lies at the origin?

K.75   PK went adventuring yesterday. At 0900 h he was cavorting in a park, 50 km due west of the campus, and two hours later, at 1100 h, he was bird watching in the forest, 50 km south of the campus. Determine the direction of PK's average velocity throughout the two-hour time interval from 0900 h to 1100 h.

K.76   PK went skiing on a fine winter's day. At 9:00 AM he was schussing 3 km south of the chalet, and at 3:00 PM he was mogul-bashing 2 km north of the chalet. Determine the magnitude and direction of PK's average velocity during those six glorious hours from 9:00 to 3:00.

K.77   On the day following his skiing exertions, PK went to the library. At 9:00 AM he was "shushing"[2] in the new periodicals area, and at 3:00 PM he was immersed in the stacks, two floors directly above the periodicals. Determine the magnitude and direction of PK's average velocity during those six illuminating hours from 9:00 to 3:00.

K.78   Whilst wrestling with a calculation, PK hoped that a change of scenery and a good cup of coffee might provide the requisite inspiration. He rode his bicycle to a nearby coffee shop, quaffed a decaf double espresso, and returned, reinvigorated, to his office. Determine the magnitude and direction of PK's average velocity from the time he got up from his desk to the time of his return.

K.79   Avenues and streets in Fargo, North Dakota, as in many cities, are laid out in an approximate rectangular grid. Streets run north–south and are each separated by 1/12 of a mile, while avenues run east–west and are spaced 1/6 of a mile apart.

(a) What distance must one walk to get from the corner of Fifth Street West and Eleventh Avenue North to the corner of Fourteenth Street West and Thirteenth Avenue North?

(b) What is the distance between these two corners "as the crow flies?"

K.80   A ship sails due north-west at $20\sqrt{2}$ km/h for two hours and then north at 20 km/h for an additional two hours.

(a) Through what distance did the ship travel during those four hours?

(b) What displacement did the ship undergo in the four-hour time interval?

(c) Determine the average speed of the ship during those four hours.

(d) What was the average velocity of the ship throughout its four-hour voyage?

K.81   A ship sails due north at 20 km/h for two hours and then at 36.87° west of north at 25 km/h for three hours.

(a) Through what distance did the ship travel during those five hours?

(b) What displacement did the ship undergo in the five-hour time interval?

(c) Determine the average speed of the ship throughout the five hours.

(d) What was the average velocity of the ship throughout its five-hour voyage?

K.82   A ship sails due north-east at 30 km/h for two hours and then north at 20 km/h for an additional two hours.

(a) Through what distance did the ship travel during those four hours?

(b) What displacement did the ship undergo in the four-hour time interval?

(c) Determine the average speed of the ship during those four hours.

(d) What was the average velocity of the ship throughout its four-hour voyage?

---

[2]Being shushed is more likely!

K.83 A particle moves in 2-d with constant velocity $\vec{v}_0 = (1, -2)$ m/s. At time $t = 0$ s the particle is located at $(5, 10)$ m.

(a) Express the position of the particle as a function of time, t.

(b) Plot   (i) $r_x(t)$ vs. t and $r_y(t)$ vs. t.

(c) Plot the trajectory of the particle in xy-space.

(d) Does the particle's trajectory pass through the point $(100, -190)$?

(e) Whereabouts along the line $x = 250$ ought one stand to intercept the particle, and when will this occur?

K.84 A particle moves in 3-d with constant velocity $\vec{v}_0 = (2, -3, 4)$ m/s. At $t = 0$ the particle is found at the origin, $\vec{r}_0 = (0, 0, 0)$ m.

(a) Determine the speed (magnitude of the velocity) of the particle.

(b) Determine the position of the particle at $t = 2$ s.

(c) Determine where the particle crosses the plane $z = 10$ m.

K.85 A particle moves in 3-d with constant acceleration $\vec{a}_0 = (12, -12, 6)$ cm/s^2. At $t = 0$, the particle is instantaneously at rest, $\vec{v}_0 = (0, 0, 0)$ cm/s, at $\vec{r}_0 = (0, 24, 17)$ cm.

(a) Determine the magnitude of the acceleration.

(b) Determine the   (i) velocity and   (ii) position   of the particle at $t = 1$ s.

(c) Determine where the particle crosses the xz-plane, *i.e.*, $y = 0$, after $t = 0$.

K.86
Brad Pitt (B), his sisters Olive (O) and Peach (P), and his brother Mosh (M), attended a gala film screening and cast party at the Toronto International Film Festival. Post-party, each returned to his or her hotel as indicated on the map.

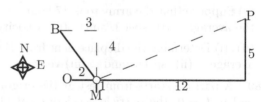

(a) Peach set out by taxi. Upon arrival at her destination, the taxi's trip odometer/timer read 17 km distance and 39 minutes duration. Determine Peach's   (i) average speed, (ii) displacement, and   (iii) average velocity.

(b) Olive walked the 2 km distance in one half-hour. Determine Olive's   (i) average speed,   (ii) displacement, and   (iii) average velocity.

(c) Mosh got lost several times, taking five hours to reach his hotel located 1 km from the cinema. All told, he wandered a distance of 5 km. Determine Mosh's   (i) average speed,   (ii) displacement, and   (iii) average velocity.

(d) Brad's helicopter flew high above the buildings, traffic, and crowds, going directly to his hotel located 4 km north and 3 km west of the gala venue. His short hop took five minutes. Determine Brad's   (i) average speed,   (ii) displacement, and   (iii) average velocity.

K.87
The three Spears sisters: Asparagus (A), Brittany (B), and Shake (S) met in Las Vegas for a celebratory singing gig. A map showing the concert hall and their respective hotels is shown nearby.

(a) Asparagus set out by taxi to the concert hall. The trip (through heavy traffic) took 35 minutes. Upon arrival, the taxi's odometer read 7 km. Determine Asparagus'   (i) average speed,   (ii) displacement, and   (iii) average velocity.

(b) Brittany took a helicopter and travelled directly to the concert hall, a distance of 4 km, in two minutes. Determine Brittany's   (i) average speed,   (ii) displacement, and   (iii) average velocity.

(c) Shake walked to the hall in four hours, getting lost several times along the way. All told, she walked a distance of 10 km, though her hotel was only 1 km away. Determine Shake's   (i) average speed,   (ii) displacement, and   (iii) average velocity.

K.88   *Macbeth shall never vanquish'd be until, / Great Birnam wood to high Dunsinane hill, / Shall come against him*   Macbeth Act IV, scene 1

The nearby figure shows the positions of Birnam forest (B), the castle Dunsinane (D), and a river ford (F). The forces arrayed against Macbeth collected boughs at Birnam and thence advanced to attack Dunsinane. The road from Birnam to the ford runs straight north for 4 km and then due east for 3 km. Dunsinane is 12 km from the ford at a heading of 36.87° south of east.

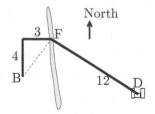

(a) Determine the   (i) distance travelled by the soldiers,   (ii) magnitude of the displacement, and   (iii) direction,   from Birnam to the ford.

(b) Suppose that the army took 35 hours to travel from Birnam to the ford. Compute the average   (i) speed, and   (ii) velocity   of the army during this time interval.

(c) Suppose that the army took 17 hours to travel from the ford to the castle. Compute the average   (i) speed, and   (ii) velocity   of the army during these 17 hours.

(d) (i) Determine the displacement from Birnam to Dunsinane.   Compute the army's average   (ii) speed and   (iii) velocity   as it marched from Birnam to Dunsinane.

K.89   A particle starts from rest at the origin at time $t = -2$. In the time interval from $t = -2$ to $t = 0$, the particle accelerates at $(12\,t^2\,,\,1)$ m/s^2, while from $t = 0$ onward, it accelerates at $(-10\,,\,-1)$ m/s^2. Determine the   (a) velocity, and   (b) position   of the particle at times   (i) $t = -2$,   (ii) $t = 0$, and   (iii) $t = 2$.

K.90   A particle is at rest at the origin on a flat horizontal surface until time $t = 0$ s when it suddenly begins to accelerate at 4 m/s^2 [north]. This acceleration persists until $t = 4$ s, at which time the acceleration abruptly changes to 5 m/s^2 [at 36.87° east of south]. At $t = 8$ s the acceleration ceases, $\vec{a} = \vec{0}$ m/s^2.

(a) Consider the time interval $0 < t < 4$ seconds.   (i) Write down formulae expressing the components of the velocity as functions of time, valid in this interval.   (ii) Determine the velocity of the particle at $t = 4$ s.   (iii) Write down formulae expressing the components of the position as functions of time, valid in this interval.   (iv) Determine the position of the particle at $t = 4$ s.

(b) Consider the time interval $4 < t < 8$ seconds, and repeat the steps outlined in part(a), with attention to the final time, $t = 8$ s.

(c) Consider the time interval $t > 8$ s and repeat the analyses outlined in parts (a) and (b), with explicit concern for the velocity and position at $t = 12$ s.

K.91   A particle remained at rest at the origin on a horizontal frictionless plane until time $t = 0$, whereupon it suddenly experienced a constant acceleration $\vec{a} = (12\,,\,-5)$ m/s^2, lasting until $t = 4$ s, at which time it suddenly changed to $\vec{a} = 0$.

(a) Sketch graphs representing the magnitude of the acceleration, and its x and y components, at time t.

(b) (i) Determine an expression for the velocity of the particle valid for times $0 \leq t \leq 4$. (ii) Sketch graphs showing the x and y components of the velocity, and the speed [magnitude of the velocity], at times $0 \leq t \leq 4$.

(c) (i) Determine an expression for the position of the particle valid for times $0 \leq t \leq 4$. (ii) Sketch graphs representing the x and y components of the position of the particle at time t. (iii) Sketch the trajectory followed by the particle in the xy-plane.

(d) Describe in words, or in formulae, the motion of the particle for times $t > 4$.

K.92 A particle is initially at rest at the origin on a frictionless plane. At $t = 0$, the particle experiences an acceleration, $\vec{a} = (2\,\hat{\imath} + 1\,\hat{\jmath})$. At $t = 2\,\text{s}$, the acceleration suddenly changes to $\vec{a} = (2\,\hat{\imath} - 1\,\hat{\jmath})$, and at $t = 4\,\text{s}$, it instantly becomes zero.

(a) Determine the (i) velocity and (ii) position of the particle at $t = 2\,\text{s}$.

(b) Determine the (i) velocity and (ii) position of the particle at $t = 4\,\text{s}$.

(c) Determine the particle's (i) velocity and (ii) position for $t > 4\,\text{s}$.

K.93 A particle moving on a Cartesian plane is found at $(1\,,\,2)$ at time $t = 2\,\text{s}$. Earlier, at $t = 0$, the particle was seen to be moving with velocity $(-1\,,\,2)\,\text{m/s}$. It is believed that the particle undergoes constant acceleration, $\vec{a} = (1\,,\,1/2)\,\text{m/s}^2$, throughout the time interval $-1 \leq t \leq 4$ s. Determine formulae expressing the velocity and position of the particle as functions of time.

K.94 A projectile is launched from ground level with an initial speed of $v_0 = 11\,\text{m/s}$.

(a) Determine how far away from its initial position the particle returns to the ground when it is launched at initial angle (i) $\theta_0 = 30°$ and (ii) $\theta_0 = 60°$.

(b) Determine the two horizontal downfield distances at which the projectile is at a height of $2\,\text{m}$ above the ground when it is launched at an initial angle of $60°$.

K.95 PK is practicing American-style football kick-offs,[3] hoping [in vain] to be scouted by a professional team. In all of the following cases, the ball starts and ends at field level.

(a) PK boots the ball at an initial speed of $15\,\text{m/s}$ and initial angle of $30°$ above the horizontal. (i) Write down the constant acceleration kinematical equations describing the trajectory of the ball. Determine (i) how long the ball is in the air, and (ii) how far away the ball lands.

(b) A coach suggests that PK kick higher. With some additional effort, PK boots the ball at an initial speed of $15\,\text{m/s}$ and initial angle of $60°$. Repeat (i–iii) above in this instance.

(c) Another coach suggests that PK increase the strength of his kicking leg. After several months of diligent weight training, PK is able to propel the ball at an initial speed of $25\,\text{m/s}$ and initial angle of $60°$. Repeat (i–iii) for these kicks.

K.96 Determine, in the standard setting of projectile motion, the launch angle(s) at which the range of a projected particle is maximised. [Assume that all parameters other than the initial angle are held fixed, and show all relevant details in your analysis.]

K.97 A snowball is fired with initial speed v_0, at initial angle above the horizontal θ_0, and from an initial height H. Assuming that the snowball passes over level ground, the acceleration due to gravity is constant, and air resistance is negligible, determine the point at which the snowball lands on the ground.

---

[3]These are not unlike corner kicks in the other style of football.

K.98   Recently, PK clambered up the exterior of the Student Union Building with the intention of affixing a banner reading FIZZIKS RULZ to the roof. At $10\,\mathrm{m}$ above the ground, a gust of wind slipped his knapsack off his shoulder, imparting to it a velocity of $3\,\mathrm{m/s}$ [horizontally outward from the building]. Effects of air resistance are assumed to be negligible.

(a) Determine how long the knapsack was in flight [*i.e.*, from the time it escaped PK's grasp to the time it hit the ground].

(b) At what distance away from the base of the building did the knapsack land?

(c) With what velocity did the knapsack hid the ground?

K.99
A cannon is mounted on a battlement on a cliff face overlooking a harbour as shown in the nearby figure. The cannon fires shells horizontally with an initial speed of $v_0 = 60\,\mathrm{m/s}$, and is $H = 80\,\mathrm{m}$ above sea level. [Neglect drag, wind, and other effects.]

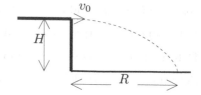

(a) At what distance R do the canoes belonging to the invading forces come into range of the cannon?

(b) At what angle should the cannon be aimed so as to maximise its range?

K.100   During the 2008 financial crisis, "Helicopter Ben" Bernanke flew throughout the United States dropping suitcases, valises, and satchels of cash in an attempt to "re[in]flate" the economy. Passing above a playing field nearby the campus, at an altitude of $1135\,\mathrm{m}$ above ground, Ben gently released a small bag of bills. PK saw this from the vantage point of his office, located $10\,\mathrm{m}$ above ground in a building $750\,\mathrm{m}$ from the playing field.

(a) With what speed must Ben's helicopter be flying in order that his falling currency crash through PK's office window? [Neglect the effects of financial friction.]

(b) Determine the velocity of the money just as it hits the window.

(c) Are the results for (a–b) affected by how much money Ben throws at the problem?

K.101
A stone is thrown at an initial angle of $45°$ from a point $5\,\mathrm{m}$ above the surface of a pond, as shown in the figure. The initial speed of the stone is $10\sqrt{2}\ \mathrm{m/s}$.

Determine how long the stone is in the air, and where it hits the water.

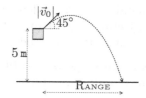

K.102   *I will not be afraid of death and bane, Til Birnam forest come to Dunsinane*
Macbeth ACT V, scene 3

The soldiers approaching Macbeth's castle bear longbows capable of launching arrows with an initial speed of $40\ \mathrm{m/s}$ at any angle of initial elevation. The soldiers garrisoned within the castle, equipped with the same type of longbows, fire arrows from narrow windows located $15\,\mathrm{m}$ above the ground.

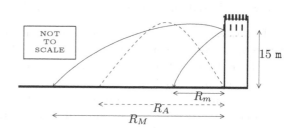

The maximum and minimum angles at which the castle archers may fire are $+30°$ and $-30°$ with respect to the horizontal.

(a) Determine the distance R_A at which the attacking soldier's arrows are just barely able to strike the base of the tower. [Assume that the attacking soldiers fire their arrows at 45° to maximise their range.]

(b) Determine the maximum distance R_M at which the defender's arrows strike the ground. [The garrison have the same type of bows as the attackers.]

(c) Determine the minimum distance R_m at which the defender's arrows land.

(d) Estimate the width of the zone through which the attackers must pass with their shields overhead (to ward off defensive arrows) before undertaking any attempt to return fire. Suggest a tactic to ameliorate this situation.

K.103 *Lay on, Macduff* Macbeth ACT V, scene 8
In problem K.102, the window design is an impediment to efficient defense of the castle. Were the castle archers to fire from the rooftop, 20 m above ground, and without restriction on the initial angle, their maximum range would be increased.

(a) Adapt the constant acceleration kinematical formulae to the case in which the archers fire at [unknown] angle θ from the rooftop at initial height 20 m.

(b) Develop an expression for the range, in terms of the initial angle, at which the arrow hits the level ground.

(c) Differentiate the expression in found in (b) and ascertain the value of θ, the initial angle for which the range is maximised.

(d) Compute the maximum range in this case and compare with the results of the preceding problem.

K.104
In "shotgun" formation, the centre (C) and the wide-out (WO) stand at the line of scrimmage [in this instance 20 m apart], while the quarterback (QB) stands 5 m behind the centre. At the instant that the ball is snapped to the QB, the wide-out runs a slant pattern downfield, with the intent of receiving a forward pass. Exactly 2 s from the instant that the ball is snapped, the QB throws it directly downfield with initial speed 20 m/s, at 15° above the horizontal.

(a) How long is the ball in flight, from the instant it is thrown to the time it returns to shoulder height?

(b) Where (with respect to the QB and with respect to the line of scrimmage) does the football return to shoulder height?

(c) How far must the receiver run in order to get to the position determined in (b)?

(d) At what angle, with respect to straight-ahead, must the receiver run?

(e) With what average speed must the receiver run in order to meet the ball?

K.105 PK has always wanted to to kick a field goal from midfield. [This would be roughly equivalent to a free-kick sailing over the net in the *other* form of football.] One way of realising his ambition involves a great deal of training effort and coaching [with little prospect of success]. His preferred approach is to compensate for his many and variegated deficiencies by raising the kicking tee—*to an absurd height*.
In our model for this event, the horizontal distance to the goal posts is 50 m, and the cross-bar, over which the ball must pass, is 3 m above the playing field. Air resistance is negligible. The maximum speed which he is able to consistently impart to the ball (throughout a wide range of angles) is $v_0 = 7$ m/s.

(a) Suppose that the ball is placed at midfield, on a tee at ground level, and the ball is kicked at an initial angle of 45° above the horizontal. Determine the maximum height attained by the ball and the range of the kick, *i.e.*, the distance from the tee at which the ball hits the ground.   [This attempt falls well short of the goal!]

(b) Determine the minimum height, H_{min}, of the tower that must be built at midfield, along with the particular initial angle, $\theta_{0,\mathrm{min}}$, at which the ball must be kicked with $v_0 = 7\,\mathrm{m/s}$ in order that it barely sails through the uprights.

[HINTS: (1) Parameterise the height of the tower in terms of its height above the crossbar, but don't forget that the base of the tower is at field level. (2) The constraint that the ball pass through the uprights, $50\,\mathrm{m}$ away, relates H to θ_0. Find the extrema of this function to ascertain H_{min} and $\theta_{0,\mathrm{min}}$.]

K.106

A C130 "Hercules" transport plane is attempting to drop emergency paper and pencil supplies to a group of physics students who are stuck in the middle of a large calculation. The plane is flying $320\,\mathrm{m}$ (*i.e.*, about $1000\,\mathrm{ft}$) above level ground at $40\,\mathrm{m/s}$ as illustrated in the figure. Neglect the effects of air resistance (no parachute, too).

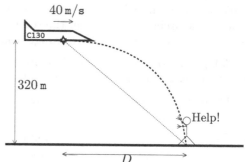

(a) How long is the package in flight from its release to its reaching the students?

(b) (i) At what horizontal distance D from the target should the bombardier release the package?   (ii) At what angle will the target be in his line of sight when it is time to release the package?

(c) Suppose that the package collides with the ground and is stopped within 0.5 seconds. (i) Determine the horizontal and vertical components of the average acceleration of the package during this half-second time interval.   (ii) Compute the magnitude and direction of the average acceleration as the particle comes to rest.

K.107   *Oh Romeo, Romeo, wherefore art thou Romeo?*

In one of the lesser-known drafts of *Romeo and Juliet*, Romeo, Benvolio, and Mercutio are students at the Verona Institute of Technology. In this version of the famous balcony scene [ACT II, scene 2], Romeo vaults (in a sprightly manner) over the railing with the aid of a trebuchet that he and his friends had built as a class project. Our task, in aid of the star-cross'd lovers, is to calculate the launch parameters within the context of the model presented below.

♡   The balcony railing is $8\,\mathrm{m}$ above ground and along the wall's edge. The trebuchet arm releases its projectile at an initial height of $2\,\mathrm{m}$. There are no obstacles, and the ground is level adjacent to the structure.

♡   The effects of friction and drag are negligible.

♡   Romeo is moving horizontally, with speed $1\,\mathrm{m/s}$, as he clears the railing.[4]

(a) Determine the launch speed and the initial angle that satisfy the design constraints, and how far, horizontally, away from the wall Romeo's launch point should be.

(b) Romeo is accelerated from rest to his launch speed by the trebuchet arm. Let's say that the distance through which he moves while accelerating is $2\,\mathrm{m}$ (the height at

---

[4]He'd rather it not be said (of him): *But soft! What bloke through yonder window breaks?*

which he is released). Estimate the average acceleration that Romeo must endure to be launched at the speed computed in (a).

K.108 On his way to his office the other day, PK was ensnarled in traffic. Seeking to better his mode of commuting, he pondered the possibility of projectile motion, under the model assumptions listed below.

o   Neither friction nor drag forces are operative.

o   The terrain is sufficiently flat [neglecting the curvature of the Earth's surface, too], with no obstacles between his home and his office 49 km due west.

o   The launch height and the landing height are effectively the same, since the rail gun with which PK shall be impelled will be constructed partially below ground level, and a large, soft cushion is to be placed at the landing site.

o   The launch angle is set at 45° above horizontal so as to maximise the range.

(a) Determine the launch speed that PK must attain in order to fly to the beach, and his time spent in flight.   [Derive and/or explain any formulae that you use.]

(b) Suppose that the rail gun launcher is 8 m in length and that PK is at rest just before launch. What average acceleration must PK endure to attain his needed launch speed? Comment.

K.109 Albert and Bertal construct a cannon to lob water balloons at unsuspecting passersby. The launch speed is set at 12 m/s. The target is 10 m away, at the same height as the cannon. There is a fence of height 5 m halfway between the cannon and the target. At what angle should the cannon be fired?

K.110
A car moved along a flat circular track of radius 500 m. At an initial time, $t = 0$, the car was located 30° north of east with respect to the centre of the track.

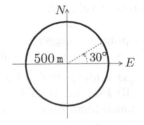

(a) Two minutes later, the car had moved one-third of the way around the circle.   (i) Through what distance did the car move in these two minutes?   Determine the   (ii) average speed, (iii) displacement, and   (iv) average velocity   of the car during this two minute interval.

(b) At five minutes the car reached the two-thirds mark.   (i) Through what distance did the car move from $t = 2$ min to $t = 5$ min?   Determine the   (ii) average speed, (iii) displacement, and   (iv) average velocity   of the car during this three minute time span.

(c) After a total of eight minutes had elapsed, the car was back at its starting point. (i) Through what distance did the car move in the eight minute interval?   Determine the   (ii) average speed,   (iii) displacement, and   (iv) average velocity   of the car during these eight minutes.

K.111 Determine the conversion factors   (a) from radians per second [rad/s] to revolutions per minute [RPM], and   (b) from revolutions per minute to radians per second.

K.112 The orbit of the Earth about the Sun is approximately circular, with [mean] radius $R_\oplus = 1.496 \times 10^{11}$ metres. There are about 3.156×10^7 seconds in a year. Estimate the magnitude of the centripetal acceleration experienced by the Earth in its orbit about the Sun. Is this acceleration constant in time?

K.113 A carousel spins at a constant rate, each complete revolution taking 10 seconds. There are two rows of horses on the carousel, at radii 4 m and 5 m, respectively. Compute the magnitude of the centripetal acceleration experienced by riders on the   (a) inner and (b) outer   horses.

K.114 A carousel spins at a constant rate. Each revolution takes τ seconds to complete. There are two rows of horses on the carousel. The inner horses are at radius R_i, while the outer ones are at R_o, with $R_i < R_o$. Compute the magnitude of the centripetal acceleration experienced by riders on the   (a) inner and   (b) outer   horses.

K.115 A compact disk spins at variable speed so as to maintain a (roughly) constant data-reading speed of $1.3\,\mathrm{m/s}$ throughout its radius. The information-carrying part of the disc is annular, with inner radius $R_i = 25\,\mathrm{mm}$ and outer radius $R_o = 115\,\mathrm{mm}$. At what angular speed does the disc rotate when the player is reading information stored near the   (a) inner and   (b) outer   radius?

K.116 A particular compact disk rotates at 400 RPM while it is in operation.

(a) Determine the angular speed of the disk in radians per second.

(b) Suppose that the data-carrying portion of the disk has outer radius 6 cm. Determine the linear speed of a point on the disk   (i) at the outer limit of the data region, and   (ii) midway from the centre to the outer limit (*i.e.*, at 3 cm).

(c) The compact disk spins clockwise when viewed from above. A microbe is "standing" on the disk near the midway point between the centre and the edge. Describe what happens when the microbe attempts to "walk" radially   (i) outward toward the rim, and   (ii) inward toward the centre   of the spinning disk.

K.117   *Throw physic[s] to the dogs*   Macbeth ACT V scene 3

Lady Macbeth took her dog, Spot [of course], to a wide level field for a game of fetch. Keeping the dog at her side, Lady Macbeth threw a dagger, which landed 30 m away. Spot raced directly toward the dagger, accelerating at $5\,\mathrm{m/s^2}$ for two seconds and thenceforth maintaining a constant velocity.

Grasping the dagger triumphantly in his jaws, Spot quickly turned and trotted back to his mistress along a semicircular arc at a constant speed of $2\pi \simeq 6.283\,\mathrm{m/s}$.

(a) While pursuing the dagger: (i) What was Spot's speed after 2 seconds of running? (ii) How far from Lady Macbeth was Spot when he ceased accelerating?   (iii) How much more time elapsed before Spot reached the dagger?

(b) While returning to Lady Macbeth: (i) What distance did Spot travel?   (ii) For how long did Spot trot?   What were the   (iii) magnitude and   (iv) direction   of Spot's acceleration?

K.118 A particle moves along a circle of radius R according to the parametric description:

$$x(t) = R \cos\left(\omega t + \phi\right) \qquad\qquad y(t) = R \sin\left(\omega t + \phi\right).$$

(a) Compute the velocity of this particle and verify that it moves with constant speed.

(b) Compute the acceleration of this particle and verify that it is both constant in magnitude and directed centripetally.

K.119 The trajectory of a certain particle moving in a straight line at constant speed in flat two-d space is given, parametrically, by $\vec{r}(t) = \left(x(t),\, y(t)\right) = (3t,\, 4t)$.

(a) Verify that the trajectory   (i) is straight [has constant direction],   (ii) is uniform [constant speed], and   (iii) passes through the origin.

(b) (i) Verify that the trajectory may be re-expressed as $5t$ at an angle of $53.13°$ with respect to the x-axis,   (ii) infer from this re-expression that the angular speed is zero, and   (iii) compute the velocity of the particle.

K.120 The trajectory of a certain particle (for $t > 0$) is

$$\vec{r}(t) = \left(x(t),\, y(t)\right) = a_0 \left(\frac{t}{\sqrt{1+\gamma^2}},\, \frac{\gamma\, t}{\sqrt{1+\gamma^2}}\right), \quad \text{where } a_0 \text{ and } \gamma \text{ are positive constants.}$$

(a) Verify that the trajectory   (i) is straight [has constant direction],   (ii) is uniform [constant speed], and   (iii) passes through the origin.

(b)   (i)   Verify that the trajectory may be re-expressed as $\vec{r}(t) = \left(r(t),\, \theta(t)\right) = \left(a_0\, t,\, \tan^{-1}(\gamma)\right)$,   (ii) compute the velocity and the acceleration of the particle in both of its parametric representations, and   (iii) verify that the results are consistent.

K.121 [See also K.122.] Consider a specific parabolic trajectory passing through the origin:

$$\vec{r}(t) = \left(x(t),\, y(t)\right) = \left(t,\, t^2\right), \quad \text{for } t > 0.$$

(a) Re-express the trajectory in terms of polar coordinates: $\left(r(t),\, \theta(t)\right)$.

(b) Compute the velocity of the particle, at time t, in both systems of coordinates.

(c) Compute the acceleration, at time t, in both systems of coordinates.

K.122 [See also K.121.] Consider a somewhat more general parabolic trajectory passing through the origin:

$$\vec{r}(t) = \left(x(t),\, y(t)\right) = \left(V_0\, t,\, \frac{A_0}{2}\, t^2\right), \quad \text{for } t > 0.$$

(a) Re-express the trajectory in terms of polar coordinates: $\left(r(t),\, \theta(t)\right)$.

(b) Compute the velocity of the particle, at time t, in both systems of coordinates.

(c) Compute the acceleration, at time t, in both systems of coordinates.

K.123 [See also K.124, R.32.] Consider a particle trajectory, parallel to the Cartesian y-axis, described by $\vec{r}(t) = (1,\, t)$, for times $t \geq 0$.

(a) Re-express the trajectory in terms of polar coordinates: $\left(r(t),\, \theta(t)\right)$.

(b) Compute the velocity of the particle, at time t, in both systems of coordinates.

(c) Compute the acceleration, at time t, in both systems of coordinates.

K.124 [See also K.123, R.33.] Consider a particle trajectory parallel to the Cartesian y-axis described by $\vec{r}(t) = (X_0,\, V_0\, t)$, for times $t \geq 0$.

(a) Re-express the trajectory in terms of polar coordinates: $\left(r(t),\, \theta(t)\right)$.

(b) Compute the velocity of the particle, at time t, in both systems of coordinates.

(c) Compute the acceleration, at time t, in both systems of coordinates.

K.125 A particle moves in a logarithmic spiral: $r(t) = R_0 \ln(\gamma\, t)$, and $\theta(t) = \omega\, t$, for constants R_0, γ, and ω. Determine the velocity and acceleration of the particle as functions of time.

K.126 Show, using the orthonormality of $\{\hat{\imath}, \hat{\jmath}\}$, that the polar basis vectors, $\{\hat{r}, \hat{\theta}\}$, are orthonormal.

D

Dynamics Problems

D.1 Let's rework the "Constant Acceleration: PK and the Police Car" example found in Chapter 3 from the perspective of Mr. John Q. Motoring-Public [hereafter known as Q], whose car is moving at a constant speed of $36\,\text{km/h}$, or $10\,\text{m/s}$, in the direction opposite to that in which PK is travelling. [The analysis simplifies if Q happens to pass by PK at the same instant that PK initially passes by the police car. Furthermore, Q will agree to start his watch at this common instant of passing, if asked nicely.]

(a) Sketch a spacetime diagram showing trajectories for both PK and the police car from Q's perspective.

(b) At what time does Q observe the police car overtaking PK?

(c) Where, from his perspective, does Q observe the police car overtaking PK?

D.2 The climax of the book *The Cat in the Hat Comes Back* [by Dr. Seuss] occurs when THE CAT initiates the launch of a succession of cats out of a succession of hats, culminating in a VOOM which sets all the previously described chaos aright. In our model description of this event, we suppose that The Cat's hat is at rest with respect to the children, and that LITTLE CAT A emerges at $1\,\text{m/s}$ in the forward direction. Little Cat A launches LITTLE CAT B from his hat at a relative velocity of $1\,\text{m/s}$ [fwd]. All of the remaining cats emerge from their progenitors' hats at $1\,\text{m/s}$ [fwd]. Finally, the VOOM emerges from LITTLE CAT Z's hat at $1\,\text{m/s}$ [fwd] with respect to Little Cat Z. Determine the speed at which the children, SALLY AND ME, see the VOOM emerge.

D.3 Albert, A, and Bertal, B, are again sliding particles across a table whilst riding on the *Inertial Express*. A is facing forward, B faces the rear. The train moves forward at a steady $10\,\text{m/s}$ along its track. Two keen-eyed observers watch the happenings. One of these, T, stands alongside, and at rest with respect to, the tracks. The other, C, pilots an airplane moving parallel to the tracks, in the same direction as the train, with speed $50\,\text{m/s}$. Describe the motion of the particle as seen by each of the observers, T and C, in each of the scenarios listed below.

The particle is slid at $\begin{bmatrix} \text{(a)} & 5\,\text{m/s} \\ \text{(b)} & 10\,\text{m/s} \\ \text{(c)} & 15\,\text{m/s} \end{bmatrix}$ WRT the table, $\left(\begin{array}{ll} \text{(i)} & \text{from A to B} \\ \text{(ii)} & \text{from B to A} \end{array} \right)$.

D.4 Albert, A, and Bertal, B, are again sliding particles across a table whilst riding on the *Inertial Express*. The fellas are sitting transversely, facing sideways, while the train moves forward at $10\,\text{m/s}$ along its track. Two keen-eyed observers watch the happenings. One of these, T, stands alongside, and at rest with respect to, the tracks. The other, C, pilots an airplane moving parallel to the tracks, in the same direction as the train, with speed $50\,\text{m/s}$. Describe the motion of the particle as seen by each of the observers, T and C, in each of the scenarios listed below.

(a) The particle is slid at $5\,\text{m/s}$ WRT the table, from A toward B.

(b) The particle is slid at $15\,\text{m/s}$ WRT the table, from B toward A.

D.5   [See also M.53.]

Three inertial observers, Larry (L), Curly (C), and Moe (M), find themselves in relative motion on a horizontal frictionless plane. From the perspective of a fourth observer, at rest on the surface, their velocities are: $\vec{v}_L = 8$ m/s $[\rightarrow]$, $\vec{v}_C = 4.5$ m/s $[\leftarrow]$, and $\vec{v}_M = 6$ m/s $[\uparrow]$. Their positions at the instant $t = -2$ s are recorded in the adjacent figure.

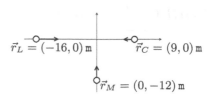

(a) With what velocity does Larry observe   (i) Curly and   (ii) Moe   to be moving?

(b) With what velocity does Curly observe   (i) Larry and   (ii) Moe   to be moving?

(c) With what velocity does Moe observe   (i) Larry and   (ii) Curly   to be moving?

(d) For each of the three fellows, sketch the observed trajectories (spatial) of the other two.

D.6   Analyse the following sports scenario in terms of Newton's Laws of Motion.

A hockey puck is (momentarily) at rest on the icy surface of a hockey rink. A player, with a graceful and smooth wrist shot taking one-fifth of a second, fires the puck at 20 m/s along the ice toward the opposing team's goal.

D.7   Two skydivers, A and B, emerge from their respective planes at altitudes $H_A = 3080$ m and $H_B = 3000$ m. Their jumps are coordinated such that B exits his airplane just as A passes him by. [Assume that both airplanes are moving horizontally with the same velocity, and that the skydivers do not propel themselves out of the plane. Instead, they simply drop from a wing strut. Also, one can consistently choose to neglect the (inertial) forward motion of the skydivers. Most importantly, we neglect air resistance.]

(a) (i) Verify that A accelerates at a constant rate, and   (ii) adapt the constant acceleration kinematical formulae to describe his motion, with $t = 0$ at the instant he starts to freefall.

(b) (i) Verify that B experiences constant acceleration, and   (ii) determine his trajectory using split time, starting when he begins to freefall.

(c) Ascertain   (i) the time at which A passes B and   (ii) A's velocity at this time.

(d) Recast B's trajectory in terms of A's time, t, and sketch the vertical components of both trajectories on velocity *vs.* t and position *vs.* t axes.

(e) At the instant that A passes by B, he throws a ball directly behind him at 20 m/s. Determine the trajectory of the ball in terms of   (i) B's split time and   (ii) time, t.

(f) (i) Plot the trajectory of the ball on the graphs sketched in part (d).   (ii) Comment.

D.8   Two small children are engaged in a tug-of-war competition. Each pulls with a force of 125 N on opposite ends of an ideal rope, and it does not move. What is the magnitude of the tension force found in the ideal rope?

D.9   Two physics students pull on opposite ends of a rope, each with a force of 500 N. The rope does not move. What is the force of tension in the rope?

D.10   Two teams of students are engaged in competition to drive a 100 kg block across the opponent's goal. The block, of width 2 m, is initially at rest at the centre of a 10 m long frictionless playing field [*i.e.*, equidistant from each goal line].

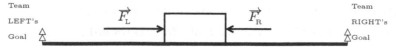

Each team sustains a constant force, $\vec{F}_L = 110$ N and $\vec{F}_R = 90$ N, on the block.

(a) Which team is bound to win the contest?

(b) What is the acceleration of the block?

(c) How long does the competition last?

(d) With what speed does the block cross the losing team's goal line?

D.11 An elevator riding in a frictionless vertical shaft has total mass of 1000 kg.

(a) The elevator cable exerts a force of 9000 N in the upward direction on the elevator. Determine (i) the acceleration of the elevator, and (ii) how long it takes for the elevator, starting from rest, to attain a speed of 5 m/s.

(b) The cable exerts a force of 12000 N [up] on the elevator. Determine (i) the acceleration of the elevator, and (ii) how long it takes for it to attain a speed of 5 m/s, starting from rest.

D.12 A block of mass $M = 5$ kg is at the origin of a horizontal frictionless plane. At the instant $t = 0$, it is moving in the negative x-direction with speed 4 m/s. A constant force of 10 N is applied to the block in the positive x-direction starting at $t = 0$ and lasting until a later time, t_f, when the block comes to rest.

(a) Draw a free body diagram and enumerate the forces which act on the block.

(b) Compute the acceleration of the block while the force is applied.

(c) Compute the (i) time, t_f, and (ii) position, x_f, at which the block stops.

D.13 A particle of mass $M = 3$ kg is at rest at the origin on a horizontal frictionless plane until $t = 0$ s. At that time, two forces, $\vec{F}_1 = (20, -20)$ N and $\vec{F}_2 = (16, 5)$ N, are applied to the particle. Determine the (a) velocity and (b) position of the particle at $t = 4$ s.

D.14 A block of mass 4 kg is moving along a frictionless horizontal plane. At time $t = 0$, the particle is found at the origin and is moving with velocity 10 m/s at a heading of 30° north of east. In the time interval $0 \le t \le 1$ s, the block is acted upon by a force with magnitude 40 N acting due south. From $t = 1$ s to $t = 6$ s, the block experiences a force of 8 N [north]. After $t = 6$ s the block moves inertially.

(a) Express the components of the acceleration (with respect to the natural basis: east and north) in the time intervals: (i) $0 \le t \le 1$, (ii) $1 \le t \le 6$, and (iii) $6 \le t$.

(b) Determine the velocity for: (i) $0 \le t \le 1$, (ii) $1 \le t \le 6$, and (iii) $6 \le t$.

(c) Determine the position for: (i) $0 \le t \le 1$, (ii) $1 \le t \le 6$, and (iii) $6 \le t$.

D.15 A block of mass $m = 10$ kg riding on a frictionless horizontal plane is subjected to three forces: its weight, the normal force of contact with the plane, and an applied force. The information that we have been able to gather about these forces is:
$$\vec{W} = 100 \text{ N} [\downarrow] \qquad \vec{N} = ? \qquad \vec{F}_A = 20 \text{ N} [\rightarrow]$$
The initial velocity of the block is $\vec{v}_0 = 2$ m/s $[\leftarrow]$, while its initial position is $\vec{r}_0 = (0, 0)$.

(a) Sketch the force vectors on a FBD. Note the direction of the normal force.

(b) Ascertain the magnitude of the normal force.

(c) Determine the net force acting on the block.

(d) Determine the acceleration of the block. Is this acceleration constant?

(e) Determine the time at which the block returns to its starting point (*i.e.*, the origin).

(f) Determine the velocity of the block at the instant that it returns to the origin.

D.16 [See also E.4, M.5.] A block of mass $m = 10\,\text{kg}$ riding on a frictionless horizontal plane is subjected to five forces. The weight of the block and the normal force of contact with the plane both act in the vertical direction, cancel, and may be henceforth ignored. Three applied forces, $\{\vec{F_1}, \vec{F_2}, \vec{F_3}\}$, act horizontally on the block. The information that we have been able to gather about the applied forces is summarised as follows:

$$\vec{F_1} = 10\,\text{N}\ [\rightarrow] \qquad \vec{F_2} = 40\,\text{N}\ [\downarrow] \qquad \vec{F_3} = 50\,\text{N}\ [+53.13°\ \text{WRT}\ x\text{-axis}]$$

The initial velocity of the block is $\vec{v_0} = 2\ \text{m/s}\ [\leftarrow]$, while its initial position is $\vec{r} = (0\,,0)$.

(a) Sketch the applied force vectors on a FBD.

(b) Determine the net applied force.

(c) Determine the acceleration of the block. Is this acceleration constant?

(d) Determine the   (i) velocity, and   (ii) position   of the block at $t = 4\,\text{s}$.

D.17 [See also E.14, M.7.] A block of mass $2.5\,\text{kg}$ riding on a horizontal frictionless plane is subjected to four applied forces: $\vec{F_1} = 13\ \text{N}\ [22.62°\ \text{WRT positive}\ x\text{-axis}]$, $\vec{F_2} = 5\ \text{N}\ [53.13°$ WRT positive y-axis], $\vec{F_3} = 8\ \text{N}\ [\leftarrow]$, and $\vec{F_4} = 3\ \text{N}\ [\downarrow]$. The weight and normal force of contact act perpendicular to the plane, and they cancel everywhere and always. At the instant $t = 0\,\text{s}$ the block is at the origin, $\vec{r_0} = (0\,,0)$ m, and moving with velocity $\vec{v_0} = 3\,\text{m/s}\ [\rightarrow]$. Our goal is to determine the speed of the block at the time $t = 2\,\text{s}$, when it reaches the point $\vec{r_2} = (6\,,4)$ m.

(a) Draw a free body diagram for the block and determine the net force.

(b) Determine the acceleration response of the block.

(c) Verify that the block reaches $\vec{r_2}$ at $t = 2\,\text{s}$.

(d) Determine the   (i) velocity and   (ii) speed   of the block at $t = 2\,\text{s}$.

D.18 A block of mass $M = 32\,\text{kg}$ is at rest at $(1,2)$ on a frictionless horizontal plane. At time $t = 0\,\text{s}$, the particle experiences a force $\vec{F}_{0\rightarrow4} = \left(6\,t\,,\ \frac{15}{4}\,\sqrt{t}\right)$ [acting within the horizontal plane]. During the time interval $4 \le t \le 8\,\text{s}$, the force ceases to vary and instead remains constant at $\vec{F}_{4\rightarrow8} = \left(24\,,\ \frac{15}{2}\right)$ N. At $t = 8\,\text{s}$, the force on the block suddenly drops to zero. Derive general formulae for the   (i) velocity and   (ii) position   of the block, along with the particular values of the   (iii) velocity and   (iv) position   at the final instant, for each of the time intervals:

    (a) $0 \le t \le 4$,   (b) $4 \le t \le 8$, and   (c) $8 \le t \le 12$.

D.19 A particle of mass $5\,\text{kg}$ is at rest at $(-1,\,-1)$ on a frictionless xy–plane. At $t = 0\,\text{s}$, a constant force $\vec{F} = 10\,\hat{\imath} - 5\,\hat{\jmath}$ N acts on the particle. At $t = 1\,\text{s}$, the force suddenly changes to $\vec{F} = -10\,\hat{\imath} + 10\,\hat{\jmath}$ N. At $t = 2\,\text{s}$, the force suddenly becomes, and afterward remains, zero. Determine the   (i) velocity and   (ii) position   of the particle at the following times: (a) $t = 1\,\text{s}$,   (b) $t = 2\,\text{s}$, and   (c) $t = 4\,\text{s}$.

D.20 A $3\,\text{kg}$ block lies at rest at the point $(1\,,1\,,0)$ m on the surface of a large frictionless horizontal plane [the xy-plane]. At $t = 0\,\text{s}$, a constant applied force, $\vec{F}_{A1} = (6\,,9\,,6)$ N, acts on the block. At $5\,\text{s}$ an additional force, $\vec{F}_{A2} = (-12\,,-15\,,6)$ N, begins to act. At the instant $t = 10\,\text{s}$, both forces cease to operate on the block.

(a) Determine the net force on the block for times: (i) $0 < t < 5$,   (ii) $5 < t < 10$, and   (iii) $10 < t < 15$. [Remember to include both the weight of the block and the normal force of contact exerted on the block by the horizontal plane on which it rides.]

(b) Compute the acceleration of the block for times: (i) $0 < t < 5$,   (ii) $5 < t < 10$, and   (iii) $10 < t < 15$.

(c) Determine the block's velocity at times: (i) $t = 0$,   (ii) $t = 5$,   (iii) $t = 10$, and   (iv) $t = 15$.

(d) Determine the block's position at times: (i) $t = 0$, (ii) $t = 5$, (iii) $t = 10$, and (iv) $t = 15$.

D.21 A particle of mass $M = 6\,\text{kg}$ rides on a horizontal frictionless plane. At $t = 0$, the particle is observed to be at $(x, y) = (0, 0)\,\text{m}$, and is moving with velocity $(v_x, v_y) = (12, 5)\,\text{m/s}$. A force $\vec{F}_{03} = (-18, -18)\,\text{N}$ acts on the particle from $t = 0$ until $t = 3\,\text{s}$. From $t = 3\,\text{s}$ to $t = 7\,\text{s}$, a force $\vec{F}_{37} = (-9/2, 6)\,\text{N}$ acts. Subsequent to $t = 7\,\text{s}$ the particle moves inertially. Determine the position and the velocity of the particle at $t = 8\,\text{s}$.

D.22 A particle of mass M is confined to move in one dimension and subjected to a time-dependent applied force, $F_{\text{A}}(t) = F_0 \sin(2\pi t)$, where F_0 is the maximum magnitude of the force. At $t = 0$, the particle is observed to be at the origin, $x = 0$, and moving with speed $\frac{F_0}{2\pi M}$ in the negative x direction. Determine the following kinematical properties of the particle (as functions of time): (a) acceleration, (b) velocity, and (c) position.

D.23 A particle of mass M is confined to move in one dimension. Prior to $t = 0$, the particle is observed to be at rest at the origin, $x = 0$. At $t = 0$, the particle is subjected to a time-dependent applied force, $F_{\text{A}}(t) = F_0 (1 - e^{-\gamma t})$, where F_0 is the maximum magnitude of the force, and γ is a constant. Determine the following kinematical properties of the particle (as functions of time): (a) acceleration, (b) velocity, and (c) position.

D.24 At time $t = 0\,\text{s}$, a block of mass $M_1 = 1/2\,\text{kg}$ lies at rest at the origin, $(0, 0)$, of a horizontal frictionless plane. A second block, with mass $M_2 = 2\,\text{kg}$, moving with velocity $(2, 1)\,\text{m/s}$, is beside the first block, at the point $(3, 0)$, at the same instant. The blocks experience constant forces $\vec{F}_1 = (\frac{1}{4}, \frac{1}{6})$ and $\vec{F}_2 = (-\frac{1}{3}, 0)$.

(a) Determine the trajectory of Block 1.

(b) Determine the trajectory of Block 2.

(c) Ascertain whether the particles collide. Provide details.

D.25 [See also D.26, D.76]
Two blocks, with mass m_1 and m_2, are able to move in a straight line along a frictionless horizontal plane surface. The blocks are linked by an ideal rope lying horizontally. In addition, an external agent applies a force on another length of ideal rope which is affixed to Block 1.

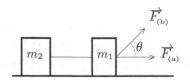

(a) Suppose that the rope which pulls the blocks lies parallel to the plane. Determine the ratio of the tensions in each of the two ropes in terms of the masses of the blocks.

(b) Suppose that the rope which pulls the system is inclined at angle θ above the horizontal [and that Block 1 remains on the surface]. Determine the ratio of the tensions in the ropes.

D.26 [See also D.25, D.77, E.13, M.6] Suppose that the system in D.25 is at $x_0 = 0$, and moving forward with speed v_0 at $t = 0$. Determine the speed of the blocks at the instant that the system has displaced a distance D, for cases (a) and (b), described just above.

D.27
Two equal-length pieces of ideal rope hold a block of mass M suspended from a ceiling, as shown in the figure. The block is at rest and each rope makes the same angle, θ, with respect to the vertical. Compute the tension in each of the ropes.

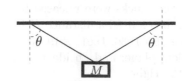

D.28

A dune buggy is stuck in some sand a short distance away from a palm tree. A bevy of physics students attach one end of a stout [read "ideal"] rope to the towing hook on the buggy, and the other end to the palm tree. The rope is pulled as tight as they can manage, and the students push *sideways* on the middle of the rope, as illustrated, with a total force of 2500 newtons, deflecting it by 10°.

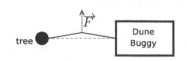

Determine the magnitude of the tension force in the rope.

D.29

Three ideal ropes are arranged, with two ceiling anchors and a block of mass M, as in the figure. The ropes and the block remain at rest.   (a) Ascertain which of the two ropes connected to the ceiling has the greater tension.   (b) Which of the three ropes has the greatest tension?

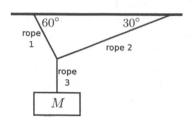

D.30   [See also D.31, D.57, D.73, E.16, R.18.]

Two blocks, with mass $m_1 = 2\,\text{kg}$ and $m_2 = 4\,\text{kg}$, are joined by a length of ideal rope passing over an ideal pulley, as shown. Block 1 slides along a frictionless plane, while Block 2 falls. Neither block is affected by drag.

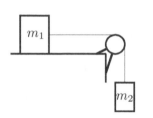

(a) Compute the acceleration of the system and the tension in the rope.

(b) Suppose that the blocks are interchanged, *i.e.*, 1 ↔ 2. Compute the acceleration of the system and the tension in the rope in this case.

D.31   [See also D.30, D.72, E.17.] Repeat the previous problem with arbitrary m_1 and m_2.

D.32   [See also D.30, D.72, E.17.] Repeat the previous problem with $m_1 = M = m_2$.

D.33   [See also D.34, D.59, E.18, M.9.]

Three blocks, each of mass M; two ideal ropes; a frictionless plane; and an ideal pulley comprise the dynamical system illustrated in the adjacent figure.

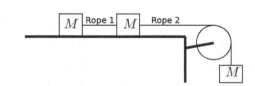

(a) Draw a free body diagram for each of the blocks and the pulley. Enumerate the forces acting on each object.

(b) Solve for the acceleration of the system and the tensions in the ropes.

D.34   [See also E.18, M.9.] Consider the three-block system in the previous problem. Suppose that the blocks were released from rest, and determine the speed of the system once each block has moved a distance H from its initial location.

D.35   [See also D.36, D.61, E.19, M.10.]   Three blocks, each of mass M; two ideal ropes; a frictionless plane; and an ideal pulley comprise the dynamical system illustrated in the figure to the right.

(a) (i) Draw a free body diagram for each of the blocks and the pulley.   (ii) Enumerate the forces acting on each object. (iii) Choose coordinates and express N2 for each constituent.

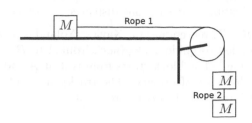

(b) Solve for   (i) the acceleration of the system,   and   (ii) the tensions in the ropes.

D.36 [See also E.19, M.10.]   Consider the three-block system in the previous problem. Suppose that the blocks were released from rest, and determine the speed of the system once each block has moved a distance H from its initial location.

D.37 [See also D.62, E.20, M.11.] Four blocks, each of mass M; three ideal ropes; an ideal pulley; and a frictionless plane comprise the dynamical system illustrated in the nearby figure.

(a) Draw a free body diagram for each of the blocks and the pulley. Enumerate the forces acting on each object.

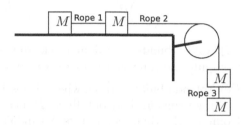

(b) Ascertain the acceleration of the system.

(c) Determine the tension in each of the ropes.

D.38 [See also E.20, M.11.] Consider the four-block system in the previous problem. Suppose that the blocks were released from rest, and determine the speed of the system once each block has moved a distance H from its initial location.

D.39 [See also E.36, M.47, R.24.]

Two blocks are joined by a length of ideal rope which is slung over an ideal pulley of radius R, as illustrated in the figure to the right. The system was held at rest until time $t = 0$, at which time it was released and given a short sharp tap, (instantaneously) setting the system in motion with speed v_0. WLOG, the left block may be assumed to rise while the right block falls. The mass of the left block is M_L, and that of the right block is M_R. The local acceleration due to gravity is constant and directed along the negative y-axis.

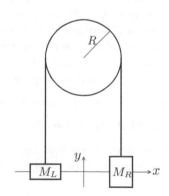

(a) Draw separate FBDs for the left and right blocks and for the ideal pulley.

(b) Suppose that the blocks have equal mass, $M_L = M = M_R$, and determine:
(i) the acceleration of the system,   (ii) the tension in the rope,   (iii) the trajectory of each of the blocks, and   (iv) the velocities of the blocks at the instant that each has been displaced a distance H from its initial position.

(c) Suppose that $M_L = M - m$, while $M_R = M + m$. [This case could be considered a modification of the equal-mass block case by the transfer of mass m from one block to the other.] Determine   (i)   the acceleration of the system,   (ii)   the tension in the rope, (iii)   the trajectory of each of the blocks, and   (iv) the velocities of the blocks at the instant that each has displaced a distance H from its initial position.

D.40   Three engineers, Xavier, Yolanda, and Zoe, were commissioned to design a track intended to guide a $1\,\mathrm{kg}$ block from A to B [shown in the figures below] in the least amount of time.[1] The block starts from rest at A and may not be propelled in any way other than by the force of gravity. The tracks are made of frictionless material. The velocity of the block as it reaches B is irrelevant.

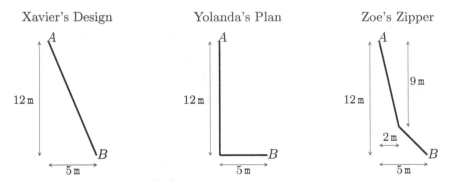

Xavier's Design   Yolanda's Plan   Zoe's Zipper

(X) Xavier builds a straight track from A to B, because this is the shortest distance. Compute the time that it takes Xavier's block to move from A to B.

(Y) Yolanda builds a track which drops $12\,\mathrm{m}$ vertically, has a very small curved section, and then runs $5\,\mathrm{m}$ horizontally to B, because the average speed is greatest in this case. Compute the time required for Yolanda's block to travel from A to B.

(Z) Zoe builds a two-stage track. The first stage drops $9\,\mathrm{m}$ over a horizontal run of $2\,\mathrm{m}$. The second stage drops the final $3\,\mathrm{m}$ over a horizontal run of $3\,\mathrm{m}$. How long does Zoe's block take to zip from A to B?

D.41   [See also E.24.] A block of mass $M = 5\,\mathrm{kg}$ is held at rest on a frictionless plane with angle of inclination $\theta = 30° = \frac{\pi}{6}\,\mathrm{rad}$. At $t = 0\,\mathrm{s}$, it is released. With what speed does the block pass by an observer who is at rest with respect to the surface and located a distance $D = 5\,\mathrm{m}$ from the initial position of the block?

D.42   [See also E.25.] A block of mass $M = 70\,\mathrm{kg}$ is launched, with speed $v_0 = 6\,\mathrm{m/s}$, up a frictionless plane inclined at angle $\theta = 45° = \frac{\pi}{4}\,\mathrm{rad}$.

(a) What maximum distance upward along the incline is attained by the block?

(b) With what speed does the block pass by its initial position on its way back down?

D.43   Various blocks, each of mass M, are held in place on frictionless inclined planes ($\theta < 45°$) by lengths of ideal rope. The ropes differ in their points of attachment to fixed anchors.

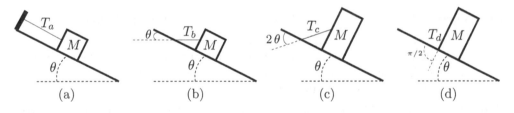

    (a)   (b)   (c)   (d)

---

[1] The optimal shape for the track was determined by J. Bernoulli, *circa* 1696!

(a) One rope, shown in the figure labelled (a), is tied in such a way that it lies parallel to the incline. Compute the tension in this rope.

(b) Another rope is tied to a hook on the incline such that the rope is horizontal, as shown in figure (b). Compute the tension in this rope.

(c) This rope is attached to the incline, making an angle of 60° with respect to the surface. Compute the tension in the rope in this case.

(d) Suppose that the angle between the rope and incline is increased until it approaches $\pi/2$. What happens then to the tension in the rope?

D.44 [See also D.80, E.21.] Various blocks of mass M are moving upward along frictionless planes with angle of inclination $\theta < 45°$. Each block is acted upon by an applied force, with magnitude F_A, transmitted through a length of ideal rope.

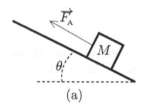
(a)
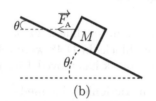
(b)
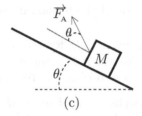
(c)

(a) In this case, the ideal rope extends parallel to the incline. (i) Determine the acceleration of the block for arbitrary F_A. (ii) Ascertain the particular value of F_A for which the block coasts up the inclined plane. Compute the acceleration of the block when $M = 5\,\text{kg}$, $\theta = 20°$, and (iii) $F_A = 10\,\text{N}$, and (iv) $F_A = 40\,\text{N}$.

(b) Here, the rope lies horizontally and remains so as the block moves. Repeat the steps (i–iv) above, in this instance.

(c) The third block is pulled by a rope which remains elevated at twice the ramp angle. Repeat steps (i–iv).

D.45
Three blocks, with respective masses M_1, M_2, and M_3, lie in contact with one another on a horizontal frictionless plane. An applied force, F_A, acts horizontally on the leftmost block as illustrated in the figure. Determine the acceleration of the system of blocks and the magnitudes of the internal forces acting between the blocks.

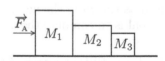

D.46 [See also D.82.]

Two blocks, of mass m_1 and m_2 respectively, together ride on a frictionless plane inclined at angle θ. A constant force F_A is applied parallel to the incline.

For what value of the magnitude of the applied force does the system of blocks coast along the incline, and what is the magnitude of the force of contact acting between the blocks in this case?

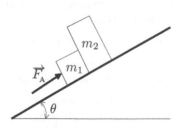

D.47 [See also D.83.]

Two blocks, of mass m_1 and m_2 respectively, together ride on a frictionless plane inclined at angle θ. A constant force, $\vec{F}_A$, is applied horizontally.

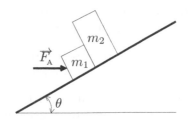

For what value of the magnitude of the applied force does the system of blocks coast along the incline, and what is the magnitude of the force of contact acting between the blocks in this case?

D.48 [See also D.49, D.50, D.64.]

Two blocks of mass $m = 25\,\text{kg}$ are joined by a length of ideal rope which passes over an ideal pulley, as shown in the figure. One of the blocks rides along a frictionless plane which is inclined at $15°$, while the other drops vertically. Compute the acceleration of the system of blocks and the magnitude of the tension in the ideal rope.

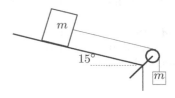

D.49 Repeat D.48 for arbitrary mass m and incline angle θ.

D.50 [See also E.34.] Suppose that the blocks in D.48 were released from rest, and determine the speed of the system once each block has moved $1.5\,\text{m}$ from its initial location.

D.51

Two blocks, each of mass M, ride on frictionless inclined planes and are connected by a piece of ideal rope which passes over an ideal pulley, as shown in the figure. The inclination angles of the planes are $\theta_l = 30°$ and $\theta_r = 60°$, respectively.

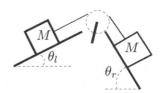

(a) Determine the acceleration of the system.

(b) Suppose that at time $t = 0$ the block on the right is at postion x_0 on its plane and is moving upward along the incline with speed $v_0 = 2\,\text{m/s}$. Determine the time at which the right block returns to x_0.

D.52

Three blocks with masses $M_1 = 15\,\text{kg}$, $M_2 = 2\,\text{kg}$, and $M_3 = 8\,\text{kg}$, respectively; an ideal pulley; some ideal rope; and a frictionless plane, inclined at $\theta = 30°$, are arranged as shown in the nearby figure. Our goal is to compute the acceleration of this system.

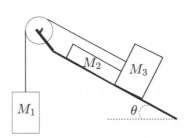

(a) Draw FBDs for each of the blocks (and the pulley, too).

(b) Choose incline-oriented coordinates and express N2 for each of the blocks. Link together the dynamical equations, using the assumptions and constraints inherent in the idealisations.

(c) Solve for the acceleration of the system.

D.53 A block of mass $m = 5\,\text{kg}$, riding on a frictionless plane, traversing a circular path of radius $R = 0.8\,\text{m}$, is held by an ideal rope fastened at the centre of the circle.

(a) Compute the forces of tension in the rope   (i) when the speed of the block is $v_{(a)} = 2\,\text{m/s}$, and (ii) when the block goes around the circle in $T_{(b)} = 3\,\pi \approx 9.425$ s.

(b) Suppose that the magnitude of the tension is $T = 6.25\,\text{N}$. Compute   (i) the speed of the block, and   (ii) the time it takes the block to complete one revolution.

D.54 [See also D.84.] Two blocks, of mass $M_1 = 1\,\text{kg}$ and $M_2 = 2\,\text{kg}$ respectively, riding on a frictionless horizontal plane, are joined by an ideal spring with force constant $k = 3\,\text{N/m}$. The system is pulled by a horizontal applied force, $F_A = 5\,\text{N}$, which is transmitted by an ideal rope as shown in the figures below. The system on the left[right] is in motion to the right[left], and both blocks are moving with precisely the same velocity.

(a) In this case, the ideal rope is attached to Block 2. Determine (i) the acceleration of the system, and (ii) the amount by which the spring is stretched.

(b) The ideal rope is attached to Block 1 in this instance. Determine (i) the acceleration of the system, and (ii) the amount by which the spring is stretched.

(c) Comment on the results obtained for parts (a) and (b).

D.55 A hockey mom passed the puck with an initial speed of $3\,\text{m/s}$ along a horizontal ice surface to her child, who inadvertently failed to intercept it. Kinetic friction brought the puck to a stop $20\,\text{m}$ away from where it was launched. Estimate the coefficient of kinetic friction for the rubber puck (of mass $170\,\text{g}$) sliding along the ice. [Hint: Attribute the deceleration of the puck entirely to the force of kinetic friction.]

D.56
Three blocks are stacked as in the adjacent figure. The respective masses of the blocks are: $M_1 = M$, $M_2 = 4\,M$, and $M_3 = 9\,M$. The blocks are at rest with respect to one another, while they accelerate along a horizontal frictionless surface under the influence of an external horizontal force of magnitude F_A acting on Block 1. All of the surfaces of contact between the various blocks possess the same coefficient of static friction.

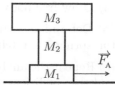

Explain what happens to the stack when the magnitude of the applied force is increased without bound. In particular, address whether the structure falls apart or remains intact. If it falls apart, which interface is the first to slide?

D.57 [See also D.30.] Reconsider Problem D.30 with kinetic friction acting. Neglect the fact that at the instant $t = 0$ the block is (momentarily) at rest, and take $\mu_k = 0.15$.

D.58 [See also D.32.] Repeat the previous problem with $m_1 = M = m_2$.

D.59 [See also D.33.] Let's reanalyse the problem with three blocks, each of mass M. Two of them ride on a horizontal plane, while the third hangs from a pulley as shown in the figure with D.33. Assume that the system is in motion (to the right and down). Determine the (i) acceleration of the system and (ii) tension in each of the ropes in the following cases.

(a) Kinetic friction, with coefficient μ_k, acts on the leftmost block only.

(b) Kinetic friction acts on both blocks with precisely the same coefficient, μ_k.

(c) Kinetic friction acts on both blocks with differing coefficients: μ_{kl} and μ_{kr}.

D.60 [See also D.34, E.18.] Consider case (a) examined in the previous problem. Suppose that the system was released from rest [instantaneously passing from static to kinetic friction], and determine its speed at the instant the blocks have each moved through a distance D.

D.61 [See also D.35.] Let's reanalyse the problem with three blocks, each of mass M. One of them rides on a horizontal plane, while the other two hang from a pulley as shown in the figure with D.35. Assume that the system is in motion (to the right and down). Suppose that kinetic friction, with coefficient μ_k, acts on the leftmost block. Determine the
(a) acceleration of the system and (b) tension in each of the ropes.

D.62 [See also D.37.] Let's reanalyse the problem with four blocks, each of mass M. Two of them ride on a horizontal plane, while the other two hang from a pulley as shown in the figure near D.37. Assume that the system is in motion (to the right and down). Determine the   (i) acceleration of the system and   (ii) tension in each of the ropes   in the following cases.

(a) Kinetic friction, with coefficient μ_{k}, acts on the leftmost block only.

(b) Kinetic friction acts on both blocks with precisely the same coefficient, μ_{k}.

(c) Kinetic friction acts on both blocks with differing coefficients: $\mu_{\text{k}1}$ and $\mu_{\text{k}r}$.

D.63 [See also D.38, E.18.] Consider case (a) examined in the previous problem. Suppose that the system was released from rest [instantaneously passing from static to kinetic friction], and determine its speed at the instant each block has moved through a distance D.

D.64 [See also D.48.] Repeat Problem D.48 with kinetic friction (coefficient $\mu_{\text{k}} = 0.1553$) acting between the block and the incline.

D.65 [See also D.49.] Repeat the previous problem for unspecified block mass M (keeping $\theta = 15°$).

D.66
Two blocks, of mass M and m respectively, are joined by a spring with force constant k, and pulled along a horizontal frictionless plane by a constant applied force. In this system the blocks move together with precisely the same velocity.

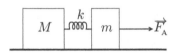

(a) Draw free body diagrams and enumerate the forces which act on each block.

(b) Determine   (i) the acceleration of the system and   (ii) the amount by which the spring is stretched.

(c) Repeat the analysis, supposing that the blocks are moving to the right, friction acts, and the coefficient of kinetic friction, μ_{k}, is the same for both blocks.

D.67 [See also E.26.] A block of mass $M = 5\,\text{kg}$ rides on a horizontal frictionless plane. Until $t = 0\,\text{s}$, the block was moving with speed $6\,\text{m/s}$ in the negative y-direction, *i.e.*, $v_0 = -6\,\text{m/s}\,[\hat{\jmath}]$. At $t = 0$, the block is at the origin, and it suddenly becomes subject to a constant force of kinetic friction, with magnitude $f_{\text{K}} = 10\,\text{N}$. Eventually the block stops.

(a) Draw a free body diagram and enumerate the forces which act on the block.

(b) Determine the acceleration of the block.

(c) Ascertain when and where the block comes to rest.

D.68 [See also E.27.] A block of mass $M = 5\,\text{kg}$ is located at the origin on a horizontal frictionless plane and is moving with speed $4\,\text{m/s}$ in the direction specified by $\frac{1}{\sqrt{5}}(2\,\hat{\imath} + \hat{\jmath})$ at an initial time $t = 0\,\text{s}$. At this same instant, $t = 0$, the block is suddenly subject to a constant force of kinetic friction, with magnitude $f_K = 10\,\text{N}$, acting in opposition to the block's motion. Eventually the block comes to rest.

(a) Draw a free body diagram and enumerate the forces which act on the block.

(b) Determine the acceleration of the block.

(c) Ascertain the   (i) time, t_f, and   (ii) position, x_f,   at which the block stops.

D.69 A block of mass $M = 10\,\text{kg}$ slides on a horizontal frictional surface. The coefficient of the kinetic friction operative between the block and the surface is $\mu_{\text{k}} = 1/2$. At $t = 0\,\text{s}$, the block is at the origin and moving with velocity $5\,\text{m/s}$ in the positive x-direction. Also at $t = 0\,\text{s}$, a time-dependent force with magnitude $F_{\text{A}} = 50\left(1 + e^{-t/3}\right)\,\text{N}$, where t is measured in seconds, begins to act in the positive x-direction. Confine attention to times $t \geq 0\,\text{s}$.

(a) Sketch curves showing the magnitudes of the applied and kinetic frictional forces acting on the block as functions of time.

(b) Draw an FBD for the block.

(c) Determine the (i) acceleration, (ii) velocity, and (iii) position of the block as functions of time.

(d) Discuss the asymptotic behaviour of the block, *i.e.*, its acceleration, velocity, and position, in the limit $t \to \infty$.

D.70

A block of mass M is at rest on a frictional plane inclined at angle θ. The coefficient of static friction acting between the block and the surface is μ_s. An ideal rope passing over an ideal pulley attaches a second block, of mass m, to the first as shown in the figure to the right.

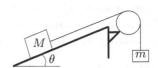

(a) Suppose that θ is less than the angle of repose. Determine the maximum possible value of m for which the system will remain at rest.

(b) Suppose that θ is much greater than the angle of repose. Determine the minimum value of m needed for the system to remain at rest.

D.71 A block of mass $m = 50\,\text{kg}$ is at rest on a frictional plane inclined at $15°$. The coefficient of static friction acting between the block and the surface on which it rides is $\mu_s = 0.8$.

(a) Show that the block remains at rest on the incline.

(b) Determine the magnitude of the minimum applied force, (i) directed parallel to the incline, $\vec{F}_{\|,\text{m}}$, and (ii) directed horizontally, $\vec{F}_{\text{h,m}}$, required to set the block in motion.

(c) Determine the magnitude and direction of the minimum applied force, $\vec{F}_{\text{A,m}}$, needed to set the block in motion.

D.72 [See also D.30.] Two blocks, with mass $m_1 = 2\,\text{kg}$ and $m_2 = 4\,\text{kg}$, are joined by a length of ideal rope passing over an ideal pulley, as shown in the figure accompanying D.30. Block 1 slides along a frictional plane, while Block 2 falls. The blocks are in forward motion and the coefficient of kinetic friction [acting between Block 1 and the plane] is $\mu_k = 0.3$.

(a) Compute the acceleration of the system and the tension in the rope.

(b) Suppose that the blocks are interchanged, *i.e.*, $1 \leftrightarrow 2$, and that Block 2 has the same frictional coefficient (*i.e.*, 0.3). Compute the acceleration of the system and the tension in the rope.

D.73 [See also D.31.] Repeat D.72 with the two masses, m_1 and m_2, and the common coefficient of friction, μ_k, left unspecified. [Assume that the system is in motion in the forward direction.]

D.74 [See also D.73.]

Two blocks of equal mass, M, are joined by a length of ideal rope which passes over an ideal pulley, as shown in the figure. One of the blocks rides along a horizontal frictional plane with coefficient of kinetic friction $\mu_k = 0.2$, while the other drops vertically down. At the time $t = 0\,\text{s}$, the blocks are in motion at $2\,\text{m/s}$ to the right and down, respectively.

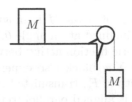

(a) Determine the acceleration of the system of blocks and the magnitude of the tension in the ideal rope.

(b) Compute the speed of the system when the blocks have moved through a distance of 1.5 m.

D.75 Repeat D.74 with the blocks initially in motion to the left and upward at 4 m/s, and the distance through which they move equal to 1.0 m.

D.76 [See also D.25.]

Two blocks of equal mass, M, joined by a length of ideal rope, are pulled along a horizontal frictional plane by a constant force, $\vec{F_A}$, as shown in the nearby figure. The two blocks have the same coefficient of kinetic friction, μ_k, and are moving with the same velocity, $\vec{v_0}$.

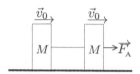

Determine the acceleration of the system of blocks and the magnitude of the tension in the ideal rope.

D.77 [See also D.26.] Suppose that the system in D.76 is at $x_0 = 0$, and moving forward with speed v_0 at $t = 0$. Determine the speed of the blocks at the instant that the system has displaced a distance D.

D.78 An IDEAL ROD [*viz.*, a massless and perfectly rigid device for transmitting a force] is employed to push a sled of mass M riding upon a horizontal frictional surface. The coefficients of static and kinetic friction are μ_s and μ_k, with $\mu_s > \mu_k$, as usual. The rod is horizontal in parts (a) and (b), and tilted upwards in part (c).

(a) What is the magnitude of the minimum horizontal applied force needed to set the sled in motion [starting from rest]?

(b) What is the magnitude of the horizontal force needed to keep the sled in motion, once it has just begun to move?

(c) At what angles should the rod by held in order to minimise the forces required to (i) initiate   and   (ii) sustain   the motion of the sled?

D.79
Three blocks are sliding down a frictional incline with slope $\pi/6$. Block 1 is attached to Block 2 by means of an ideal rod (massless, with fixed length), while Blocks 2 and 3 are in contact. The respective masses of the blocks and their coefficients of kinetic friction with respect to the incline are quoted below.

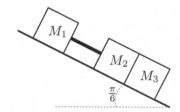

$M_1 = 1\,\text{kg}, \quad M_2 = 2\,\text{kg}, \quad M_3 = 3\,\text{kg}, \quad \text{and} \quad \mu_{K1} = \frac{1}{10\sqrt{3}}, \quad \mu_{K2} = \frac{1}{5\sqrt{3}}, \quad \mu_{K3} = \frac{3}{10\sqrt{3}}.$

(a) Determine the acceleration of the system.

(b) Ascertain whether the rod is under tension or compression.

D.80 [See also D.44, E.22.]

A block of mass M is moving upward along a frictional plane inclined at an angle θ, with $\theta < 45°$. The coefficient of kinetic friction acting between the block and the incline is μ_k. The block also experiences an applied force, with magnitude F_A, transmitted through a length of ideal rope which is aligned parallel to the incline.

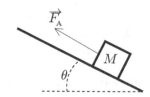

(a) Compute the acceleration of the block for arbitrary F_A.

(b) Determine the particular value of F_A for which the block coasts up the plane.

(c) Suppose that $M = 5\,\text{kg}$, $\theta = 20°$ and $\mu_k = 0.2$. Determine the acceleration of the block when (i) $F_A = 10\,\text{N}$, and (ii) $F_A = 40\,\text{N}$.

D.81 [See also E.46.]

Two blocks, $M_1 = 10\,\text{kg}$ and $M_2 = 5\,\text{kg}$, one ideal pulley, some ideal rope, and an inclined plane ($\theta = 30°$) are arranged as shown in the figure. The coefficient of kinetic friction between block 1 and the plane is $\mu_k = 1/\sqrt{3}$. At $t_i = 0$, Block 1 is in motion up the incline, while 2 is moving downward, with speed $v = 20\,\text{m/s}$. At $t_f = 6\,\text{s}$, each block has moved 60 m from its starting point. Determine the speed of the blocks at t_f.

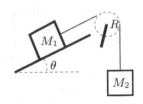

D.82 [See also D.46.] Two blocks, of mass m_1 and m_2 respectively, together ride on a frictional plane inclined at angle θ. The coefficients of kinetic friction for these blocks are μ_{k1} and μ_{k2}, respectively. A constant force, $\vec{F}_A$, acting parallel to the incline, is applied to block 1, which remains in contact with block 2.

For what value of the magnitude of the applied force does the system of blocks coast along the incline, and what is the magnitude of the force of contact acting between the blocks in this case?

D.83 [See also D.47.] Two blocks, of mass m_1 and m_2 respectively, together ride on a frictional plane inclined at angle θ. The coefficients of kinetic friction for the two blocks are μ_{k1} and μ_{k2}, respectively. A constant force, $\vec{F}_A$, is applied horizontally to block 1, which remains in contact with block 2.

For what value of the magnitude of the applied force does the system of blocks coast along the incline, and what is the magnitude of the force of contact acting between the blocks in this case?

D.84 [See also D.54.] Two blocks, of mass $M_1 = 1\,\text{kg}$ and $M_2 = 2\,\text{kg}$ respectively, riding on a frictional horizontal plane, are joined by an ideal spring with force constant $k = 3\,\text{N/m}$. The relevant coefficients of kinetic friction are $\mu_{k1} = 0.15$ for Block 1 and $\mu_{k2} = 0.3$ for Block 2. The system is pulled by a horizontal applied force, $F_A = 5\,\text{N}$, which is transmitted via an ideal rope as shown in the figures below. The system on the left[right] is in motion to the right[left], and both blocks are moving with precisely the same velocity.

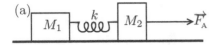 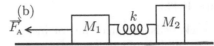

(a) In this case, the ideal rope is attached to Block 2. Determine (i) the acceleration of the system and (ii) the amount by which the spring is stretched.

(b) The ideal rope is attached to Block 1 in this instance. Determine (i) the acceleration of the system and (ii) the amount by which the spring is stretched.

(c) Repeat (a) with $\{\mu_{k1} = 0.3, \mu_{k2} = 0.15\}$.

(d) Repeat (b) with $\{\mu_{k1} = 0.3, \mu_{k2} = 0.15\}$.

(e) Repeat (a) with $\{\mu_{k1} = 0.3 = \mu_{k2}\}$.

(f) Repeat (b) with $\{\mu_{k1} = 0.3 = \mu_{k2}\}$.

D.85 [See also E.10, M.8.] A sled of mass 40 kg is pulled along a horizontal frictional surface with an applied force of 50 N acting at an angle of $\pi/6$ rad with respect to horizontal. The sled is moving in the forward direction, and the coefficient of kinetic friction acting between the sled and the surface on which it rides is $\mu_k = 0.04$. It so happens that in a 4-second time interval, the sled advances a distance of 10 m. Determine the speed of the sled at the (a) start and (b) end of the 4-second interval.

D.86 [See also D.87.]

Two parents, walking side-by-side on a horizontal snowy plane, pull a sled on which their child rides. Together the child and the sled have mass $M = 25$ kg. The coefficient of kinetic friction acting between the sled and the ground is $\mu_k = 0.2$. The mom applies a constant force, $F_m = 40$ N, at an angle of $\pi/6$ rad above the horizontal. The dad, who is somewhat taller, pulls with constant force F_d at $\pi/4$ rad above the horizontal.

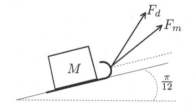

(a) Determine the force that the dad must apply in order that the sled move forward uniformly.

(b) Determine how hard the dad must pull to get the sled "airborne" for an instant. [HINT: The normal force of contact between the ground and the sled vanishes when they separate.]

D.87 [See also D.86.]

The family in the previous problem encounters a snowy slope inclined at angle $\pi/12$ upward with respect to the horizontal. The mom maintains a constant force, $F_m = 40$ N, at an angle of $\pi/6$ rad above the slope, while the dad pulls with constant force F_d at $\pi/4$ rad above the slope.

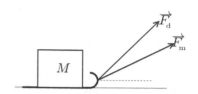

(a) Determine the force that the dad must apply in order that the sled move uniformly up the slope.

(b) Determine how hard the dad must pull to get the sled "airborne" for an instant.

D.88 [See also E.43.]

Run-away truck beds are stretches of increasingly soft sand found at the bottom of steeply inclined roadways, in which vehicles whose brakes have failed may be safely brought to rest.

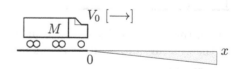

In this case, the truck, of mass M, enters the bed at $x = 0$, with speed V_0 at some initial instant. Our model shall attribute the effect of the truck bogging down in the increasingly soft bed to a force of kinetic friction with increasing coefficient, *viz.*, $\mu_k = \frac{x}{L}$, where L is a constant with dimensions of length. Once the truck comes to rest, it remains at rest, as the frictional force vanishes from that time onward. Ascertain the distance (into the bed) at which the truck stops.

D.89 [See also E.37, R.19.]

Two blocks with mass m_1 and m_2, $(m_1 > m_2)$, hang by ideal rope wrapped around an ideal pulley. The system is held at rest, as shown in the figure, and then suddenly released at time t_i. We shall compute the speeds of the two masses at the instant at which they pass each other by means of the following sequence of steps.

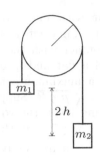

(a) Draw FBDs for each of the masses (and the pulley) and enumerate the forces.

(b) Determine the acceleration of the system.

(c) Ascertain the time required for the blocks to each move a distance h from rest.

(d) Determine the speed of the blocks at the instant they pass.

D.90   A car of mass M experiences centripetal acceleration a_c when it goes around a curve of radius R at speed V. Compare this acceleration to that of a car of mass $2M$ going around a curve of radius $3R$ at speed $2V$.

D.91   PK drove his Triumph Spitfire around a smooth, flat, and level curve at such a prodigious speed that the fuzzy dice[2] hanging from his rear-view mirror were tilted away from the vertical at 45°. Determine the centripetal acceleration that PK experienced in rounding this corner.

D.92

An ideal rope is threaded through a tiny hole in a horizontal frictionless plane. A block of mass M_1 hangs from the rope (beneath the plane), while a block of mass M_2 attached to the other end of the rope moves along a circular path of radius R, at constant speed v. Neglect air resistance, and assume that the rope slides without friction along the edge of the hole.

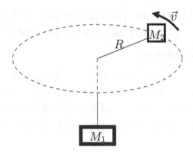

(a) Draw free body diagrams for the two masses.

(b) Express the centripetal $m\,a$ experienced by Block 2 in terms of its mass, its tangential speed, and the radius of its trajectory.

(c) Given that Block 2 moves in a circular path, determine the tension in the rope.

(d) Apply Newton's Second Law to determine an expression for the speed of the orbiting block in terms of measurable parameters: $\{g, R, M_1, M_2\}$.

(e) Discuss what happens if drag or friction forces act on the orbiting block.

---

[2]This story is clearly fictitious.

D.93
A car with mass M, equipped with tires having coefficient of static friction (in contact with the roadway) μ_s, can round a flat and level curve of radius R at a maximum safe speed given by $v_{\mathrm{MAX}} = \sqrt{\mu_s\, g\, R}$. Suppose that, instead of being level, the road is banked the "wrong way," as in the figure.

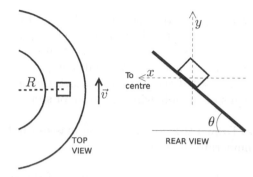

(a) Draw a free body diagram for the car, and on it indicate the net $M\,\vec{a}$.

(b) Assuming that the net force gives rise to a purely centripetal acceleration, determine the maximum safe speed in this case.

(c) Take the flat-track limit, $\lim_{\theta \to 0}$, of the result obtained in (b).   (i) What result ought one get?   (ii) Is your expectation realised?   (iii) How does the $\theta \neq 0$ case compare to its $\theta \to 0$ limit?

D.94 A block is in motion down an inclined frictionless plane surface. The angle of inclination is such that the acceleration down the incline is $2\,\mathrm{m/s^2}$. [$\theta \simeq 11.54°$ when $g = 10\,\mathrm{m/s^2}$.] As the block passes by the point $x = 0$, it happens to be moving with speed $v_0 = 4\,\mathrm{m/s}$. At this same instant, two additional blocks are launched, side by side, down the incline from $x = 0$. The initial speeds of these blocks are $v_{S0} = 2\,\mathrm{m/s}$ (slow) and $v_{F0} = 6\,\mathrm{m/s}$ (fast) respectively.

(a) Determine the trajectories of the three blocks from the perspective of an observer at rest with respect to the incline.

(b) Determine the trajectories of the slow and fast blocks from the perspective of an observer riding along with the initial block.

(c) Comment.

D.95 [See also E.48.] A block of mass M slides down a long frictional incline while immersed in a draggy medium. The angle of descent is $\theta = 30°$. The coefficient of kinetic friction between the surfaces of the block and incline is $\mu_k = \frac{4}{5\sqrt{3}}$. The linear drag coefficient is small enough to be negligible, $b_1 = 0$, while the quadratic coefficient happens to satisfy the relation
$$\frac{b_2}{M\,g} = 0.025\,\frac{\mathrm{s^2}}{\mathrm{m^2}}.$$
Employ a dynamics argument to determine the terminal velocity of the block.

D.96 Consider a mass-spring system $(M,\,k)$ riding on a horizontal frictionless plane, and subject to a linear drag force with drag coefficient b. Suppose that at $t = 0$ the system is at a turning point, a distance of $+A$ from its equilibrium position.

(a) Write the (nontrivial) equation of motion for the system (*i.e.*, $M\,a = F_{\mathrm{net}}(x, v; t)$).

(b) Suppose that there exist trajectories of the form $x(t) = A\,e^{-\gamma t}\cos(\omega t)$ which satisfy the equation of motion for unknown constant factors γ and ω. [Here we are making an *Ansatz*.]   (i) Formally compute the velocity and acceleration functions associated with the trajectory, $x(t)$.   (ii) Substitute the results from (i) into the equation of motion, and collect terms in **sine** and **cosine** separately. This ought to yield two equations involving the unknowns γ, ω.   (iii) Solve these equations to express γ and ω in terms of (M, k, b).   (iv) Write down the trajectory $x(t)$ in terms of (M, k, b).

E

Energetics Problems

E.1 A constant applied force, $\vec{F_A} = 3\,\text{N}\ [\leftarrow]$, acts on particles undergoing a variety of straight-line displacements, which are described below. For each displacement, compute the work done by $\vec{F_A}$.
(a) $\vec{S_a} = 5\,\text{m}\ [\leftarrow]$,   (b) $\vec{S_b} = 7\,\text{m}\ [\uparrow]$,   (c) $\vec{S_c} = 5\,\text{m}\ [36.87° \text{ above} \leftarrow]$,
(d) $\vec{S_d} = \sqrt{8}\,\text{m}\ [45° \text{ below} \leftarrow]$.

E.2 A constant force, $\vec{F_A} = (3, 4)\,\text{N}$, acts on particles undergoing a variety of straight-line displacements. Compute the work done by $\vec{F_A}$ as the particle moves from the initial to the final positions enumerated below.
(a) $(0, 0) \rightarrow (2, 0)$,   (b) $(2, 2) \rightarrow (2, 4)$,   (c) $(5, 3) \rightarrow (2, 2)$,
(d) $(-4, -3) \rightarrow (2, 5)$.

E.3 A variable force, $\vec{F_A}(x, y) = (2x^2, y)\,\text{N}$, acts on particles $\{a, b, c, d\}$. Compute the work done by $\vec{F_A}$ acting through each of the straight-line displacements listed below. [The positions are expressed in metres.]
(a) $(0, 0) \rightarrow (0, 6)$,   (b) $(0, 0) \rightarrow (6, 0)$,   (c) $(6, 0) \rightarrow (12, 0)$,
(d) $(-6, -6) \rightarrow (12, 12)$.

E.4 [See also D.16, M.5.] A block of mass $m = 10\,\text{kg}$ rides on a frictionless horizontal plane. Three applied forces, $\vec{F_1} = 10\,\text{N}\ [\rightarrow]$, $\vec{F_2} = 40\,\text{N}\ [\downarrow]$, and $\vec{F_3} = 50\,\text{N}\ [+53.13° \text{ WRT } x\text{-axis}]$, act on the block. The block's weight and the normal force both act vertically, and cancel. The initial velocity of the block, at $t = 0\,\text{s}$, is $\vec{v_0} = 2\,\text{m/s}\ [\leftarrow]$, while its initial position is $\vec{r_i} = (0, 0)$. Subsequently, the block is seen to move in a straight line within the horizontal plane to the point $\vec{r_f} = (24, 0)$.

(a) Determine the net (applied) force acting on the block.

(b) Compute the mechanical work done by the net applied force on the block as it travelled from its initial to its final position.

(c) State briefly why the work done by the normal force as well as that done by the weight is 0 J.

(d) Separately compute the work done by each of the three applied forces. Sum these to get the total work performed on the block. Comment.

(e) Determine the speed of the block when it reaches $\vec{r_f}$.

E.5 A block of mass $m = 10\,\text{kg}$ is pushed along a frictional horizontal plane by an applied force, $\vec{F_A}$, in such a way that it moves at a constant speed, $3\,\text{m/s}$, in the positive x-direction. The coefficient of kinetic friction acting between the block and the plane is $\mu_k = 1/4$. The initial position of the block is $\vec{r_i} = (0, 0)\,\text{m}$, and its final position is $\vec{r_f} = (5, 0)\,\text{m}$.

(a) Determine the force of kinetic friction acting on the block.

(b) Infer the magnitude and direction of the applied force acting on the block.

(c) Compute the mechanical work done by friction on the block as it moved from its initial to its final position.

(d) Compute the mechanical work done by the applied force acting on the block as it is displaced from its initial to its final position.

(e) Comment on the results obtained for (c) and (d).

(f) Compute the power input to the block by   (i) each of the forces individually and (ii) the net force. Comment.

E.6

A particle of mass $4\,\text{kg}$ is constrained to move on a horizontal plane while subjected to a number of forces. These include an applied force, $\vec{F}_A = (8,\,0)\,\text{N}$, which happens to be constant in both magnitude and direction; the force of kinetic friction, with constant magnitude $f_K = \mu_K N$, where $\mu_K = 0.2$; and a variable force, $\vec{F}_V(x, y) = (3\,x^2 + y,\,0)\,\text{N}$. In all cases below, the particle moves from the origin, $\vec{r}_i = (0,\,0)$, to a common final destination, $\vec{r}_f = (5,\,4)$, via a path comprised of straight line segments.

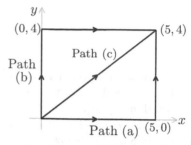

Common Figure for E.6–9

(a) This path consists of two straight line segments, with intermediate point $\vec{r}_a = (5,\,0)$. Compute the mechanical work done by each of the three forces acting through this two-stage displacement.

(b) Two straight line segments make up this path. The corner lies at $\vec{r}_b = (0,\,4)$. Compute the mechanical work done by each of the forces acting through this two-stage displacement.

(c) This path consists of one straight (diagonal) segment. Compute the work done by each of the three forces acting through this displacement.

(d) Compare the results obtained in parts (a–c). Under what circumstances is the work done by the forces path-independent?

E.7   A constant applied force, $\vec{F}_A = (8,\,0)\,\text{N}$, acts on a particle of mass $4\,\text{kg}$ as the particle moves on a horizontal plane surface. Compute the total amount of work done by $\vec{F}_A$ acting on the particle as it traverses the following paths leading to a net displacement from its initial position at the origin, $\vec{r}_i = (0,\,0)$, to its final position at $\vec{r}_f = (5,\,4)$.

(a) Two straight line steps with corner at $\vec{r}_a = (5,\,0)$, i.e., $(0,\,0) \to (5,\,0) \to (5,\,4)$.

(b) Two straight line steps with corner at $\vec{r}_b = (0,\,4)$, i.e., $(0,\,0) \to (0,\,4) \to (5,\,4)$.

(c) One direct straight line diagonal path: $(0,\,0) \to (5,\,4)$.

(d) Compare the results obtained for parts (a–c), above.

E.8   Kinetic friction, $\vec{f}_K$, acts on a block of mass $M = 4\,\text{kg}$ which is sliding along on horizontal plane surface. The coefficient of kinetic friction is taken to be $\mu_K = 0.20$. Compute the total amount of work done by kinetic friction as the block moves from the origin, $\vec{r}_i = (0,\,0)$, to $\vec{r}_f = (5,\,4)$, via a number of different paths.

(a) Two straight line steps with corner at $\vec{r}_a = (5,\,0)$, i.e., $(0,\,0) \to (5,\,0) \to (5,\,4)$.

(b) Two straight line steps with corner at $\vec{r}_b = (0,\,4)$, i.e., $(0,\,0) \to (0,\,4) \to (5,\,4)$.

(c) One direct straight line diagonal path: $(0,\,0) \to (5,\,4)$.

(d) Compare the results obtained for parts (a–c), above.

E.9   A variable force, $\vec{F}_V(x, y) = (3\,x^2 + y,\,0)\,\text{N}$, acts on a $4\,\text{kg}$ particle constrained to lie on the surface of a horizontal xy-plane. Compute the total net amount of work performed on the particle by this variable force while the particle travels from $\vec{r}_i = (0,\,0)$, the origin, to $\vec{r}_f = (5,\,4)$, via a variety of distinct paths.

(a) Two straight line steps with corner at $\vec{r}_a = (5, 0)$, *i.e.*, $(0, 0) \rightarrow (5, 0) \rightarrow (5, 4)$.

(b) Two straight line steps with corner at $\vec{r}_b = (0, 4)$, *i.e.*, $(0, 0) \rightarrow (0, 4) \rightarrow (5, 4)$.

(c) One direct straight line diagonal path: $(0, 0) \rightarrow (5, 4)$.

(d) Compare the results obtained for parts (a–c), above.

E.10 [See also D.85, E.42, and M.8.] A sled of mass 40 kg, is pulled along a horizontal frictional surface with an applied force of 50 N acting at an angle of $\pi/6$ rad with respect to horizontal. The sled is moving in the forward direction, and the coefficient of kinetic friction acting between the sled and the surface on which it rides is $\mu_k = 0.04$.

(a) Draw a FBD for the sled and enumerate the forces which act.

(b) Determine the net force acting on the sled.

(c) Suppose that the sled is pulled through a distance $\Delta s = 10$ m in a time interval $\Delta t = 4$ s. Compute the amount of work performed by the   (i) weight,   (ii) normal, (iii) applied, and   (iv) friction   forces as the sled underwent its displacement.

(d) Determine the net amount of work done on the sled. Comment.

E.11 A block of mass M was accelerated from rest, 0 m/s, to 10 m/s. This took an amount of work $W_{0 \rightarrow 10}$. Then the same block was accelerated from 10 m/s to 20 m/s with the addition of work $W_{10 \rightarrow 20}$. Compare the relative sizes of $W_{0 \rightarrow 10}$ and $W_{10 \rightarrow 20}$.

E.12 [See also M.19.] Superman observes a fast passenger train of mass 200 metric tonnes travelling at 180 km/h toward a bridge which spans a yawning chasm. The bridge has been damaged, and the train will derail as soon as the engine reaches it. The train is but 200 m from the beginning of the bridge when Superman arrives on the scene. We shall determine, via two different means, the minimum [average] force that Superman must exert on the train to stop it before it hurtles disastrously into the chasm.

(a) Invoking the WORK–ENERGY THEOREM: Compute the train's   (i) initial and (ii) final [when it is safely stopped]   kinetic energy.   (iii) How much work must Superman do on the train to stop it?   (iv) What minimum constant force must Superman exert to do this much work within the 200 m distance remaining before the bridge?   (v) Determine the average acceleration of the train.

(b) Employing standard dynamics methods: (i) Use a kinematic argument to determine the constant acceleration of the train that will bring it to a stop just at the edge of the bridge.   (ii) What constant force must Superman exert on the train so as to give rise to the necessary acceleration?

E.13 [See also, D.26, M.6.] Two blocks, with masses m_1 and m_2, are able to move in a straight line along a frictionless horizontal plane surface. The blocks are linked by an ideal rope lying horizontally. In addition, an external agent applies a force on another length of ideal rope which is affixed to Block 1. The situation is illustrated in the figure alongside D.25. We assume that the system is in forward motion with speed v_0 at the time $t = 0$. We wish to determine properties of the system at the instant each of the blocks has displaced a forward distance D from its position at $t = 0$. For the two cases,   (a) $\theta = 0$ [the external force is applied parallel to the plane] and   (b) $\theta \neq 0$,   compute   (i) the net work done on each of the blocks [after each undergoes a displacement D],   (ii) the net work input to the system, and   (iii) the final speed of the system.

E.14 [See also D.17, M.7.] A block of mass 2.5 kg riding on a horizontal frictionless plane is subjected to four applied forces: $\vec{F}_1 = 13$ N [22.62° WRT positive x-axis], $\vec{F}_2 = 5$ N [53.13° WRT positive y-axis], $\vec{F}_3 = 8$ N [←], and $\vec{F}_4 = 3$ N [↓]. The weight and normal force of contact act perpendicular to the plane, and they cancel everywhere and always.

At the instant $t = 0\,\mathrm{s}$, the block is at the origin, $\vec{r}_0 = (0, 0)\,\mathrm{m}$, and moving with velocity $\vec{v}_0 = 3\,\mathrm{m/s}\,[\to]$. Our goal is to determine the speed of the block at the time $t = 2\,\mathrm{s}$, when it reaches the point $\vec{r}_2 = (6, 4)\,\mathrm{m}$.

(a) Determine the net force acting on the block.

(b) (i) Argue that the particle's trajectory is NOT a straight line through space, and (ii) determine the net displacement from $\vec{r}_0$ to $\vec{r}_2$, nonetheless.

(c) (i) Argue that the work done by each of the forces along the trajectory of the particle is precisely equal to the work done by the same force acting through a straight-line trajectory yielding the same displacement. (ii) Compute the work done by each of the forces separately along the straight-line trajectory.

(d) Determine the net work done on the block.

(e) (i) Determine the initial kinetic energy of the block. (ii) Ascertain the value of the kinetic energy of the block when it is at $\vec{r}_2$. (iii) Thence, compute the speed of the block when it is at $\vec{r}_2$.

E.15 [See also D.74.] Two blocks of equal mass, M, are joined by a length of ideal rope which passes over an ideal pulley, as shown in the figure alongside D.74. One of the blocks rides along a horizontal frictional plane with coefficient of kinetic friction $\mu_k = 0.2$, while the other drops vertically down. At the time $t = 0$ s, the blocks are in motion at $2\,\mathrm{m/s}$ to the right and down, respectively. At a later time, t, each of the blocks has displaced $1.5\,\mathrm{m}$. Determine the speed of the system at time t.

E.16 [See also D.30, D.73, R.18.] Two blocks, with masses $m_1 = 2\,\mathrm{kg}$ and $m_2 = 4\,\mathrm{kg}$, are joined by a length of ideal rope passing over an ideal pulley, as shown in the figure alongside D.30. Block 1 slides along a frictionless plane, while Block 2 falls. Neither block is affected by drag. The system is released from rest at $t = 0$. Determine the speed of the blocks once each has moved $1.2\,\mathrm{m}$.

E.17 [See also D.31, D.73.] Repeat the previous problem for arbitrary m_1 and m_2, and unspecified distance D.

E.18 [See also D.34, M.9.] Three blocks, each of mass M, two ideal ropes, a frictionless plane, and an ideal pulley comprise the dynamical system studied in D.33. At the instant $t = 0\,\mathrm{s}$, the system is at rest. At a later time, $t_f > 0$, the two blocks on the plane have each moved a distance H to the right. Use energy methods to determine the speed of the system of blocks at time t_f.

E.19 [See also D.36, M.10.] Three blocks, each of mass M, two ideal ropes, a frictionless plane, and an ideal pulley comprise the dynamical system illustrated in D.35. At the instant $t = 0\,\mathrm{s}$, the system is at rest. At a later time, $t_f > 0$, the two blocks to the right have each dropped a distance H. Use energy methods to determine the speeds of the blocks at t_f.

E.20 [See also D.37, M.11.] Four blocks, each of mass M, three ideal ropes, a frictionless plane, and an ideal pulley comprise the dynamical system considered in D.37. At the instant $t = 0\,\mathrm{s}$, the system is at rest. At a later time, $t_f > 0$, the two blocks to the right have each dropped a distance H. Use energy methods to determine the speeds of the blocks at t_f.

E.21 [See also D.44, D.80, E.22.] Various blocks of mass M are moving upward along frictionless planes with angle of inclination $\theta < 45°$. Each block is acted upon by an applied force, with magnitude F_{A}, transmitted through a length of ideal rope.

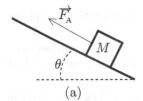

(a)

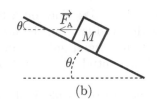

(b)

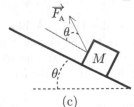

(c)

(a) In this case, the ideal rope extends parallel to the incline. (i) Compute the net work done on the block as it displaces a distance D upward along the plane. Compute the change in the kinetic energy of the block when $M = 5\,\text{kg}$, $\theta = 20°$, $D = 3\,\text{m}$, and (ii) $F_A = 10\,\text{N}$ and (iii) $F_A = 40\,\text{N}$.

(b) Here, the rope lies horizontally and remains so as the block moves. Repeat steps (i–iii) above, in this instance.

(c) The third block is pulled by a rope which remains elevated at twice the ramp angle. Repeat steps (i–iii).

E.22 [See also D.44, D.80, E.21.] Various blocks of mass M are moving upward along frictional planes with angle of inclination $\theta < 45°$ and common coefficient of kinetic friction μ_k. Each block is acted upon by an applied force, with magnitude F_A, transmitted through a length of ideal rope, as displayed in the figures accompanying E.21.

(a) In this case, the ideal rope extends parallel to the incline. (i) Compute the net work done on the block as it displaces a distance D upward along the plane. Compute the change in the kinetic energy of the block when $M = 5\,\text{kg}$, $\theta = 20°$, $\mu_k = 0.2$, and (ii) $F_A = 10\,\text{N}$ and (iii) $F_A = 40\,\text{N}$.

(b) Here, the rope lies horizontally and remains so as the block moves. Repeat steps (i–iii) above, in this instance.

(c) The third block is pulled by a rope which remains elevated at twice the ramp angle. Repeat steps (i–iii).

E.23 A body of mass M experiences a straight-line displacement from an initial position, $\vec{r}_i = (x_i, y_i)$, to a final position, $\vec{r}_f = (x_f, y_f)$, in a time interval lasting from t_i to t_f. The local acceleration due to gravity is $\vec{g} = (0, -g)$, for g, a positive constant with appropriate dimensions.

(a) How much work is done on the body, as it undergoes its displacement, by the weight force?

(b) By what amount does the gravitational potential energy of the body change during the time interval?

(c) Discuss the manner in which your results for (a) and (b) would be changed if the body proceeded from $\vec{r}_i = (x_i, y_i)$, to $\vec{r}_f = (x_f, y_f)$, via a circuitous route.

E.24 [See also D.41.] A block of mass $M = 5\,\text{kg}$ is held at rest on a frictionless plane with angle of inclination $\theta = 30° = \frac{\pi}{6}\,\text{rad}$. At $t = 0\,\text{s}$, it is released. With what speed does the block pass by an observer at rest with respect to the surface and located a distance $D = 5\,\text{m}$ from the initial position of the block?

E.25 [See also D.42.] A block of mass $M = 70\,\text{kg}$ is launched, with speed $v_0 = 6\,\text{m/s}$, up a very large frictionless plane inclined at angle $\theta = 45° = \frac{\pi}{4}\,\text{rad}$.

(a) What maximum distance upward along the incline is attained by the block?

(b) With what speed does the block pass by its initial position on its way back down?

E.26 [See also D.67.] A block of mass $M = 5\,\text{kg}$ rides on a horizontal frictionless plane. Until $t = 0\,\text{s}$ the block was moving with speed $6\,\text{m/s}$ in the negative y-direction, *i.e.*, $v_0 = -6\,\text{m/s}\,[\hat{\jmath}]$. At $t = 0$, the block is at the origin, and it suddenly becomes subject to a constant force of kinetic friction with magnitude $f_{\text{K}} = 10\,\text{N}$. Determine where the block comes to rest.

E.27 [See also D.68.] At an initial time $t = 0\,\text{s}$, a block of mass $M = 5\,\text{kg}$ is located at the origin on a horizontal frictionless plane and is moving with speed $4\,\text{m/s}$ in the direction specified by $\frac{1}{\sqrt{5}}(2\hat{\imath} + \hat{\jmath})$. At this same instant, the block is suddenly subject to a constant force of kinetic friction, with magnitude $f_K = 10\,\text{N}$, acting in opposition to the block's motion. Ascertain where the block comes to rest.

E.28 A spring with force constant $k = 12\,\text{N/m}$ is anchored at one end. A block of mass $M = 25\,\text{kg}$ is attached to the free end of the spring. At $t_i = 2\,\text{s}$, the spring is compressed by $0.25\,\text{m}$, and later, at $t_f = 22\,\text{s}$, it is extended $0.5\,\text{m}$ from its natural length. Compute the change in the potential energy of the spring in the twenty-second time interval.

E.29 A block–spring system with mass M and force constant k is aligned with the x-direction. With one end of the spring securely fixed, the equilibrium position of the block is at $x_0 \neq 0$. Determine   (a) the spring force exerted on the block and   (b) the spring potential energy of the system   when the block is at x.

E.30 A spring with characteristics k and L is anchored at the base of a frictionless plane which is inclined at angle θ, as illustrated below and to the left. A block of mass M is brought quasi-statically down the incline until it makes contact with the free end of the spring at $X_0 = 0$.

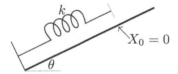
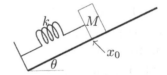

(a) Suppose that the block is allowed to slowly slide further down the incline until it is supported by the spring. At what position on the incline [with respect to the free equilibrium position, X_0] is the equilibrium position, x_0, of the mass–spring–incline system?

(b) Suppose instead that the block is simply let go at the point at which it makes contact with the spring. Determine the farthest point down the incline reached by the block in its subsequent motion.

E.31
A toy car with mass $M = 25\,\text{g}$ rides on a flexible track as illustrated in the adjacent figure. Suppose that the start of the track is 0.3 metres above the finish line, and that the car starts from rest.

(a) Estimate the speed of the car at the finish line using energy principles.

(b) Does the track shape affect the speed with which the car reaches the finish line?

(c) Does the shape of the track affect the time at which the car reaches the finish line?

E.32 A toy car of mass $40\,\text{g}$ rides along a frictionless track of length $5\,\text{m}$. One end of the track is held $2\,\text{m}$ above the floor. The end on the floor meets a tiny wedge inclined upward at $30°$, which acts as a ramp. If the car is let go from rest at the top of the incline, how far from [the ramp at the end of] the track does it land?

E.33

A toy car with mass $M = 25\,\text{g}$ rides on a flexible track containing a loop-the-loop, as shown in the figure. Compute the minimum height [above the base of the loop] at which the car must start in order that it may pass safely through the loop-the-loop.

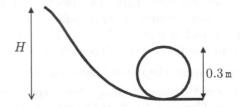

N.B.: The minimum height is not $0.3\,\text{m}$, because, if this were the case, the car would stop at the top of the loop, and then fall off the track. To barely stay on the track, the car must instead experience ZERO normal force of contact with the track only at the instant at which it is upside-down at the top of the loop-the-loop. As a consequence, the centripetal $m\,a$ of the car, at that particular instant, is entirely attributable to its weight.

E.34 [See also D.50.] Two blocks of mass $m = 25\,\text{kg}$ are joined by a length of ideal rope which passes over an ideal pulley, as shown in the figure accompanying D.48. One of the blocks rides along a frictionless plane which is inclined at $15°$, while the other drops vertically down. The blocks were released from rest. Determine the speed of the system at the instant that each block has moved $1.5\,\text{m}$ from its original location.

E.35 In an analog pinball game, the balls are initially launched up a long shallow ramp by means of a spring-loaded plunger system. The player chooses the degree of compression of the spring. Before and after the spring is compressed [from its equilibrium length], the ball is at rest against the plunger. When the spring is released, the ball is propelled up the ramp. Here we shall model the rolling pinball as though it were a particle sliding along a frictionless incline. For definiteness, take the length of the ramped channel in which the ball travels before entering into the "game region" to be $L = 1.2\,\text{m}$, the slope angle $\theta = \pi/12 = 15°$, the mass of the ball $M = 25\,\text{g}$, and the spring constant $k = 100\,\text{N/m}$.

(a) The spring is compressed by an amount, Δx, which we shall [for now] assume is negligible insofar as the ball is concerned. The ball remains in contact with the plunger and is at rest prior to its explosive launch into the game. Determine whether the ball escapes the channel when (i) $\Delta x = 1\,\text{cm}$, (ii) $\Delta x = 4\,\text{cm}$, and (iii) $\Delta x = 10\,\text{cm}$.

(b) Repeat the analyses, (i–iii), taking into account that the ramp length is increased by the amount by which the plunger is compressed.

E.36 [See also D.39, M.47, R.24.] Consider the two blocks joined by a length of ideal rope which is slung over an ideal pulley of radius R, illustrated in the figure alongside D.39. The system was held at rest until time $t = 0$, at which time it was released and given a short sharp tap, (instantaneously) setting the system in motion with speed v_0. WLOG, the left block may be assumed to rise while the right block falls. The mass of the left block is M_L, and that of the right block is M_R.

(a) Determine the initial [*i.e.*, just after the tap at $t = 0$] kinetic energies of (i) Block L, (ii) Block R, and (iii) the $(L + R)$-block system.

(b) Assign initial potential energies to (i) Block L, (ii) Block R, and (iii) the $(L + R)$-block system.

(c) At time t_H, $t_H > 0$, the left block has risen a distance H, while the right block has dropped by the same amount. Determine the amounts by which the potential energies of (i) Block L, (ii) Block R, and (iii) the $(L + R)$-block system change.

(d) Determine the change in the kinetic energy of the system throughout the time interval $t = 0 \to t_H$, and hence infer the final speed of the system of blocks.

E.37 [See also D.89, R.19.] Two blocks with masses m_1 and m_2 ($m_1 > m_2$) hang by ideal rope wrapped around an ideal pulley, as shown in the figure accompanying D.89. The system is held at rest, as shown in the figure, and then suddenly released at time t_i. We shall compute the speeds of the two masses at the instant at which they pass each other.

(a) Compute the initial kinetic energies of   (i) Block 1 and   (ii) Block 2.

(b) Say that m_2 is rising with speed v at the instant, t_f, that the masses pass each other. Compute the final kinetic energies of   (i) Block 1 and   (ii) Block 2.

(c) Determine the net change in gravitational potential energy for the system during the time interval t_i to t_f.

(d) Use an energy argument to determine the speed, v.

E.38 A block of mass $M = 10\,\text{kg}$ is at rest at the origin of a horizontal frictionless plane. At time $t = 0$ a constant applied force, $\vec{F}_A = 5\,\text{N}$ [in the positive x-direction], begins to act on the block. Once the block reaches a distance of $4\,\text{m}$ from the origin, the applied force is suddenly reduced to zero.

(a) Draw a free body diagram and enumerate the forces which act on the block while it is in motion from $x = 0$ to $x = 4\,\text{m}$.

(b) Compute the work done by the   (i) applied and   (ii) weight   forces as the block translates from the origin to $x = 4\,\text{m}$.

(c) Compute the net work done on the block as it translates from the origin to $x = 4\,\text{m}$.

(d) (i) Apply the WORK–ENERGY THEOREM to determine the kinetic energy of the block when it reaches $x = 4\,\text{m}$.   (ii) Compute the speed of the block when it reaches $x = 4\,\text{m}$.

(e) (i) Determine the instantaneous power input by the applied force just prior to the instant at which the block reaches $x = 4\,\text{m}$.   (ii) Compare the instantaneous power, obtained in (i), with the average power input by the applied force throughout the time it took for the block to reach $x = 4\,\text{m}$.

E.39 A block of mass M is held at rest on an inclined plane. The angle of inclination is θ, and the block lies a distance L_1 from the bottom. At the bottom of the incline there is a smooth transition to a horizontal plane. At distance L_2 away from the base of the incline lies the free end of a spring with force constant k. The other end of the spring is anchored. The situation is illustrated in the figure on the left below.

(a) Suppose that the surfaces of the inclined and horizontal planes are frictionless and that the block is released from rest. Determine the maximum compression of the spring (illustrated in the sketch on the right, just below).

(b) Repeat the analysis, incorporating kinetic friction with coefficient μ_k along the ramp and the part of the horizontal plane up to, but not including, that beneath the [uncompressed] spring.

(c) Repeat the analysis, incorporating kinetic friction with coefficient μ_k along the ramp and the entire horizontal plane including that part to the right of the equilibrium position of the spring.

E.40 The net external force acting on a block of mass $M = 2\,\text{kg}$ does a total of 36 J work on the block during the time interval from initial time t_i, to final time $t_f = t_i + 3\,\text{s}$.

(a) Suppose that the initial speed of the particle is zero. Determine the (i) final speed of the block, and (ii) average net power input to the block during the time interval.

(b) Suppose that the initial speed of the particle is $8\,\text{m/s}$. Determine the (i) final speed of the block, and (ii) average net power input to the block during the time interval.

E.41 A constant horizontal applied force of $15\,\text{N}$ acts on a $3\,\text{kg}$ block which rides on a horizontal frictionless plane. At the instant $t = 0$ the block is at rest at the origin.

(a) Employing dynamics methods, compute the position and velocity of the block at $t = 1, 3$, and 5 s.

(b) Compute the net amounts of work done on the block from $t = 0$ to $t = 1, 3$, and 5 s.

(c) Compute the instantaneous power input to the block at $t = 1, 3$, and 5 s.

(d) Express the instantaneous power input to the block at any time $t \geq 0$, and integrate it from $t = 0$ to t. Evaluate your result for $t = 1, 3$, and 5 s. Comment.

E.42 [See also D.85, E.10.] A sled of mass $40\,\text{kg}$ is pulled along a horizontal frictional surface with an applied force of $50\,\text{N}$ acting at an angle of $\pi/6\,\text{rad}$ with respect to horizontal. As is usual in this sort of situation, the applied force is transmitted via a short length of ideal rope. The coefficient of kinetic friction acting between the sled and the surface on which it rides is $\mu_{\text{k}} = 0.04$. The sled moves forward $10\,\text{m}$ in 4 s.

(a) Compute the average power input to the sled by the (i) weight, (ii) normal, (iii) applied, and (iv) friction forces during the $4\,\text{s}$ time interal.

(b) Compute the net average power input to the sled in this interval.

(c) Discuss why it is the case that, even though the forces which act are constant, the instantaneous power [evaluated at times during the interval] is not equal to the average power [throughout the interval].

E.43 [See also D.88.] Consider the run-away truck bed in D.88, modelled with a position-dependent coefficient of kinetic friction, $\mu_{\text{k}} = x/L$, for a length parameter, L. Suppose that the truck, with total mass M, has initial speed V_0 when it leaves the road, at $x = 0$.

(a) Compute the work done by the frictional force as the truck moves from $x = 0$ to $x = a$, an arbitrary distance, while remaining within the bed.

(b) Compute the initial kinetic energy of the truck.

(c) Determine the distance through which the frictional force must act so as to cancel all of the initial kinetic energy.

E.44 PK is considering two driving options for a $D = 900\,\text{km}$ car trip he has planned.

Tortoise	proceed sedately at 90 km/h for 10 h
Hare	race along at 150 km/h for 6 h

Here our intention is to model the relative amounts of gasoline consumed in each of these scenarios under the assumptions of linear $[F_{D1} = -b_1\,v]$, quadratic $[F_{D2} = -b_2\,v^2]$, and mixed $[F_{D1,D2} = -b_1\,v - b_2\,v^2 = -b_1\,v - \xi\,b_1\,v^2 = -b_1(v - \xi\,v^2)]$ drag forces. [The mechanical work done by the drag force(s) would reduce the kinetic energy of the car, if not for the compensatory energy input derived from combustion of fuel. The expression for the combined force has been simplified through the introduction of a constant parameter, ξ, relating the quadratic and linear drag coefficients of PK's car.]

(a) Determine the work done by the   (i) linear,   (ii) quadratic, and   (iii) combined drag forces acting on the car through distance D as it travels at speed v.

(b) (i–iii) Compute the ratio of the work done by the drag force at $v = 150\,\text{km/h}$ vs. that at $v = 90\,\text{km/h}$ in each of the three cases.

(c) As a test of consistency, take the limits   (i) $\xi \to 0$ and   (ii) $\xi \to \infty$ in your results for the combined case. Comment.

E.45   Two blocks, with masses $M_1 = 2\,\text{kg}$ and $M_2 = 4\,\text{kg}$ respectively; one inertialess pulley; some ideal rope; and a horizontal frictional plane are arranged as shown in the figure alongside D.30. The coefficient of kinetic friction acting between Block 1 and the surface on which it rides is $\mu_k = 0.5$. The effects of drag are negligible. The system is held in place at its initial position as shown and released at $t = 0\,\text{s}$. As soon as it is released, it begins to move, *i.e.*, static friction is immediately overcome. Two seconds later, $t_f = 2\,\text{s}$, each block has displaced 10 m from its initial position. In parts (b) and (d), the final speed of the system shall be derived by two independent methods.

(a) Draw free body diagrams for each of the blocks and the pulley. Enumerate the forces acting on each object.

(b) Apply Newton's Second Law to solve for the acceleration of the system and determine the speed of the blocks at $t_f = 2\,\text{s}$.

(c) Ascertain the initial kinetic energy of the system.

(d) Determine   (i) the change in the gravitational potential energy of each of the blocks, and   (ii) the work done by kinetic friction   during the time interval from 0 to 2 seconds.

(e) Infer the kinetic energy at t_f, along with the speed of the blocks.

E.46   [See also D.81.] Two blocks, with masses $M_1 = 10\,\text{kg}$ and $M_2 = 5\,\text{kg}$ respectively, one inertialess pulley, some ideal rope, and a plane inclined at angle $30°$ are arranged as shown in the figure alongside D.81. The coefficient of kinetic friction acting between block 1 and the surface on which it rides is $\mu_k = 1/\sqrt{3}$. The effects of drag are considered to be negligible. The system is in motion (upward along the incline) at $t = 0\,\text{s}$, with initial speed $v_0 = 20\,\text{m/s}$. Six seconds later, $t_f = 6\,\text{s}$, each block has displaced 60 m from its initial position.

(a) Ascertain the initial kinetic energy of the system.

(b) Determine   (i) the change in the gravitational potential energy of each of the blocks, and   (ii) the work done by kinetic friction   during the time interval from 0 to 6 seconds.

(c) Infer the   (i) kinetic energy and   (ii) speed   of the blocks at t_f.

E.47   A block is in motion down an inclined plane located on the surface of the moon. The angle of inclination is $60°$. The mass of the block is $5\,\text{kg}$. The coefficient of kinetic friction between the block and the incline is $\mu_k = 0.1$. Take the lunar acceleration due to gravity to be $\frac{5}{3}\,\text{m/s}^2$.

(a) Draw a free body diagram, enumerate the forces, and choose a coordinate system.

(b) Compute the work done by each of the external forces that act on the block as the block moves down the incline a distance of 2 m.

(c) Determine the change in the   (i) gravitational potential energy and   (ii) kinetic energy   of the block after it has moved 2 m.

(d) (i) Repeat the calculation for the work done by friction when the incline gets progressively rougher, in such a way, as the block moves along, that $\mu_k(x) = 0.1 + 0.1x$ for $0\,\text{m} \le x \le 2$ m. (ii) Determine the change in the kinetic energy of the block in this case.

E.48 [See also D.95.] A block of mass M slides down a long frictional incline while immersed in a draggy medium. The angle of descent is $\theta = 30°$, and the coefficient of kinetic friction between the surfaces of the block and incline is $\mu_k = \frac{4}{5\sqrt{3}}$. The linear drag coefficient is small enough to be negligible, $b_1 = 0$, while the quadratic coefficient happens to satisfy the relation

$$\frac{b_2}{M\,g} = 0.025\,\frac{\text{s}^2}{\text{m}^2}\,.$$

Determine the terminal velocity of the block, employing an energetics (power) argument.

E.49 A block of mass M slides along a horizontal planar surface toward a spring with force constant k. At an initial time t_i, the block is at a distance D from the end of the spring, and is moving with speed v_i.

(a) Suppose that the plane is frictionless. (i) By how much will the spring be compressed when the block is stopped? (ii) What speed will the block have if/when it returns to distance D from the end of the spring?

(b) Suppose that the coefficient of kinetic friction acting between the block and the plane is μ_k. (i) By how much will the spring be compressed when the block is stopped? (ii) What speed will the block have if/when it returns to distance D?

E.50 [See also E.49.] A block of mass $2\,\text{kg}$ slides along a horizontal planar surface towards a spring with force constant $5\,\text{N/m}$. At initial time $t_i = 0$, the block is $1\,\text{m}$ from the end of the spring and moving at $3\,\text{m/s}$.

(a) Suppose that the plane is frictionless. (i) By how much will the spring be compressed when the block is stopped? (ii) What speed will the block have if/when it returns to distance D from the end of the spring?

(b) Suppose that the coefficient of kinetic friction acting between the block and the plane is $\mu_k = 1/20$. (i) By how much will the spring be compressed when the block is stopped? (ii) What speed will the block have if/when it returns to distance D?

E.51 [See also M.18.] At $t = 0$, a block of mass $10\,\text{kg}$ is at the origin, and moving with velocity v_{a-e}. Also at $t = 0$, a constant force, $F_A = 50\,\text{N}$, acting in the positive x-direction, is applied. We are interested in the state of motion of the block at the final time, $t = 2\,\text{s}$.

(a - e) For each of the initial velocities,
$$v_a = 0, \qquad v_b = 5, \qquad v_c = -10, \qquad v_d = -15, \text{ and} \qquad v_e = -5,$$
compute, or otherwise determine, the (i) velocity and position of the block at $t = 2\text{s}$, (ii) initial kinetic energy of the block, (iii) amount of mechanical work performed by the applied force, (iv) final kinetic energy of the block.

(f) Comment on the consistency of the Work–Energy Theorem in the analyses (a–e).

(g) Show, with the aid of a simple diagram, how the five cases (a–e) might be regarded as descriptions of the same particle dynamics from the perspectives of different inertial observers. Determine how these other observers are in motion with respect to the observer who sees the block initially moving at v_a.

E.52 A particle of mass m, confined to move in one-d, is subject to a single conservative force. The potential energy function associated with this force is taken to be $U(x) = \frac{1}{2}k\,x^2$, for a fixed parameter, k. The total mechanical energy of the particle is E_0, which we may express as $E_0 = \frac{1}{2}k\,A^2$, where A is a constant [with dimension of length]. We shall perform the analysis mentioned at the end of Chapter 27 for this specific form of potential energy function and total mechanical energy.

(a) Start with $\sqrt{\frac{2}{m}}\, dt = \frac{dx}{\sqrt{E_0 - U(x)}}$. Specialise to the present case and simplify.

(b) Rewrite these relations in terms of the parameter $\omega = \sqrt{k/m}$, and the scaled variable $u = x/A$. Ensure that the t- and u-dependences are gracefully separated.

(c) Integrate self-consistently [employing a trigonometric substitution, $u = \sin(\theta)$, to effect the u-integration] from $t = 0$ to an arbitrary later time t and from a corresponding initial angle θ_0 to the final angle θ [occurring at the final time].

(d) Transform the relation $t(\theta)$, obtained in (c), into an expression for the position of the particle as a function of time, *i.e.*, $x(t)$. Comment.

E.53   Consider a force field in one dimension, $F(x) = -\kappa\, x^3 + k\, x$, where both κ and k are positive constants.

(a) Ascertain the points at which the force vanishes.

(b) Determine the potential energy function associated with this force, whose reference value is $U(0) = 0$.

(c) Sketch the potential energy function *vs.* position.

(d) Comment on the nature of $U(x)$ in the vicinity of each of the points listed in (a).

E.54   Consider a force in one-d with magnitude and direction given by $F(x) = -4\,x^3 + 2\,x$.

(a) Ascertain the points at which the force vanishes.

(b) Determine the potential energy function associated with this force, whose reference value is $U(0) = 0$.

(c) Sketch the potential energy function *vs.* position.

(d) Comment on the nature of $U(x)$ in the vicinity of each of the points listed in (a).

E.55   A potential energy function in one dimension is given by $U(x) = \tan^{-1}(x^2 - 1) - \frac{\pi}{4}$.

(a) Compute the force associated with this potential energy.

(b) Plot $U(x)$ *vs.* x.

(c) From (b) and the analysis in (a), determine the equilibrium points and characterise their stability.

M

Momentum and Systems Problems

M.1   A block of mass M, moving with initial velocity $\vec{v}_i = (a\,,\,b\,,\,c)\,\mathrm{m/s}$, where a, b, and c represent real constants, is later observed to have a final velocity of $\vec{v}_f = (2\,a\,,\,-b\,,\,0)\,\mathrm{m/s}$. The initial time was $t_i = 112\,\mathrm{s}$, and the final time was $t_f = 117\,\mathrm{s}$ [as measured on a particular clock].

(a) Infer the net impulse received by the block during the time interval $t_i \le t \le t_f$.

(b) Determine the average net force acting on the block throughout the time interval.

M.2   A block of mass $M = 10\,\mathrm{kg}$ is at rest at the origin of a horizontal frictionless plane. At time $t = 0$, a constant applied force, $\vec{F}_A = 5\,\mathrm{N}$ [in the positive x-direction], begins to act on the block. At time $t = 4\,\mathrm{s}$, the force suddenly ceases to act.

(a) Compute the impulse provided to the block by the applied force while it acted, and explain (briefly) why this is the net impulse received by the block.

(b) (i) Compute the initial momentum of the block, *i.e.*, at $t = 0$.   Determine the (ii) momentum and   (iii) velocity   of the block at $t = 4\,\mathrm{s}$.

M.3   A block of mass $M = 2\,\mathrm{kg}$ is at rest at the origin of a horizontal frictionless plane. At time $t = 0$, a constant applied force, $\vec{F}_A = 4\,\mathrm{N}$ [in the negative x-direction], begins to act on the block. At time $t = 2\,\mathrm{s}$, the force suddenly ceases to act.

(a) Compute the impulse provided to the block by the applied force while it acted, and explain (briefly) why this is the net impulse received by the block.

(b) (i) Compute the initial momentum of the block, *i.e.*, at $t = 0$.   Determine the (ii) momentum and   (iii) velocity   of the block at $t = 2\,\mathrm{s}$.

M.4   A football (soccer) player intends to redirect a 425 g ball, using her head. The initial velocity of the ball is $(8\,,\,6)\,\mathrm{m/s}$ [in the plane containing both the incoming and outgoing velocities].

(a) Determine the impulse provided to the ball when its final velocity is   (i) $(8\,,\,-6)\,\mathrm{m/s}$, (ii) $(-8\,,\,6)\,\mathrm{m/s}$, and   (iii) $(-8\,,\,-6)\,\mathrm{m/s}$.

(b) For the three cases in part (a), determine the average force exerted by the player's head on the ball if the collision occurs in a time span of 0.008 s. [Owing to N3, this is (minus) the average force experienced by the player. The peak forces are probably double the average.]

M.5   [See also D.16, E.4.] A block of mass $m = 10\,\mathrm{kg}$ rides on a frictionless horizontal plane. Three applied forces act horizontally on the block, while the block's weight and the normal force from the plane both act vertically and cancel. The applied forces are $\vec{F}_1 = 10\,\mathrm{N}\,[\,\rightarrow\,]$, $\vec{F}_2 = 40\,\mathrm{N}\,[\,\downarrow\,]$, and $\vec{F}_3 = 50\,\mathrm{N}\,[\,+53.13°\,\mathrm{WRT}\ x\text{-axis}]$. The initial velocity of the block, at $t = 0\,\mathrm{s}$, is $\vec{v}_0 = 2\,\mathrm{m/s}\,[\,\leftarrow\,]$, while its initial position is $\vec{r}_i = (0\,,\,0)$. We are interested in determining the velocity of the block at the time $t = 4\,\mathrm{s}$.

(a) Determine the net (applied) force acting on the block.

(b) Compute the impulse produced by the net applied force acting on the block during the time interval $t = 0 \rightarrow 4\,\text{s}$.

(c) State briefly why the impulses produced by the normal force and weight cancel when added.

(d) Compute the impulse produced by each of the three applied forces. Sum these to get the total impulse received by the block. Comment.

(e) Determine the speed of the block at the time $t = 4\,\text{s}$.

M.6   [See also D.26, E.13.] Reconsider the situation, previously studied in D.26 and illustrated in D.25, in which two blocks, m_1 and m_2, are pulled along a straight-line trajectory upon a frictionless plane. Consider a time interval from $t = 0$, when the blocks are moving forward with speed v_0, to generic time t, and examine two cases: (a) in which $\vec{F}_{(a)}$ is exerted parallel to the plane, and   (b) in which $\vec{F}_{(b)}$ is applied at fixed angle θ with respect to the plane. Compute   (i) the net impulse provided to each of the blocks,   (ii) the net impulse provided to the system, and   (iii) the final speed of the system   in both cases.

M.7   [See also D.17, E.14.] A block of mass 2.5 kg riding on a horizontal frictionless plane is subjected to four applied forces: $\vec{F}_1 = 13\,\text{N}$ [$22.62°$ WRT positive x-axis], $\vec{F}_2 = 5\,\text{N}$ [$53.13°$ WRT positive y-axis], $\vec{F}_3 = 8\,\text{N}\ [\leftarrow]$,   and $\vec{F}_4 = 3\,\text{N}\ [\downarrow]$. The weight and normal force of contact act perpendicular to the plane, and they cancel everywhere and always. At the instant $t = 0\,\text{s}$, the block is at the origin, $\vec{r}_0 = (0,0)\,\text{m}$, and is moving with velocity $\vec{v}_0 = 3\,\text{m/s}\ [\rightarrow]$. Our goal is to determine the speed of the block at the time $t = 2\,\text{s}$, when it reaches the point $\vec{r}_2 = (6,4)\,\text{m}$.

(a) (i) Determine the net force acting on the block.   (ii) Ascertain whether the net force is constant.

(b) Compute the impulse provided to the block by the net force acting through the time interval $0 \rightarrow 2$ s.

(c) (i) Compute the impulse provided to the block by each of the four applied forces, and   (ii) sum to obtain the net impulse. Comment.

(d) Determine the initial momentum of the block, *i.e.*, at $t = 0$.

(e) Infer the   (i) momentum,   (ii) velocity, and   (iii) speed   of the block at $t = 2\,\text{s}$.

M.8   A sled of mass 40 kg is pulled along a horizontal frictional surface with an applied force of 50 N acting at an angle of $\pi/6\,\text{rad}$ with respect to horizontal. The sled is moving in the forward direction, and the coefficient of kinetic friction acting between the sled and the surface on which it rides is $\mu_k = 0.04$. It so happens that in a 4-second time interval, the sled advances a distance of 10 m.

(a) (i) Determine the net force acting on the sled.   (ii) Ascertain whether the net force is constant.

(b) Compute the impulse provided to the block by the net force during the 4 s time interval.

(c) (i) Compute the impulse provided to the block by each of the four forces that it experiences, and   (ii) sum to obtain the net impulse. Comment.

(d) Determine the change in the momentum of the block throughout the time interval.

M.9   [See also D.33, E.18.] Three blocks, each of mass M, two ideal ropes, a frictionless plane, and an ideal pulley comprise the dynamical system illustrated in the figure accompanying problem D.33. At the instant $t = 0\,\text{s}$, the system is at rest. At a later time, $t_f > 0$, the block to the right has dropped a distance H.

(a) Draw free body diagrams for each of the blocks and the pulley. Enumerate the forces which act on each object.

(b)   (i) Invoke $\frac{\partial \vec{p}}{\partial t} = \vec{F}_{\text{net ext'l}}$ for the momentum of each block and [formally] of the [massless] pulley.   (ii) Sum these expressions to get the time rate of change of the total momentum of the three-block and pulley system expressed in terms of the forces which act on the system.

(c) (i) Given that the acceleration experienced by the blocks has magnitude $a = \frac{1}{3}\, g$, compute the change in momentum experienced by each block throughout the time interval from $t = 0$ to $t = t_f$.   (ii) Sum these to get the total momentum change for the system.

(d) Employ the results from parts (b) and (c), above, to determine the magnitude and direction of the force which is holding the pulley in place.

M.10   [See also D.35, E.19.] Three blocks, each of mass M, two ideal ropes, a frictionless plane, and an ideal pulley comprise the dynamical system illustrated in the figure associated with D.35.
At the instant $t = 0\,\text{s}$, the system is at rest. At a later time, $t_f > 0$, the two blocks to the right have each dropped a distance H.

(a) Draw free body diagrams for each of the blocks and the pulley. Enumerate the forces which act on each object.

(b)   (i) Invoke $\frac{\partial \vec{p}}{\partial t} = \vec{F}_{\text{net ext'l}}$ for the momentum of each block and [formally] of the [massless] pulley.   (ii) Sum these expressions to get the time rate of change of the total momentum of the three-block and pulley system expressed in terms of the forces which act on the system.

(c) (i) Given that the acceleration experienced by the blocks has magnitude $a = \frac{2}{3}\, g$, compute the change in momentum experienced by each block throughout the time interval from $t = 0$ to $t = t_f$.   (ii) Sum these to get the total momentum change for the system.

(d) Employ the results from parts (b) and (c), above, to determine the magnitude and direction of the force which is holding the pulley in place.

M.11   [See also D.37, E.20.] Four blocks, each of mass M, three ideal ropes, a frictionless plane, and an ideal pulley comprise the dynamical system illustrated in the figure alongside D.37. At the instant $t = 0\,\text{s}$, the system is at rest. At a later time, $t_f > 0$, the two blocks to the right have each dropped a distance H.

(a) Draw free body diagrams for each of the blocks and the pulley. Enumerate the forces which act on each object.

(b)   (i) Invoke $\frac{\partial \vec{p}}{\partial t} = \vec{F}_{\text{net ext'l}}$ for the momentum of each block and [formally] of the [massless] pulley.   (ii) Sum these expressions to get the time rate of change of the total momentum of the four-block and pulley system expressed in terms of the forces which act on the system.

(c) (i) Given that the acceleration experienced by the blocks has magnitude $a = \frac{1}{2}\, g$, compute the change in momentum experienced by each block throughout the time interval from $t = 0$ to $t = t_f$.   (ii) Sum these to get the total momentum change for the system.

(d) Employ the results from parts (b) and (c), above, to determine the magnitude and direction of the force which is holding the pulley in place.

M.12

A time-varying force, $F(t)$, which acts on a particle confined to move in one dimension, is plotted in the nearby graph.

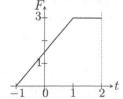

Estimate the average value of the force throughout the following time intervals:

(a) $-1 < t < 1$,   (b) $1 < t < 2$, and   (c) $-1 < t < 2$.

M.13   A force acting in one dimension oscillates according to $F(t) = 1 + \sin(\omega t)$, where ω, the angular frequency, has units of $\mathtt{rad/s}$. Compute the value of the average force throughout the time intervals: (a) $t \in [0, 2\pi/\omega]$, *i.e.,* one complete cycle,   (b) $t \in [0, \pi/(2\omega)]$, the first quarter-cycle,   (c) $t \in [0, 3\pi/(2\omega)]$, the first three-quarters of a cycle, and   (d) a generic time interval, $t \in [t_i, t_f]$.

M.14   A force acting in one dimension oscillates according to $F(t) = 1 + \sin(\omega t)$, where ω, the angular frequency, is equal to $\pi/2$ $\mathtt{rad/s}$. Compute the value of the average force throughout the time intervals: (a) $t \in [0, 4]$ $\mathtt{s}$, *i.e.,* one complete cycle,   (b) $t \in [0, 1]$ $\mathtt{s}$, the first quarter-cycle,   (c) $t \in [0, 3]$ $\mathtt{s}$, the first three-quarters of a cycle, and   (d) a generic time interval, $t \in [t_i, t_f]$.

M.15   A force acting in one dimension is described by $F(t) = t\,(T_f - t)\,F_0\,\exp(-\gamma t)$ during the time interval lasting from $t = 0$ to $t = T_f > 0$. The force vanishes otherwise. The constant γ is positive and has the dimensions of reciprocal time, while F_0 has the appropriate units. Ascertain the value of the average force acting throughout the time intervals:

(a) $t \in [0, T_f]$,   (b) $t \in [0, T_f/2]$,   (c) $t \in [T_f/2, T_f]$, and   (d) $t \in [t_i, t_f]$, with $t_i > 0$ and $t_f < T_f$ (*i.e.,* any generic time interval).

M.16   A force acting in one dimension is described by $F(t) = t\,(4 - t)\,F_0\,\exp(-\gamma t)$ during the time interval from $t = 0 \to 4$. The force vanishes otherwise. The constant $\gamma = 1/2$ is measured in $\mathtt{s^{-1}}$, while $F_0 = 3$ has units $\mathtt{N/s^2}$. Compute the value of the average force acting throughout the time intervals: (a) $t \in [0, 4]$,   (b) $t \in [0, 2]$, and   (c) $t \in [2, 4]$.

M.17   A force acting in one dimension is described by $F(t) = (t - T_i)\,(T_f - t)\,F_0\,\exp(-\gamma t)$ during the time interval lasting from $t = T_i$ to $t = T_f > 0$. The force vanishes otherwise. The constant γ is positive and has the dimensions of reciprocal time, while F_0 has the appropriate units. Adopt a regimes approach to explicitly recast this force into the form considered in the previous problem.

M.18   [See also E.51.]   At $t = 0$, a block of mass $10\,\mathrm{kg}$ is at the origin and is moving with velocity v_{a-e}. At this same time, a constant force, $F_A = 50\,\mathrm{N}$, acting in the positive x-direction, is applied. We are interested in the state of motion of the block at $t = 2\,\mathrm{s}$.

(a–e)   For each of the initial velocities,

$$v_a = 0, \qquad v_b = 5, \qquad v_c = -10, \qquad v_d = -15, \text{ and } \qquad v_e = -5,$$

compute, or otherwise determine, the   (i) velocity and position of the block at $t = 2\mathrm{s}$, (ii) initial momentum of the block,   (iii) impulse provided to the block by the applied force,   (iv) final momentum of the block.

(f) Comment on the consistency of the Impulse–Momentum Theorem in (a–e).

(g) Show, with the aid of a simple diagram, how the five cases (a–e) might be regarded as descriptions of the same particle dynamics from the perspectives of different inertial observers. Determine how these other observers are in motion with respect to the observer who sees the block initially moving with velocity v_a.

M.19   [See also E.12.]   Superman observes a fast passenger train of mass 200 metric tonnes travelling at $180\,\mathrm{km/h}$ toward a bridge which spans a yawning chasm. The bridge has been damaged, and the train will derail as soon as the engine reaches it. There are but 8 seconds remaining in the episode [before the commercials commence] when Superman arrives on the scene. We shall determine, via application of the IMPULSE–MOMENTUM THEOREM, the minimum [average] force that Superman must exert on the train to stop it before it hurtles disastrously into the chasm. Compute the train's (a) initial and (b) final [*i.e.,* when it is safely stopped] momentum. (c) What impulse must Superman provide to the train to stop it? (d) What minimum constant force must Superman exert throughout the 8 s remaining to yield the required impulse? (e) Determine the average acceleration of the train.

M.20 The Sun has mass $M_\odot \simeq 2.0 \times 10^{30}$ kg and radius $R_\odot \simeq 7.0 \times 10^8$ m, while the Earth's mass and radius are $M_\oplus \simeq 6 \times 10^{24}$ kg and $R_\oplus \simeq 6.4 \times 10^6$ m, respectively. The mean distance between the Earth and the Sun is called an astronomical unit (AU), where 1 AU $\simeq 1.5 \times 10^{11}$ m. Determine the location of the centre of mass of the Earth–Sun system. Comment.

M.21 A "coxed fours" crew team consists of four oarsmen, and a coxswain (C) who steers the boat and sets the cadence. On one particular team the crew, as enumerated in the figure, have masses (all in kg) $m_4 = 100$, $m_3 = 95$, $m_2 = 90$, $m_1 = 85$, and $M_C = 50$. The boat has mass $M = 80$ kg and is roughly symmetric fore and aft, as well as port to starboard.

$$4 \quad 3 \quad 2 \quad 1 \quad C$$

Identify the rower who is nearest to the CofM of the system comprised of boat and oarsmen.

M.22 A thin rod of length $2\,L$ lies along the x-axis and is centred on the origin. The parts which extend to the left, $x < 0$, and the right, $x > 0$, have uniform lineal mass densities λ_l and λ_r, respectively, with $\lambda_r > \lambda_l$. Determine (a) the total mass of the rod, and (b) the location of its CofM.

M.23 A thin rod of length L lies along the x-axis, stretching from the origin, $x = 0$, to $x = L$. The lineal mass density of the rod is $\lambda = \lambda_0\, x^2$, where λ_0 is a constant with dimensions of mass per unit volume.

(a) Determine the (i) total mass of the rod, and (ii) location of its CofM.

(b) Propose a simple physical situation which might correspond to this model.

M.24 A thin rod of length 3 m, situated along the positive x-axis, with one end at the origin, so tapers that its lineal density is $\lambda = 2\,x$ kg/m. Determine (a) the total mass, and (b) the location of the centre of mass of the rod.

M.25 A thin rod, extending from $x = 0$ to $x = L$, has lineal mass density given by $\lambda = \lambda_0 \left(1 - \frac{x^3}{L^3}\right)$, where λ_0 is a constant with appropriate units. Determine the (a) total mass, and (b) location of the centre of mass of the rod.

M.26 Xavier, Yolanda, and Zoe skip physics class to hang out at a local playground, where they espy a seesaw. The three friends have masses $M_X = 75$ kg, $M_Y = 50$ kg $= M_Z$. The seesaw consists of a plank with mass $M_P = 15$ kg and length $L_P = 4$ m supported by a very narrow fulcrum set at its midpoint.

(a) Yolanda sits at the very end of the plank on one side of the seesaw. Where with respect to the fulcrum should Xavier position himself so as to balance the seesaw?

(b) To accomodate Zoe, Xavier and Yolanda sit as far apart as possible. Where with respect to the fulcrum should Zoe sit so as to balance the seesaw?

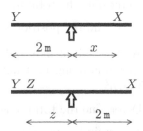

M.27
Six point-like particles are arranged on the perimeter of a
2 metre × 1 metre rectangle as in the nearby figure. The
mass of each of the upper particles is 2 kg, while that of
each lower one is 4 kg. Determine the location of $\vec{R}_{\text{CofM}}$.

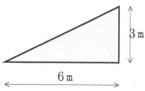

M.28
A right-triangular thin sheet of aluminium, with sides of
length 3, 6, and $\sqrt{45}$ m, has uniform areal density $\sigma_0 =$
$100\,\text{g/m}^2$.

(a) Determine the total mass of aluminium in the sheet.

(b) Explicitly compute the location of the CofM of the
sheet. Comment.

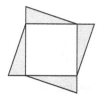

M.29   The CofM of a triangularly shaped slab of material with uniform density is located
at its **centroid**, *i.e.*, the unique point at which the three straight lines from the midpoint
of each side to the opposite vertex intersect. In Cartesian coordinates, the position of the
centroid is obtained by averaging the positions of the vertices. Furthermore, when the
triangle under consideration is a right triangle, the centroid is located one third of the
distance to the non-right vertices as illustrated in the figure below and to the left.

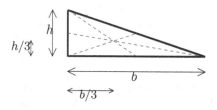

(a) Verify that, for a right-triangular shaped homogeneous slab, the mass-weighted
average position of the slab is coincident with the centroid.

(b) Determine the position of the centre of mass of the system comprised of four identical
right-triangular slabs arranged as indicated in the figure above and to the right.

M.30   A thin slab of material has length 2 m and width $1 - \frac{1}{3}\sin(\pi\,x/2)$, where $x \in [0,2]$
denotes position along its length. Take the areal density of the material to be constant,
$\sigma_0 = 0.5\,\text{kg/m}^2$.

(a) Determine the total mass of the slab.

(b) Ascertain the position of the CofM of the slab.

M.31   A thin slab of material has length L and width $A - B\sin(\pi\,x/L)$, where $x \in [0, L]$
denotes position along its length and $A > B$. Take σ_0, the areal density of the material, to
be constant.

(a) Determine the total mass of the slab.

(b) Ascertain the position of the CofM of the slab.

M.32   A thin slab of material has length 2 m and width $1 + \frac{1}{3}\cos\left(\frac{\pi x}{4}\right)$, where $x \in [0,2]$
denotes position along its length. Take the areal density of the material to be constant,
$\sigma_0 = 0.5\,\text{kg/m}^2$.

(a) Determine the total mass of the slab.

(b) Ascertain the position of the CofM of the slab.

M.33 A thin slab of material has length L and width $A + B \cos\left(\frac{\pi x}{2L}\right)$, where $x \in [0, L]$ denotes position along its length and $A > B$. Take the areal density of the material to be σ_0, a constant.

(a) Determine the total mass of the slab.

(b) Ascertain the position of the CofM of the slab.

M.34
Stacking blocks or tiles can be an amusing pastime. PK shall attempt to stack homogeneous rectangular parallelepipeds beyond the edge of a table, using the strategy of successive 1/4 overhangs, as illustrated in the adjacent figure. How many tiles can PK place before the pile comes tumbling down?

M.35 [See also R.13.] A pizza may be modelled as a thin uniform circular disk of radius R and mass-per-unit-area σ. Determine the (a) mass and (b) location of the CofM of the pizza.

M.36 [See also R.14.] In the ONE-HANDED PIZZA GRAB [a common foraging technique, recognised by anthropologists], one grasps a pizza slice directly beneath its CofM so as to reduce the likelihood of the toppings sliding off. Our goal is to ascertain the location of the CofM of a pizza slice, employing the following model.

◁ The entire pizza has total mass M_T, radius R, area $A = \pi R^2$, and an effectively constant areal mass density, σ_0.

◁ The slice is a circular wedge with angular width $\Theta_0 = 2 \times \theta_0$ and radius R.

◁ By symmetry, the CofM must lie on the angular bisector of the slice.

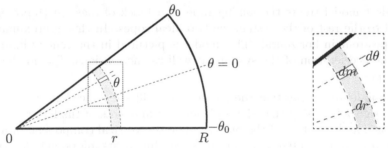

(a) Compute the mass of the slice of the pizza. [HINT: $dm = \sigma_0\,dA = \sigma_0\,r\,dr\,d\theta$]

(b) Re-express the mass of the slice in terms of M_T.

(c) Compute the mass-weighted distance of the pizza elements projected onto the angular bisector of the slice. [HINT: Mass elements with the same r, but different θ, do NOT contribute equally to the mass-weighted sum.]

(d) Determine the mass-weighted average position of the pizza elements by dividing the result in (c) by the mass of the slice, determined in (a).

(e) Confirm your result by taking the limits: $\Theta_0 \to$ (i) 0, (ii) π, and (iii) 2π.

M.37 The density of the Earth's atmosphere decreases exponentially with height above the surface. At the surface, the density is approximately $1.2\,\text{kg/m}^3$. The density decreases by one-half for every $5.9\,\text{km}$ of elevation. The radius of the Earth is 6.37×10^6 m.

(a) Estimate the total mass of the Earth's atmosphere.

(b) Determine the location of the CofM of the Earth's atmosphere.

M.38
Consider a length of uniform rope[a] with constant
mass per unit length λ_0, suspending a block of
mass M a distance L below an anchor point, as
shown. We will adopt two models, one discrete
and rather crude, and the other continuous, to
ascertain the tension at various positions in the
rope.

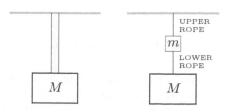

------

[a]One might *snark* that such ropes are "less than ideal."

(a) The crude model treats the real rope as a block of mass $m = \lambda_0 L$, joined to the
rest of the system by two ideal ropes. This model is represented in the adjacent figure.
Determine the tensions, $T_{l,u}$, in the lower and upper ideal ropes.

(b) A continuous model yields a rope tension which is a function of position. Setting
$y = 0$ at the point at which the rope is tied to the block and supposing that the rope
extends a distance L from the block to the ceiling, determine the tension $T(y)$ for all
$y \in [0, L]$. [HINTS: $dm = \lambda_0 \, dy$ and $T(y + dy) = T(y) + g \, dm$]

M.39   Consider a length of uniform rope with constant mass per unit length λ_0, joining
two blocks riding on a horizontal frictionless plane. The masses of the blocks are M_1 and
M_2. The system is subjected to an external applied force, F_A. Two discrete models are
developed, below, to describe this system. The second is a refinement of the first.

(a) The crudest model treats the joining rope as a block of mass m (representing its
mass) joined to the rest of the system via two ideal ropes. In this approximation, the
two ideal ropes remain horizontal. This model is pictured in the central figure above.
Determine the acceleration of the system, as well as the tensions, $T_{l,r}$, in the left and
right ideal ropes.

(b) A better model recognises that the joining rope will sag somewhat, as shown in the
figure on the right. Realising that the horizontal components of the left/right tensions
are responsible for the motion of the blocks, while the vertical component supports the
weight of the rope (represented as a point mass residing at its midpoint), determine the
acceleration of the system, the tensions, and the sag angle [assuming that it is the same
for both ideal ropes].

M.40   PK reads a thick Russian novel while sitting in the bow of his frictionless rowboat,
which happens to lie a distance of 0.5 m from the corner of the dock.

(a) PK walks to the
stern of the boat in
an attempt to clamber
onto the dock. Describe
what happens, why, and
where he ends up (quali-
tatively).

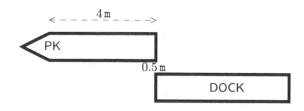

(b) PK stalks back to his original spot in the bow of the boat. Describe what happens,
why, and where he ends up.

(c) In a fit of *pique*, PK heaves the Russian novel forward over the bow of the boat. Describe what happens, why, and where he ends up (qualitatively).

M.41   Let's attempt to model the situation described in M.40. Suppose that *The Idiot*, with mass $0.5\,\text{kg}$ [we mean Dostoyevsky's book, not PK!], is tossed forward at $5\,\text{m/s}$. With what speed does the PK and boat system with aggregate mass $250\,\text{kg}$ recoil? [Assume that no drag forces act on the boat as it moves through the water.]

M.42   [See also M.41] Let's refine the quantitative model of the tossed book to include linear drag, with coefficient $b_1 = 2\,\text{N·s/m}$, acting on the boat. How far does the boat move (provided that PK is willing to wait a while[1])?

M.43   In this problem, we shall develop a crude model describing a rocket. Suppose that a composite particle of total mass $M + \Delta M$ ejects a portion of its mass, ΔM, in the rearward direction with relative speed v_e during a brief time interval, Δt.

(a) Compute the change in the momentum of the ejected mass as measured in the CofM frame of the composite particle.

(b) Infer the corresponding change in the momentum of the remainder of the particle, and the change in its speed.

(c) Take the limit $\Delta t \to 0$, realising that $\Delta M \to dM$ in this limit, and write a formal expression for the instantaneous change in the particle's speed.

(d) Self-consistently integrate the expression in (c) from the initial mass M_i, when the particle was moving with speed v_i according to an inertial observer, to its final mass, M_f, when the particle is moving with speed v_f.

(e) Suppose that $v_i = 0$ and $M_f = M_i/e^3$. Determine the final velocity of the remainder of the particle in this case. Are you surprised by this result?

M.44   Two particles, of mass M and $2\,M$ respectively, are initially at rest at $x = -1\,\text{m}$ and $x = 2\,\text{m}$. Subsequent to $t = 0$, each particle experiences a constant acceleration with magnitude a.

(a) Determine the location of the centre of mass of the two particles at $t = 0\,\text{s}$.

(b) Suppose that both particles accelerate in the positive x-direction. Determine the CofM   (i) acceleration,   (ii) velocity, and   (iii) position   at times $t > 0\,\text{s}$.

(c) Instead, suppose that the particles accelerate in opposite directions: $2\,M$ in the positive x-direction, and M in the negative x-direction. Determine the   (i) acceleration, (ii) velocity, and   (iii) position   of the CofM at times $t > 0\,\text{s}$.

M.45   Three particles, A, B, and C, with masses $M_A = 2$, $M_B = 3$, and $M_C = 5$ kg, are constrained to move along a straight, horizontal, and level track. At $t = 0$, the particles are at $x_{A0} = -10$, $x_{B0} = 5$, and $x_{C0} = 7$ m with respect to the origin, and they have constant velocities $v_{A0} = -6$, $v_{B0} = 2$, and $v_{C0} = 4$ m/s, respectively.

(a) Determine the position of the CofM of the three-particle system at $t = 0\,\text{s}$.

(b) Determine the   (i) acceleration,   (ii) velocity, and   (iii) position   of the centre of mass of the system for $t > 0$.

(c) (i) Separately determine the trajectory of each of the three particles.   (ii) Compute the centre of mass of the system at time t, employing your knowledge of the trajectories. (iii) Compare the results obtained for (c.ii) and (b.iii).

M.46   Two particles, A and B, with masses $M_A = 3\,\text{kg}$ and $M_B = 4\,\text{kg}$, move on a horizontal plane surface with velocities $\vec{v}_A = (5, 12)\,\text{m/s}$ and $\vec{v}_B = (-5, -12)\,\text{m/s}$, respectively. Determine the   (a) total momentum   and   (b) CofM velocity   of this system.

---

[1] The wait wouldn't be problematic if only PK had a book to read.

M.47 [See also D.39, E.36, R.24.] Consider two blocks joined by a length of ideal rope which is slung over an ideal pulley, of radius $R = 0.25\,\mathrm{m}$, as shown in the figure accompanying D.39. The two blocks have exactly the same mass, $M_L = M = M_R$. At the instant $t = 0$, the two blocks are at the same height. The initial velocities of the blocks are $v_{L0} = 2\,\mathrm{m/s}\;[\uparrow]$ and $v_{R0} = 2\,\mathrm{m/s}\;[\downarrow]$.

(a) Apply Newton's Laws to ascertain the   (i) acceleration,   (ii) velocity, and (iii) trajectory   of each of the blocks.

(b) Determine the   (i) acceleration,   (ii) velocity,   and   (iii) trajectory   of the CofM of the two-block system. Comment.

M.48   Repeat the previous problem for blocks with differing masses, $M_L = 2\,\mathrm{kg}$ and $M_R = 4\,\mathrm{kg}$.

M.49

Four particles, $\{A, B, C, D\}$, with masses $M_A = 1$, $M_B = 2$, $M_C = 3$, and $M_D = 4$ kg, are all at rest on a horizontal plane surface. The coordinates of the particles (all in m) are: $\vec{r}_A = (4\,,\,1)$, $\vec{r}_B = (2\,,\,3)$, $\vec{r}_C = (5\,,\,3)$, and $\vec{r}_D = (1\,,\,1)$.

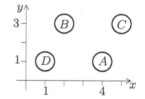

(a) Determine the location of the CofM of the four particles.

(b) Suppose that all the particles simultaneously begin to move with velocities (m/s): $\vec{v}_A = (?\,,\,?)$, $\vec{v}_B = (0\,,\,-2)$, $\vec{v}_C = (-1\,,\,-1)$, and $\vec{v}_D = (0\,,\,1)$. What must be the velocity of A, in order that the CofM of the four-particle system remain at rest?

(c) Suppose instead that the velocities of the particles are $\vec{v}_A = (3\,,\,-3)$, $\vec{v}_B = (0\,,\,-2)$, $\vec{v}_C = (-1\,,\,-1)$, and $\vec{v}_D = (0\,,\,1)$. Determine the velocity of the centre of mass.

M.50   Five particles, each of mass m, lie on a a frictionless horizontal plane. The position, velocity, and acceleration of each particle at the instant $t = 0$ are presented in the table below.

BLOCK	POSITION	VELOCITY	ACCELERATION
1	$(2\,,4)$	$(0\,,3)$	$\left(\frac{1}{2}\,,\frac{1}{2}\right)$
2	$(4\,,1)$	$(-1\,,1)$	$\left(\frac{1}{2}\,,\frac{1}{2}\right)$
3	$(6\,,4)$	$(-1\,,2)$	$(0\,,0)$
4	$(8\,,1)$	$(-1\,,1)$	$\left(\frac{1}{2}\,,\frac{1}{2}\right)$
5	$(10\,,4)$	$(-1\,,-1)$	$\left(-\frac{1}{2}\,,-\frac{1}{2}\right)$

(a) Sketch the positions of the particles on the plane and indicate with an arrow the direction in which each is moving at $t = 0$.

(b) (i) Determine the location of the centre of mass of the system at the instant $t = 0$, and   (ii) mark its position on the sketch produced in (a).

(c) (i) Determine the velocity of the centre of mass of the system at the instant $t = 0$, and   (ii) indicate its direction with an arrow at the point obtained in (b).

(d) (i) Determine the acceleration of the centre of mass of the system at the instant $t = 0$.   (ii) Comment on the net external force acting on this five-particle system.

M.51   Four blocks are constrained to move on a horizontal plane surface. The masses and trajectories of the blocks are quoted in the nearby table.

Particle	Mass	Trajectory	Particle	Mass	Trajectory
A	M	$(2+4t,0)$	B	M	$(2,6+2t)$
C	$2M$	$(-2-2t,-3-t)$	D	$4M$	$(0,0)$

(a) Compute the position of the centre of mass of the system of blocks at (i) the instant $t=0$ and (ii) generic time t.

(b) Compute (i) the velocity of each of the blocks, and (ii) the mass-weighted average velocity of the system, at time t.

(c) Compute (i) the momentum of each of the blocks, and (ii) the total momentum of the system, at any time.

(d) Compute (i) the acceleration of each of the blocks, and (ii) the mass-weighted average acceleration of the system, at t.

(e) Determine the trajectory of the CofM. Comment.

M.52 Repeat M.51 for the trajectories of the particles provided in this table.

Particle	Trajectory	Particle	Trajectory
A	$(2+4t+t^2,0+2t^2)$	B	$(2-t^2,6+2t-2t^2)$
C	$(-2-2t+4t^2,-3-t+2t^2)$	D	$(-2t^2,-t^2)$

M.53 [See also D.5.] Recall Problem D.5, in which Larry, Curly, and Moe found themselves in relative motion on a horizontal frictionless plane. Estimating their respective masses to be $M_L=75\,\mathrm{kg}$, $M_C=80\,\mathrm{kg}$, and $M_M=70\,\mathrm{kg}$, determine the velocity of the centre of mass of the three-fella system from the perspectives of (a) the observer at rest with respect to the plane, (b) Larry, (c) Curly, and (d) Moe.

M.54 A particle of mass $2\,M$, moving with speed v in the forward x direction upon a horizontal plane, collides head-on with another particle, of mass M, which is initially at rest. Suppose that the collision is perfectly elastic, and determine the speeds of the particles after the collision.

M.55 A block of mass $5\,\mathrm{kg}$, moving with a velocity of $0.9\,\mathrm{m/s}$ in the positive x direction, collides head-on and elastically with a block of mass $4\,\mathrm{kg}$ which was moving with velocity $1.8\,\mathrm{m/s}$ in the negative x direction. Determine the velocity of each of the two blocks, subsequent to the collision.

M.56 [See also M.64.] Two baseball players running for a fly ball suffer a head-on elastic collision. The first player has mass $M_1=100\,\mathrm{kg}$ and was running at $8\,\mathrm{m/s}$, while the second player has mass $M_2=80\,\mathrm{kg}$ and was running at $10\,\mathrm{m/s}$.

(a) Determine the speed of each player immediately after the collision.

(b) Compute (i) the impulse that each player receives, and (ii) the total impulse.

M.57 [See also M.65.]

Car A, of mass $M_A=500\,\mathrm{g}$, moves with speed $v=3\,\mathrm{m/s}$ along a horizontal frictionless one-dimensional track toward Car B, with $M_B=500\,\mathrm{g}$, which is originally at rest. The ensuing collision is [assumed to be] completely elastic.

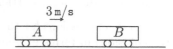

(a) Determine the initial momenta of (i) Car A, (ii) Car B, and (iii) the system.

(b) Compute the velocity of the CofM of the system prior to the collision.

(c) Ascertain the   (i) total momentum and   (ii) CofM velocity   of the two-car system just after the collision occurs.

(d) Determine the final velocity of   (i) Car A and   (ii) Car B.

(e) Ascertain the impulse delivered   (i) to Car A by Car B,   (ii) to Car B by Car A, and   (iii) in total to the two-car system,   during the elastic collision.

(f) Suppose that the collision lasted for 0.1 s. Compute the average force exerted on Car B by Car A during this time.

(g) Compute the total kinetic energy of the two-car system   (i) before and   (ii) after the collision.

M.58 [See also M.66 and M.67.] Two railway cars, with masses M and $4M$ respectively, ride without friction on a straight and horizontal section of track.

(a) The car with mass M moves with speed v and collides elastically with the heavier car, which was originally at rest. Explicitly determine the   (i) total momentum and kinetic energy of the system of railway cars prior to the collision,   (ii) velocities of the two railway cars after the collision, and   (iii) total momentum and kinetic energy of the system of railway cars after the collision.

(b) The car with mass $4M$ moves with speed v and collides elastically with the less massive car (originally at rest). Explicitly determine the   (i) total momentum and kinetic energy of the system of railway cars prior to the collision,   (ii) velocities of the two railway cars after the collision, and   (iii) total momentum and kinetic energy of the system of railway cars after the collision.

(c) Comment on the results obtained for (a) and (b) and propose a simple manner in which the systems may be related.

M.59 [See also M.69 and M.70.] Nine train cars of equal mass, M, coupled together, are at rest on a straight, flat, and level section of track. A tenth car, also of mass M, moving with speed v, collides elastically with the others.

(a) Compute the total   (i) momentum   and   (ii) kinetic energy   of the ten-car system prior to the collision.

(b) Determine the velocities of the nine- and single-car trains immediately after the collision.

(c) Compute the total   (i) momentum and   (ii) kinetic energy   of the ten-car system just after the collision.

M.60 Repeat the previous problem with the single car, initially at rest, being struck by the nine-car train moving with speed v.

M.61 An incident particle of mass $2M$, moving with speed v in the positive x-direction on a horizontal plane, undergoes a glancing collision with a target particle of mass M which is initially at rest. It happens that the incident and target particles move away at a relative angle of $\pi/2$. Let's say that the incident particle scatters at $+\theta$ with respect to its initial direction, while the target particle moves at angle $\pi/2 - \theta$ on the other side.

(a) Derive expressions for the final speeds of the incident and target particles (*i.e.*, after the collision) in terms of v and θ.

(b) Determine whether this collision is completely elastic.

M.62   A self-sustaining nuclear reaction occurs when neutrons liberated in one fission are captured by other nuclei and thence catalyse subsequent fissioning of the capturing nucleus. The kinetic energies of newly produced neutrons are typically too great to allow for ready capture by other nuclei. Reactor designs incorporate a moderating medium to slow the neutrons via elastic collisions. Graphite, composed of carbon [$^{12}_{6}C$], is commonly used as a moderator. Carbon atoms possess roughly twelve times the mass of a neutron.

(a) A neutron moving with initial speed v suffers an elastic head-on collision with a carbon nucleus which is originally at rest. Compute the fraction of the initial kinetic energy of the neutron which is given to the carbon nucleus.

(b) Estimate the minimum number of these collisions which will reduce the kinetic energy of the neutron to 4% of its initial value.

(c) To make a more efficient moderator, *i.e.*, to soak up more neutron energy per collision, should one employ lithium [$^{7}_{3}Li$] or iron [$^{56}_{26}Fe$]?

M.63   A block of mass $5\,kg$, moving with a velocity of $4.5\,m/s$ in the positive x direction, collides and sticks to a block of mass $4\,kg$ which was previously at rest. Determine the common velocity of the two blocks after the collision.

M.64   [See also M.56.] Two baseball players run for a fly ball, collide head-on, and end up entangled (stuck together). The first player has mass $M_1 = 100\,kg$ and was running at $8\,m/s$, while the second player has mass $M_2 = 80\,kg$ and was running at $10\,m/s$.

(a) Compute the speed of the tangled players just after the collision.

(b) Compute the impulse that each player receives, and the total impulse.

M.65   [See also M.57.] Car A, of mass $M_A = 500\,g$, is initially in motion with speed $v = 3\,m/s$ along a horizontal frictionless one-dimensional track toward Car B, with mass $M_B = 500\,g$, which is at rest. The two cars experience a perfectly inelastic collision, and subsequently the aggregate moves with a single unique velocity.

(a) Determine the initial momenta of Car A, Car B, and the two-car system.

(b) Compute the initial CofM velocity of the two-car system.

(c) Infer the momentum of the aggregate immediately after the collision.

(d) Determine the velocity of the aggregate after the collision.

(e) Compute the impulse   (i) given to A by B, and   (ii) given to B by A. Comment.

(f) Suppose that the collision lasted for $0.1\,s$, and infer the magnitude of the average force exerted by A on B during this time.

(g) Compute the total kinetic energy of the two-car system,   (i) before and   (ii) after the collision. Comment.

M.66   [See also M.58 and M.67.] A railway car of mass M moves with speed v along a straight horizontal section of track until it collides with and couples to another car, of mass $4\,M$, which was originally at rest.

(a) Determine the total   (i) momentum and   (ii) kinetic energy   of the system of railway cars prior to the collision.

(b) Determine the total   (i) momentum and   (ii) kinetic energy   of the aggregated rail cars after the collision.

(c) Infer the changes in the total   (i) momentum and   (ii) kinetic energy   of the system precipitated by the collision.

M.67 [See also M.58 and M.66.] A railway car of mass $4M$ moves with speed v along a straight horizontal section of track until it collides with and couples to another car, of mass M, which was originally at rest.

(a) Determine the total   (i) momentum and   (ii) kinetic energy   of the system of railway cars prior to the collision.

(b) Determine the total   (i) momentum and   (ii) kinetic energy   of the aggregated railcars after the collision.

(c) Infer the changes in the total   (i) momentum and   (ii) kinetic energy   for the system precipitated by the collision.

M.68 [See also M.66 and M.67.] Propose a simple manner in which the previous two problems may be consistently interpreted.

M.69 [See also M.59 and M.70.] Nine train cars of equal mass, M, coupled together, are at rest on a straight, flat, and level section of track. A tenth car, also of mass M, is moving with speed v toward the others. Post-collision, the tenth car is coupled to the other nine.

(a) Compute the total   (i) momentum and   (ii) kinetic energy   of the ten-car system prior to the collision.

(b) Determine the total   (i) momentum and   (ii) kinetic energy   of the ten-car system after the collision.

(c) Ascertain the amount of kinetic energy lost during the collision.

M.70 [See also M.59 and M.69.] A train car of mass M is at rest on a straight, flat, and level section of track. Nine additional joined cars, each of mass M, are moving with speed v toward the car at rest. Post-collision, the ten cars are coupled.

(a) Compute the total   (i) momentum and   (ii) kinetic energy   of the ten-car system prior to the collision.

(b) Determine the total   (i) momentum and   (ii) kinetic energy   of the ten-car system after the collision.

(c) Ascertain the amount of kinetic energy lost during the collision.

M.71 A block of mass $M + m$ is moving on a frictionless horizontal plane with speed v_i in the x-direction. At the instant $t = 0$, it explodes into two pieces! One piece, with mass M, subsequently moves with speed $13\,\text{m/s}$ at $-22.62°$ with respect to the x-axis. The other piece flies off with speed $20\,\text{m/s}$ at an angle of $+53.13°$ with respect to the x-axis.

(a) Determine the initial speed of the block (prior to the explosion).

(b) Ascertain the ratio of the masses.

M.72 Cars A and B, possessing masses $M_A = 500\,\text{g}$ and $M_B = 1.5\,\text{kg}$ respectively, are joined by an explosive release mechanism consisting of coiled ideal springs. The springs together have overall effective spring constant $k = 200\,\text{N/m}$ and are compressed by $10\,\text{cm}$. The cars move along a 1-d frictionless track.

(a) Suppose that, prior to the explosion, the two-car system is at rest. Determine the (i) velocity,   (ii) change in the momentum,   and   (iii) change in the kinetic energy of each car.   (iv) Also determine the net change in the momentum and kinetic energy of the two-car system.

(b) Suppose that, prior to the explosion, the cars were moving together with speed V in the forward direction along the track, and that, after the explosion, car A continues to move in the forward direction. Determine the   (i) velocity,   (ii) change in the momentum, and   (iii) change in the kinetic energy   of each car.   (iv) Also determine the net change in the momentum and kinetic energy of the two-car system.

M.73 Two particles, with masses M and $5\,M$ respectively, are pushed apart by a massless spring. Determine the fraction of the total potential energy (originally stored in the spring) which is borne away (as kinetic energy) by each of the particles.

M.74 At time t_0, two blocks, green with mass M and red with mass $2\,M$, both move to the right with speed v, as shown in the figure. The leading (red) block experiences a completely elastic collision with a wall at time t_1, and then collides with the trailing (green) block at time t_2.

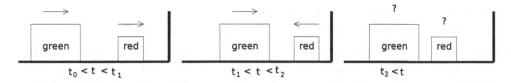

(a) Determine the velocity of the CofM prior to t_1.

(b) Compute the impulse delivered to the red block by its collision with the wall.

(c) Suppose that the collision lasted 0.01 **s**, and ascertain the average force exerted by the wall on the red block.

(d) Determine the CofM velocity during the time interval $t_1 < t < t_2$.

(e) Suppose that the collision between the two blocks is perfectly elastic. Determine the velocities of   (i) each block and   (ii) the CofM of the system   after the blocks collide.

(f) Compute the amount of kinetic energy lost by the system during the elastic collision. Comment.

(g) Suppose instead that the collision between the two blocks is completely inelastic. Determine the velocities of   (i) each block and   (ii) the CofM of the system   after the blocks collide.

(h) Maintaining the assumption of inelasticity from (g), determine the amount of kinetic energy lost by the system during the collision. Comment.

M.75 Two small balls are dropped together (one just above the other) from an initial height, H, which is large compared to the size of the balls. The lower ball has twice the mass of the upper ball. Assuming that the balls collide elastically with the ground and each other, compute the maximum height attained by the upper ball after its collision with the lower one.

M.76 A polonium nucleus $[^{210}\mathrm{Po}]$ with mass 3.49×10^{-25} **kg** is at rest. It decays into an α particle $[^{4}\mathrm{He}]$ with mass 6.64×10^{-27} **kg** and a lead nucleus $[^{206}\mathrm{Pb}]$ with mass 3.42×10^{-25} **kg**. The decay process liberates 8.65×10^{-13} **J** of kinetic energy to be shared among the products. Estimate the speeds of the α particle and the lead nucleus.

M.77 A firework of mass $4\,\text{kg}$ is launched from ground level with initial speed $V_0 = 15\sqrt{2}\,\text{m/s}$ at an initial angle of $45°$. Down-range from the launch site is a flat horizontal plain. Drag forces and wind effects are negligible. Just as the projectile reaches its maximum height, it explodes into four $1\,\text{kg}$ chunks. In the reference frame at rest with respect to the projectile at the instant it explodes, the fragments, { forward f, backward b, upward u, downward d }, move with speed $5\,\text{m/s}$ in their respective directions.

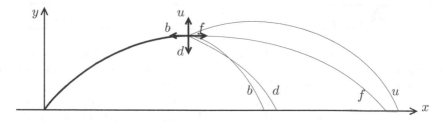

(a) Determine the trajectory of the projectile as it flies from its launch point to its maximum height.

(b) Determine the trajectory of each fragment.

(c) Ascertain the time that each fragment takes to fall to the ground.

(d) (i) Quantify the motion of the CofM of the fragments after the explosion, until the first chunk hits the ground. Comment on the motion of the CofM   (ii) subsequent to (i), while pieces are still in the air, and   (iii) after all chunks are on the ground.

M.78 Two particles, A and B, with masses $M_A = 10\,\text{kg}$ and $M_B = 20\,\text{kg}$, comprise a system which is constrained to move on a horizontal frictionless plane. At $t = 0$, the particles are at $\vec{r}_{A0} = (-3, 12)$ and $\vec{r}_{B0} = (-3, -12)$ and are moving with uniform velocities $\vec{v}_{A0} = 6\left(\cos(30°), \sin(30°)\right)$ and $\vec{v}_{B0} = 6\left(\cos(30°), -\sin(30°)\right)$ as viewed by a laboratory observer.

(a) Determine the position and velocity of the CofM of the two-particle system at $t = 0$.

(b) Determine the trajectory of each particle as viewed in the CofM IRF.

R

Rotation Problems

R.1 The angular position of a rotating body, expressed in degrees, is $\theta(t) = 30 + 5\,t$.

(a) Determine the angular (i) acceleration, (ii) speed, and (iii) position of the body at the instant $t = 3\,\text{s}$.

(b) Repeat (a) at the instant $t = 6\,\text{s}$.

(c) Ascertain the period of rotation of the body.

R.2 The angular position of a rotating body, expressed in degrees, is $\theta(t) = 30 + 5\,t - t^2$.

(a) Determine the angular (i) acceleration, (ii) speed, and (iii) position of the body at the instant $t = 3\,\text{s}$.

(b) Repeat (a) at the instant $t = 6\,\text{s}$.

(c) Determine the (i) time and (ii) angle at which the body momentarily stops.

R.3 An object rotating about a fixed axis experiences constant angular acceleration $\alpha = 5\,\text{rad/s}^2$ [clockwise]. At time $t = 0$, it is rotating at $10\,\text{rad/s}$ [anti-clockwise], and is aligned with the reference direction, so $\theta_0 = 0\,\text{rad}$.

(a) Determine the angular (i) velocity and (ii) position of the object at $t = 3\,\text{s}$.

(b) Repeat (a) at $t = 6\,\text{s}$.

R.4 A carousel starts from rest at $t = 0$ and accelerates at $0.25\,\text{rad/s}^2$ for $4\,\text{s}$. After $4\,\text{s}$, the angular acceleration suddenly drops to zero. Determine the angular (a) speed and (b) position at the following times: (i) 1s, (ii) 3s, and (ii) 5s.

R.5 A certain horse on a carousel stands at radial distance $R_H = 3\,\text{m}$ from the axis about which the carousel rotates (its centre). The carousel starts from rest at $t = 0$ and accelerates at $0.25\,\text{rad/s}^2$ for $4\,\text{s}$. Afterward, the angular acceleration suddenly drops to zero. Determine the magnitude of the (a) tangential, (b) centripetal, and (c) total, or net, acceleration of the horse at times: (i) 1 s, (ii) 3 s, and (iii) 5 s.

R.6 During the half-time show at a football game, two equal-mass members of the cheer team move in tandem as part of a dance routine. Alex coasts on a (massless) skateboard, while Bobby does backflips alongside. Which of the teammates has the greater amount of kinetic energy? Explain.

R.7 A rigid body with moment of inertia I rotates about a fixed axis with $\theta(t) = a\,t^4$, for $t \in [-10, 10]$.

(a) Determine the angular (i) velocity and (ii) acceleration of the rigid body.

(b) Infer the magnitude and direction of the net torque acting on the body (as a function of time) throughout the interval.

(c) Determine the (i) rotational kinetic energy of the rotating body and (ii) net power (the rate at which the kinetic energy is changing) at time t.

R.8   The base of an antique carousel is a disk-shaped horizontal wooden platform with radius $5\,\text{m}$ and thickness $20/\pi\,\text{cm}$, which rides on essentially frictionless bearings. The density of the wood is $900\,\text{kg/m}^3$. There are two concentric rings of six horses on the platform, with radii $3\,\text{m}$ and $4\,\text{m}$ respectively. Each horse has mass $25\,\text{kg}$. An iron railing with constant lineal mass density $10/\pi\,\text{kg/m}$ encircles the platform at its outer edge. The carousel rotates about a vertical axis through its centre.

(a) Compute the moment of inertia of   (i) the platform,   (ii) a horse at radius $3\,\text{m}$, (iii) a horse at $4\,\text{m}$, and   (iv) the iron railing.

(b) Assemble the results of (a) to compute the total moment of inertia of the carousel.

(c) Suppose that the carousel accelerates from rest to its operating [angular] speed of $0.5\,\text{rad/s}$ in $20\,\text{s}$. Ascertain the average   (i) angular acceleration of and   (ii) net torque applied to   the carousel.   Determine the   (iii) angle through which the carousel rotated and   (iv) mechanical work provided by the average torque   while the carousel started up.

(d) Compute the kinetic energy of   (i) the platform,   (ii) a horse at $3\,\text{m}$,   (iii) a horse at $4\,\text{m}$, and   (iv) the iron railing   when the carousel is at its operating speed.

(e) Assemble the results of (d) to compute the total kinetic energy of the carousel.

R.9   A carousel consists of a disk-shaped floor with mass M and radius R, along with twenty-four horses, each of mass m_h. The horses are arranged in two rings of twelve, at radii r_i [inner] and r_o [outer], respectively.

(a) Determine the moment of inertia of the carousel about its central axis of rotation (treating the floor as a uniform disk).

(b) A better model of the carousel recognises that the floor is actually an annular disk with inner radius R_i, outer radius R_o, and mass M. Compute the moment of inertia of the carousel in this model. Comment.

R.10

An object of mass M has moment of inertia I_A when rotated about axis A, and I_B about axis B. The two axes are parallel and are at perpendicular distances d_A and d_B from the centre of mass of the object, as illustrated in the adjacent figure.

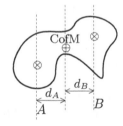

Ascertain the relation between I_A and I_B.

R.11   [See also R.10.] Redo R.10, when $d_A = d_B$. Comment.

R.12   Two point masses, M_1 and M_2, lie along the y-axis, with M_1 at the origin $(0\,,0\,,0)$ and M_2 at $(0\,,1\,,0)$.

(a) Determine the location of the CofM of this two-particle system.

(b) Determine the moment of inertia of this system about an axis parallel to the z-axis, passing through   (i) the CofM and   (ii) M_1 (*i.e.*, the origin).

(c) Determine the moment of inertia of this system about an axis parallel to the x-axis, passing through   (i) the CofM and   (ii) M_1 (*i.e.*, the origin).

(d) Determine the moment of inertia of this system about the y-axis (passing through the CofM).

(e) State which of the cases considered in (b–d) yields the   (i) smallest and (ii) greatest   value for the moment of inertia of the system.

R.13 [See also M.35.] A pizza may be modelled as a thin uniform circular disk of radius R, and constant mass-per-unit-area σ_0.

(a) Determine the total mass of the pizza.

(b) Determine the moment of inertia of the pizza about   (i) its symmetry axis and (ii) an axis through its diameter.

(c) Set up an integral for the moment of inertia of the pizza about an axis   (i) in the plane of the pizza and tangent to it, and   (ii) perpendicular to the plane containing the pizza, passing through a point on its rim.

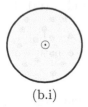
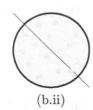
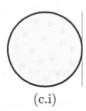
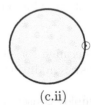

(b.i)   (b.ii)   (c.i)   (c.ii)

R.14 [See also M.36.] The moment of inertia of a pie-shaped wedge of pizza is its mass-weighted squared distance from an axis of rotation. Suppose that the slice of pizza has angular width $\Theta_0 = 2 \times \theta_0$, radius R, and constant areal mass-density σ_0.

(a) Determine the mass of the pizza slice.

(b) Determine the moment of inertia of the slice about the axis perpendicular to the slice and passing through its tip. [HINT: Here, all of the mass elements along sections of circular arc centred at the tip are the same distance from the axis.]

R.15 [See also M.36 and R.14.] Consider the slice of pizza studied in the previous problem. The location of its CofM was determined in M.36. The rotational inertia of the slice has bearing on the ultimate success or failure of the ONE-HANDED PIZZA GRAB.

(a) Employ the PARALLEL AXIS THEOREM to estimate the slice's moment of inertia about the axis perpendicular to the slice and passing through its CofM.

(b) Estimate the slice's moment of inertia about an axis perpendicular to the slice and passing through the edge of the crust along the bisector.

R.16 A thin rod has length L and lineal mass density $\lambda = \lambda_0 \left(1 + \frac{x^2}{L^2}\right)$, where λ_0 is a constant, $x = 0$ at one end of the rod, and $x = L$ at the other.

(a) Compute the total mass of the rod.

(b) Determine the location of the CofM of the rod.

(c) By explicit calculation, determine the moment of inertia of the rod about the axis perpendicular to the length of the rod and passing through   (i) $x = 0$,   (ii) $x = L$, and   (iii) the CofM.

(d) Employ the PARALLEL AXIS THEOREM to relate the results obtained in part (c).

R.17 A bolt is to be tightened to 100 N·m of torque using a 25 cm spanner. What is the theoretical   (a) minimum and   (b) maximum   force necessary to accomplish the task?

R.18 [See also D.30, E.16.] Two blocks, with masses $m_1 = 2$ kg and $m_2 = 4$ kg, are joined by a length of ideal rope passing over a pulley, as shown in the figure alongside D.30. The pulley has radius 0.5 m and moment of inertia $\frac{1}{2}$ kg · m^2. Block 1 slides along a frictionless plane, while Block 2 falls. Neither block is affected by drag. Determine   (a) the acceleration of the system,   (b) the tension in each straight section of the ideal rope, and   (c) the speed of the blocks once each has moved 1.2 m, starting from rest.

R.19 [See also D.89, E.37.] Two masses, m_1 and m_2 ($m_1 > m_2$), hang by ideal rope from a pulley with mass M and radius R, riding on frictionless bearings. The situation is as shown in the figure alongside D.89. In this instance, however, the pulley has moment of inertia $I = \frac{1}{2} M R^2$. The system is held at rest, and then suddenly released at time t_i. We shall compute the speeds of the two masses at the instant at which they pass each other.

(a) Compute the initial kinetic energies of   (i) m_1,   (ii) m_2, and   (iii) the pulley.

(b) Suppose that m_2 is rising with speed v at the instant, t_f, at which the particles pass one another. Compute the final kinetic energy of   (i) m_1,   (ii) m_2, and   (iii) the pulley.

(c) Determine the net change in the gravitational potential energy of the constituents of the system during the time interval $[t_i, t_f]$.

(d) Employ conservation of energy to ascertain the final speed of the system, v.

R.20
Two skaters, each of mass M, glide without friction on a large ice surface. An ideal rope of length $4\,l$ is held taut between them, and the skaters move in a circular path about their combined centre of mass with common speed v_i, as in the left-hand figure.

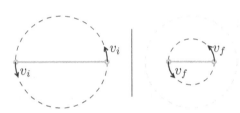

After some time has elapsed, the two skaters pull on the rope in unison, until the distance between them is decreased to $2\,l$ as shown in the right-hand figure. It took T s for the skaters to transition from the initial to the final state of motion.

(a) Compute the initial angular   (i) velocity and   (ii) momentum   of each of the skaters about the axis perpendicular to the ice and through the CofM.

(b) Compute the initial total   (i) angular momentum and   (ii) rotational kinetic energy   of the system.

(c) (i) Determine the torque exerted about the axis of rotation by each of the skaters during the time interval, T, in which they pulled inward on the rope.   (ii) By how much might one expect the total angular momentum of the skaters to change?
(iii) Explain why the total mechanical energy of the skaters might be expected to change.

(d) Infer the final   (i) angular and   (ii) tangential   speed of the skaters.

(e) Compute   (i) the final value of the kinetic energy of the skaters,   (ii) the mechanical work input to the system, and   (iii) the average power input by the skaters while they pulled on the rope.

R.21 Repeat problem R.20 when the skaters have masses $M_1 = M + m$ and $M_2 = M - m$, and initial speeds $v_{1,i}$ and $v_{2,i}$; and the centre of mass remains stationary.

R.22 A system consists of two equal-mass, M, point particles, with fixed separation $2\,L$, rotating about a common axis [the z-axis], with the same angular speed, ω.

[This system is not unlike the baton model of massive endcaps joined by a massless rod, studied in Chapter 36.]

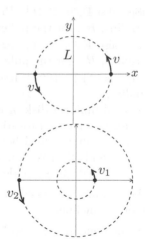

(a) Suppose that the masses rotate (symmetrically) about the axis located half-way between them and extending perpendicularly from the page. Compute the (i) moment of inertia and (ii) total angular momentum, $\vec{L}_{(a)}$, of the system about the axis.

(b) Here the masses rotate (asymmetrically) about the axis located one-quarter of the separation distance from M_1 and extending perpendicularly from the page. *I.e.*, $d_1 = L/2$ and $d_2 = 3L/2$. Compute the (i) moment of inertia and (ii) total angular momentum, $\vec{L}_{(b)}$, of the system about the axis.

(c) Compare (i) $\vec{L}_{(a)}$ to $I_{(a)}\,\vec{\omega}$, and (ii) $\vec{L}_{(b)}$ to $I_{(b)}\,\vec{\omega}$.

R.23 Repeat R.22 when the particles are rotating in the plane $z = 1$, and the point about which the angular momentum is computed is the origin.

R.24 [See also D.39, E.36, M.47.] Two blocks of the same mass, M, are joined by a length of ideal rope which is slung over a pulley. The pulley has mass $M_\mathbb{P}$, radius $R_\mathbb{P}$, and moment of inertia $I_\mathbb{P}$. The system was held at rest until $t = 0$, at which time it was released and given a short sharp tap, (instantaneously) setting the system in motion with speed v_0. WLOG, the left block may be assumed to rise while the right block falls.

(a) Determine the initial angular momenta of (i) the left block, (ii) the right block, and (iii) the pulley about the axis of rotation of the pulley. [HINT: Assume that the pulley is symmetric about its axis.]

(b) Express the total angular momentum of the system in terms of $\{M, I_\mathbb{P}, v_0, R\}$.

(c) Draw FBDs, and enumerate the forces acting on each of the blocks and on the pulley.

(d) Determine the net torque acting on the system [taken about the axis of rotation of the pulley].

(e) Comment on the time dependence of the angular momentum of the system.

R.25 [See also D.39, E.36.] Two blocks of differing mass are joined by a length of ideal rope which is slung over a pulley. The pulley has mass $M_\mathbb{P}$, radius $R_\mathbb{P}$, and moment of inertia $I_\mathbb{P}$. The system was held at rest until $t = 0$, at which time it was released and given a short sharp tap, (instantaneously) setting the system in motion with speed v_0. WLOG, the left block may be assumed to rise while the right block falls. The mass of the left block is M_L, and that of the right block is M_R.

(a) Determine the initial angular momenta of (i) each of the blocks and (ii) the pulley about the axis of rotation of the pulley. [The pulley is symmetric about its axis.]

(b) Express the total angular momentum of the system in terms of $\{M_L, M_R, I_\mathbb{P}, v_0, R\}$.

(c) Draw FBDs, and enumerate the forces acting on each of the blocks and on the pulley.

(d) Determine the net torque acting on the system [taken about the axis of rotation of the pulley].

(e) (i) Formally take the time derivative of the expression for the total angular momentum of the system. (ii) Ascertain the acceleration, $a = \frac{dv}{dt}$.

R.26 [See also D.39, R.24.] Redo D.39, incorporating a moment of inertia, I_P, and an explicit radius, R_P, for the pulley.

R.27 [See also E.36, R.24.] Redo E.36, incorporating a moment of inertia, I_P, and an explicit radius, R_P, for the pulley.

R.28 [See also M.47, R.24.] Redo M.48, incorporating a moment of inertia, $I_P = \frac{1}{9}$ kg·m^2, for the pulley.

R.29 A solid uniform disk of mass M and radius R is rolling without sliding. Compare its bulk translational and rotational kinetic energies.

R.30 A thin uniform steel hoop and a homogeneous cylindrical slab of wood both have total mass M and effective radius R. The hoop and the slab are initially at rest side-by-side on a finite inclined plane (angle θ), a distance L from the base of the incline.

(a) Determine the speed of the   (i) hoop and   (ii) cylinder   as each reaches the base of the incline.

(b) Which one arrives first?

R.31
Consider the semi-circular half-pipe with radius P shown in the figure.

(a) A snowboarder starts from rest at $\mathcal{A}$ and slides without friction. Determine   (i) his speed at the bottom of the pipe, $\mathcal{B}$, and   (ii) whether he reaches the point marked $\mathcal{C}$.

(b) A tumbler of mass M, effective radius $R \ll P$, and moment of inertia $I = M R^2/4$ rolls without sliding along the wall of the half-pipe. Determine   (i) the speed of the tumbler at the bottom of the pipe, and   (ii) whether he reaches $\mathcal{C}$.

(c) A roly-poly physics professor, with mass M, effective radius $R \ll P$, and moment of inertia $I = \frac{2}{5} M R^2$, somersaults [no sliding] along the half-pipe. Determine   (i) the speed of the professor at the bottom of the pipe, and   (ii) whether he reaches $\mathcal{C}$.

(d) A timorous tobogganist, also of mass M, digs his heels into the snow on the sides of the half-pipe in such manner as to provide a constant resistive force, F_R, in direct opposition to the motion of the toboggan. Determine   (i) the speed of the tobogganist at the bottom of the pipe, and   (ii) whether he reaches $\mathcal{C}$.

R.32 [See also K.123.] The trajectory of a particle of mass M, expressed in Cartesian coordinates, is $\vec{r}(t) = (1\,,\, t\,,\, 0)$.

(a) Determine the angular momentum [about the origin] of this particle.

(b) Ascertain the rate of change of the angular momentum in (a). Comment.

R.33 [See also K.124.] The trajectory of a particle of mass M, expressed in Cartesian coordinates, is $\vec{r}(t) = (X_0\,,\, V_0\,t\,,\, 0)$.

(a) Determine the angular momentum [about the origin] of this particle.

(b) Ascertain the rate of change of the angular momentum in (a). Comment.

S

Statics Problems

S.1

A block of mass M_1 rides on a plane. On one side, it is attached by means of a spring (with spring constant k) to an anchor. On the other side, it is connected to another block by a length of ideal rope which is slung over a pulley.

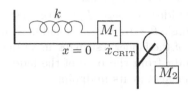

(a) Assuming that the plane upon which the block rides is frictionless, determine the position, x_{CRIT}, at which the block remains at rest.

(b) With static friction, coefficient μ_s, acting between the block and the plane, determine the range of positions, $x_- < x < x_+$, for which the block remains at rest.

S.2 A 1.0 m board of negligible mass is supported at each end by ideal ropes, forming a seat. When a child sits on the seat, the tension in the left rope is 300 N; in the right, 200 N. Determine the weight of the child, and where on the board he is sitting.

S.3 Two children, with masses 25 and 30 kg respectively, sit on a seesaw of length 4 m, which is centred on a fulcrum and held in a horizontal position. If the smaller child sits at the very end of one side, where must the other child sit in order that the seesaw not move if it is released?

S.4 A nail may be removed from a board by means of a crowbar. [The claw grasps the head of the nail to be removed; you grasp the handle.] A particular crowbar has a handle of length 50 cm, and a 10 cm long claw. Assuming that the force exerted on the nail is parallel to the nail, and that the crowbar has negligible mass, determine the maximum and minimum forces exerted on the nail when 500 N of force is applied to the handle.

S.5 A cupboard door has width W, height H, and mass M. It is symmetric, so its centre of mass is located at its geometric centre. It is supported on the door frame by two small hinges, placed at the top and bottom of the door.

(a) Relate the   (i) horizontal and   (ii) vertical   forces exerted on the door, by the hinges, to each other and to the weight of the door.

(b) Infer the forces acting at the hinges.

S.6

A block of mass M is suspended by ideal rope slung over a massless brace of length L inclined at angle θ, as shown in the figure. Determine the magnitude of the horizontal force which must be exerted on the rope to hold the mass at rest.

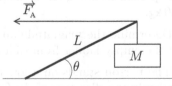

S.7 A uniform beam of length 8 m and weight 800 N extends horizontally from a pivot mounted on a vertical wall. The far end of the beam is supported by an ideal cable which is anchored at a point 6 m directly above the pivot. Furthermore, a person of weight 500 N stands on the beam at a point 2 m from the wall. Find the tension in the cable and the force provided by the pivot.

S.8
The boom of a crane supports a cable to which is attached a block of mass M, as is indicated in the nearby figure. The boom is held in place, horizontally, by a pivot at one end, and a strut joined at 45° to its midpoint.

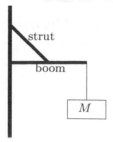

(a) Assuming that the boom is massless, determine the tension in the strut and the force which acts at the pivot.

(b) Assuming that the boom is [approximately] a uniform rod, with mass M_b, determine the tension in the strut and the force which acts at the pivot.

S.9 A ladder of mass $M_L = 50\,\text{kg}$ rests on a frictional floor and leans against a frictionless wall at angle $\theta = 67.38°$. The ladder is 6.5 m long, and the feet are 2.5 m away from the wall. [The top of the ladder rests 6.0 m above the floor.] A person with mass $M_P = 100\,\text{kg}$ stands on a rung 5.0 m from the base of the ladder [or 1.5 m from the top]. Assume that the weight of the ladder acts at its midpoint.

(a) Sketch a FBD showing all of the forces which act on the ladder and the point at which each is applied.

(b) Determine the normal force of contact acting between the ladder and the   (i) floor and   (ii) wall.

(c) Determine the minimum value of the coefficient of static friction acting between the feet of the ladder and the floor for which the ladder and person remain safely at rest.

S.10 A 5 m ladder of weight 200 N rests against a smooth wall at height 4 m. A person weighing 500 N stands 1 m from the top of the ladder. Determine the magnitude of the static friction force acting between the floor and the feet of the ladder. Comment on the value of the coefficient of static friction.

S.11 While preparing a meal, PK rests a spatula against the side of an old, much-used, tall cooking pot. The spatula is 20 cm long, and its centre of mass is located 5 cm from the end of its blade. [This is the point at which the weight force is deemed to act.] The blade touches the bottom of the pan 12 cm from the wall. The coefficient of static friction acting between the spatula and the [formerly "non-stick"] bottom surface of the pan is $\mu_s = 0.125$. Does the spatula slip?

S.12
A ladder rests on horizontal ground and leans at an angle of 60° against a vertical wall, as shown in the figures. Static friction acts between the feet of the ladder and the floor, and the wall is effectively frictionless. The ladder has mass 20 kg and length 8 m. The person standing on the ladder has mass 80 kg.

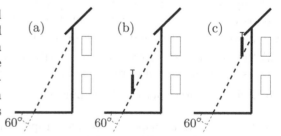

(a) Determine the magnitude and direction of the force of friction which must act to keep the empty ladder from sliding.

(b) The person stands on a low rung, 2 m from the base of the ladder. Determine the magnitude and direction of the force of friction which must act to keep the ladder-and-person system from slipping.

(c) The person stands on a high rung, 6 m from the base of the ladder. Determine the magnitude and direction of the force of friction which must act to keep the ladder-and-person system from slipping.

S.13 *Check Mate!*
A gallant knight attempts to rescue his fair queen, held captive
in a tower [rook] protected by a moat. Fortunately for the knight,
the queen is in a lower room with a window, and his bishop has
left him a ladder. The ladder is 5 m long, is symmetric, has rungs
extending all the way from the feet to the top, and weighs 200 N.
The moat is 3 m wide, so the knight can (just barely) position the
ladder to reach the window, which is 4 m up the castle wall. As
the bank is muddy, the coefficient of static friction between the
foot of the ladder and the bank is 0.5. The walls of the tower are
wet and covered in slippery vines, so the force of friction between
the ladder and the tower is effectively zero. The knight and the
queen have weights of 1000 N and 500 N, respectively.

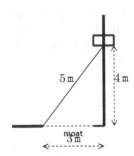

(a) Ascertain whether the knight's ladder is able to lean safely against the tower with
no one on it.

(b) In order to rescue the queen, the knight must stand on the topmost rung while
carrying her in his arms. Determine whether this configuration of the ladder–knight–
queen system is stable.

(c) The knight is accompanied by several pawns, each weighing 700 N. How many pawns
must stand on the foot of the ladder so as to not have it "kick out" while the knight is
rescuing the queen?

S.14 A radio tower stands H metres tall over a level plain. The prevailing wind, from
the west, exerts a certain force per unit length, ϕ, on the tower. A [massless] guide wire is
attached to the top of the tower and makes an angle θ with the ground at a point due west
of the base.

(a) Suppose that $\phi = \phi_c$ is constant, and determine the tension in the guide wire.

(b) Suppose that ϕ increases linearly from its ground value, ϕ_g, to its value at the top
of the tower, ϕ_t. (i) Write down an expression for the force per unit length, ϕ, as a
function of height, in this model. (ii) Determine the tension in the guide wire.

S.15 A thin uniform steel hoop and a homogeneous cylindrical slab of wood have the same
total mass M and radius R. The hoop and the slab are resting side-by-side on a flat surface,
pressed up against a small step of height h. Compare the magnitudes of the forces, applied
at corresponding sites on the hoop and slab, needed to roll the objects up the step.

S.16
A keg of Bavarian root beer, with mass 75 kg and radius
25 cm, abuts a curb of height 5 cm.
(a) Compute the force required to roll the keg up the step
when it is pushed/pulled at the point on rim directly opposite to the point of contact with the curb.
(b) Comment on the amount of force that would have been
necessary to roll the keg up the curb, had another point
been chosen for the application of force.

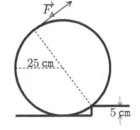

S.17 Two students, rolling a beer keg of radius 0.25 m and weight 500 N, encounter a step
of height 10 cm. One student, standing on the lower ground, lifts straight up on the vertical
side of the keg [away from the step] with a force of 100 N. The other student, standing on the
higher part, pulls the top of the keg horizontally toward himself. Compute the minimum
horizontal force that the second student must apply in order for the pair of students to
succeed in getting the keg up the step.

S.18   A sawmill has a log feeding mechanism consisting of a pair of conveyor belts arranged on the sides of a V-shaped guide as shown in the figure. Raw logs are pushed by a vertical plough until they are stopped by the lip at the edge of the guide. Once the previous log has been conveyed out the way, the force exerted by the plough is increased so as to roll the log up the lip and onto the guide. Suppose that the log has mass $M = 1100\,\text{kg}$, and that its geometry (size and shape) is well-approximated by a cylinder of radius $R = 30\,\text{cm}$ and length $L = 5\,\text{m}$. Further suppose that the lip is $h = 5\,\text{cm}$ high and that the log rolls without slipping. Determine the minimum force needed to roll the log onto the guide.

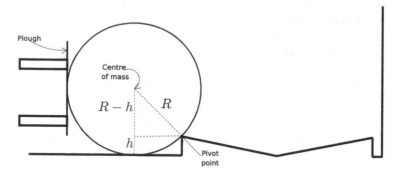

G

Gravitation Problems

Useful Data

G The gravitational force constant is [approximately] $G \simeq 6.67 \times 10^{-11} \; \frac{\text{m}^3}{\text{kg} \cdot \text{s}^2}$.

$\odot$ The Sun's mass and radius are $M_\odot \simeq 1.99 \times 10^{30}$ kg and $R_\odot \simeq 6.96 \times 10^8$ m.

$\oplus$ The Earth's mass and radius are $M_\oplus \simeq 5.98 \times 10^{24}$ kg and $R_\oplus \simeq 6.37 \times 10^6$ m.

Moon   The Moon's mass and radius are roughly 7.36×10^{22} kg and 1.74×10^6 m.

a   The mean Sun–Earth distance is 1.50×10^{11} m.

     The mean Earth–Moon distance is 3.82×10^8 m.

G.1   [See also G.5.]

Xavier, Yolanda, and Zoe have masses $M_X = 75$ kg and $M_Y = M_Z = 50$ kg. They are at rest at the locations indicated in the figure.

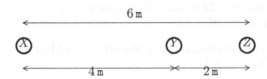

(a) Compute the magnitudes of the pairwise gravitational forces (use units of $625\,G$):
(i) Y–Z,   (ii) Z–X,   and   (iii) X–Y.

(b) Determine the net force acting on   (i) Xavier,   (ii) Yolanda,   and   (iii) Zoe.

G.2   [See also G.8.] Two point masses, M and $3\,M$, separated by a distance $2\,X$, divide the x-axis into three regions: left, middle, and right. The three field points under consideration are: $\mathcal{P}_l$, lying a distance X from M and $3\,X$ from $3\,M$; $\mathcal{P}_m$, exactly half-way between M and $3\,M$; and $\mathcal{P}_r$, at distance X from $3\,M$ and $3\,X$ from M. Determine the net gravitational field at the points   (l) $\mathcal{P}_l$,   (m) $\mathcal{P}_m$,   and   (r) $\mathcal{P}_r$.

G.3

PK's mass is $M_{\text{PK}} = 85$ kg, and Gaia's is $M_G = 6 \times 10^{24}$ kg. Each is at rest, separated by 6.37×10^6 m, as illustrated in the nearby sketch.

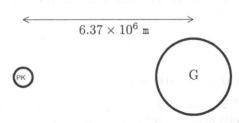

(a) Compute the magnitude and direction of the gravitational field produced by Gaia at PK's location.

(b) From the result for (a), determine the gravitational force exerted by Gaia upon PK. Comment.

G.4 Suppose that there exists a planet with one-half the mass of the Earth and one-half its radius. Determine the factor relating the weight of an astronaut on the surface of this planet to his or her weight on the surface of the Earth.

G.5 [See also G.1.] Recall the Xavier, Yolanda, and Zoe system described in G.1.

(a) Determine the pairwise gravitational potential energies (in units of $625\,G$):   (i) Y–Z, (ii) Z–X, and   (iii) X–Y.

(b) Ascertain the net, or total, potential energy of the system.

G.6 An amount of mass, M, is to be split into two parts, m_1 and m_2, which are to be held at fixed separation R. The separation scale is large enough that the two parts are both effectively point-like. The partition of the mass is determined by the factor x, representing the fraction assigned to m_1, viz., $x = \frac{m_1}{M} = \frac{m_1}{m_1+m_2}$ and $1 - x = \frac{m_2}{M} = \frac{m_2}{m_1+m_2}$.

(a) Determine the values of x which   (i) maximise and   (ii) minimise   the magnitude of the gravitational force acting between the two parts.

(b) Determine the values of x which   (i) maximise and   (ii) minimise   the gravitational potential energy of the two-mass system.

G.7 Adapt the derivation of the gravitational potential energy of two point-like masses [in Chapter 48] to show that the gravitational potential of a single point-like particle with mass m is $V_{\mathrm{G},m} = -\frac{G\,m}{r}$, where r is the distance from the source to the field point.

G.8 [See also G.2.] Recall the system of two point masses and three field points considered in G.2. Determine the net gravitational potential at the points   (l) $\mathcal{P}_l$,   (m) $\mathcal{P}_m$, and (r) $\mathcal{P}_r$.

G.9 [See also G.10.] Two concentric thin uniform spherical shells of matter are centred on the origin. They have the same mass, M. The inner shell has radius R, while the outer shell has radius $2\,R$.

(a) Determine the gravitational field in the region of space outside of both shells, *i.e.*, for $r > 2\,R$.

(b) Determine the gravitational field in the space between the two shells, *i.e.*, for $R < r < 2\,R$.

(c) Determine the gravitational field within the inner shell, *i.e.*, for $r < R$.

G.10 [See also G.9.] Consider the two equal-mass shells in G.9. Determine the gravitational potential produced by this configuration   (a) outside,   (b) between, and   (c) within the pair of shells.

G.11 Two thin spherically symmetric shells of matter are centred at the origin. The inner shell has mass M_i and radius R_i, while the outer shell has mass M_o and radius R_o.

(a) Determine the magnitude of the net gravitational field outside of the shells, *i.e.*, $r > R_o$.

(b) Determine the magnitude of the net gravitational field between the shells, *i.e.*, $R_i < r < R_o$.

(c) Determine the magnitude of the net gravitational field inside both shells, *i.e.*, $r < R_i$.

(d) By convention, the gravitational potential of any bounded system (like the two shells here) is taken to vanish at infinite distance, *i.e.*, $r \to \infty$. Starting from $V_g = 0$ as $r \to \infty$, infer the gravitational potential in the three regions studied in (a–c).

G.12

Contrary to everything that you have been told, the Earth is not solid, but instead consists of a spherically symmetric thick-shell crust supported by massless scaffolding of height 500 km above a solid interior core. The mass of the shell is $M_s = 2.99 \times 10^{24}$ kg, and its inner and outer radii are $R_{s,i} = 5000$ km and $R_{s,o} = 6370$ km respectively. The core happens to have the same amount of mass as the shell, $M_c = 2.99 \times 10^{24}$ kg, and the core radius is $R_c = 4500$ km.

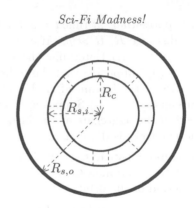

(a) Determine the magnitude and direction of the gravitational field on the outer surface of the crust, *i.e.*, where we live.

(b) Determine the strength of the gravitational field on the surface of the core, *i.e.*, where the *aliens* live.

(c) Express the magnitude of the gravitational field at distance r from the centre of the Earth, where   (i) $R_c < r < R_{s,i}$ and   (ii) $R_{s,o} < r$.

(d) What additional information or assumptions would be needed to express the gravitational field within the core and the solid shell?

G.13

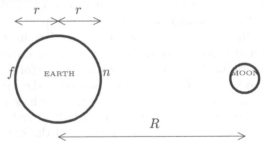

Consider the [isolated] Earth–Moon system at an instant as shown in the adjacent figure. The portion of the Earth's surface nearest to the Moon is labelled n, while the most distant point is denoted f. [The sketch is not to scale.]

(a) Compute the magnitude and direction of the gravitational field produced by the Moon at   (i) n and   (ii) f.

(b) Ascertain the difference between these two fields.

(c) Construct a Taylor series, or a binomial expansion, approximation to the gravitational field produced by the Moon in a neighbourhood centred on the centre of the Earth.

(d) Employ the approximation in (c) to estimate the difference in the gravitational fields acting at the points n and f.

(e) Comment.

G.14 A small satellite of mass m is in circular orbit about a large spherically symmetric body of mass M. The radius of the orbit is r, and the speed of the satellite is v. How does the speed of the satellite change if   (a) m,   (b) M, and   (c) both m and M are quadrupled?

G.15   Old science fiction stories proposed the existence of a *mysterious* Planet X just like Earth, but on the other side of the Sun, where we cannot see it. What effect would the discovery of *Planet X* have on our current estimate of the Sun's mass? [Recall that the solar mass is inferred from the Earth–Sun distance and the length of the year.]

G.16   Two space capsules, each of mass M, are in empty deep space. They are spherically shaped with radius ρ [to give maximum ratio of volume to surface area].

(a) Suppose that, at time $t = 0$, the two capsules are at rest with their centres separated by a distance R, $R \gg \rho$. After $t = 0$, they fall toward each other under the influence of gravity.   (i) Determine the speeds of the capsules just as they are about to collide.   (ii) Supposing that the collision is completely elastic, compute the speeds of the capsules at the instant after the collision, and determine how much mechanical energy is dissipated.   (iii) Supposing that the collision is completely inelastic, compute the speed of the aggregate at the instant following the collision, and determine how much kinetic energy is dissipated.

(b) Suppose instead that, at time $t = 0$, the capsules are moving in stable circular orbit about their common CofM.   (i) Compute the speeds of the blocks.   (ii) Determine the period of the orbit.

G.17   *The Newtonian Black Hole* [as conceived by Pierre Laplace, *circa* 1798]

(a) Derive the expression for the escape speed of a particle of mass m, at rest on the surface of the Sun, as a function of the solar radius, $R_\odot$. [Disregard for the moment the experimentally determined value of its radius.]

(b) Determine the largest value of $R_\odot$ for which the escape speed is in excess of the speed of light.

[Thus, were all of the mass of the Sun concentrated into a spherical region with radius smaller than that determined in (b), "particles of light" emitted from its surface would be unable to escape. Fortunately, the Sun's radius is far in excess of this critical value.]

G.18   Geostationary satellites maintain their position relative to points on the surface of the Earth while in orbit about its centre. For this to be so, the orbital period of the satellite must match that of the daily rotation of the Earth upon its axis. [Take this to be 24 h.]
(a) Determine the distance above the surface (of the Earth) at which geostationary satellites are found.   (b) Compare the distance of the satellite from the centre of the Earth with the distance between the centres of the Earth and Moon.

G.19   [See also G.18.] The Moon is tidally locked[1] with the Earth, and therefore its sidereal period of revolution—about the centre of the Earth—matches its period of rotation about an axis passing through the lunar poles. The duration of this common period is approximately 27.32 days.   (a) Determine the distance from the centre of the Moon at which a lunarstationary satellite would have to be placed.   (b) Compare this distance with the average radius of the lunar orbit.

G.20   [See also G.18.] The Sun does not rotate rigidly and its surface features have varying periods of rotation. For the purposes of this exercise, take the solar period to be 25.3 days.
(a) Determine the distance from the centre of the Sun at which a solarstationary satellite would have to be placed.   (b) Compare the distance you obtain with 7×10^8 m, the approximate radius of the Sun, and with 5.8×10^{10} m (roughly the semi-major axis of Mercury's rather eccentric orbit).

G.21   The International Space Station [ISS] orbits the Earth at about 370 km above the surface. Estimate the orbital   (a) speed and   (b) period   of the ISS.

G.22   The Earth's orbit around the Sun is elliptical. Given that the distance between the centres of the Earth and Sun is approximately 147.1×10^9 m at perihelion (in early January of each year) and 152.1×10^9 m at aphelion (in early July), estimate the eccentricity of the Earth's orbit.

---

[1] Tidal locking is discussed in Chapter 1 of VOLUME II.

List of Symbols

$|\vec{A}|$, $|\vec{a}|$ Magnitudes of the specified vectors $\vec{A}$ and $\vec{a}$, respectively

$\Delta\theta$ Angular displacement

$\Delta\vec{r}$ Vector displacement

ΔE The change in total mechanical energy during an explicit or implicit time interval

ΔS Arc length (especially along a segment of a circle)

Δt Time interval: a distinct period of time as measured on the synchronised clocks belonging to a set of inertial observers

Δx Displacement in 1-d; Cartesian x-component of displacement

ϵ, R, l Eccentricity, radius, and (constant) angular momentum per unit mass [in the context of conic section dynamical solutions for trajectories of small particles subjected to a central force, *e.g.*, Newtonian universal gravitation]

$F_{\parallel}$ Component of a force acting parallel to a given line or direction

$F_{\perp}$ Component(s) of a force acting perpendicular to a given line or direction

$\vec{g}$ Gravitational field: local gravitational force per unit mass

$\hat{\imath}$, $\hat{\jmath}$, $\hat{k}$ Unit vectors in the x, y, and z Cartesian basis directions

$\hat{r}$, $\hat{\theta}$ Unit vectors in the polar basis directions, r and θ

λ, σ, ρ Mass densities for lineal, areal, and volume distributions of matter

μ_{s}, μ_{k} Coefficients of static and kinetic friction, respectively

ω Angular speed; angular frequency

$r_{\perp}$, d, d Moment arm associated with a force

τ Magnitude (perhaps directed, in 1-d) of torque

a   Constant (instantaneous) acceleration in 1-d

θ, ϕ, α, β Angles

θ_0, ϕ_0 Constant (or initial) angles

$\vec{A}$, $\vec{a}$ Vectors; mathematical objects whose definition requires specification of both magnitude and direction

$\vec{A}_{\mathrm{CofM}}$ Acceleration of the CofM of a system of particles or a rigid body

$\vec{C}$ Force of contact [between two particles]

$\vec{F}_{\mathrm{NET\,ext'l}}$ Net external force acting on a particle (or system)

$\vec{F}_{\mathrm{NET}}$ Net force acting on a particle (or system)

$\vec{F}$ A generic force

$\vec{F}_{\mathrm{av}}$ Average force exerted on a particle during a specified time interval

$\vec{F_A}$ Applied force

$\vec{F_C}$ A conservative force

$\vec{F}_{D1}, \vec{F}_{D2}$ Linear and quadratic drag forces, respectively

$\vec{L}$ Angular momentum of a particle with respect to a fixed point or axis

$\vec{L}_{Total}$ Total angular momentum of a system or a rigid object

$\vec{N}$ Normal force

$\vec{P}$ Support force holding a pulley [or other dynamical constituent] in place

$\vec{P}_{Total}$ Total momentum of a system of particles

$\vec{R}_{CofM}$ Position of the CofM of a system of particles or a rigid body

$\vec{\nabla}$ The gradient operator

$\vec{T}$ Tension force

$\vec{V}_{CofM}$ Velocity of the CofM of a system of particles or a rigid body

$\vec{W}$ Weight force

$\vec{0}$ The ZERO VECTOR has zero magnitude and arbitrary (unspecified) direction; its Cartesian components are all 0.

$\vec{\alpha}$ Instantaneous angular acceleration

$\vec{\alpha}_{av}$ Average angular acceleration

$\vec{f}_S, \vec{f}_K$ Static and kinetic friction forces, respectively

$\vec{\omega}$ Instantaneous angular velocity associated with a rotating body

$\vec{\omega}_{av}$ Average angular velocity associated with a rotating object

$\vec{\tau}$ Vector torque

$\vec{a} \cdot \vec{b}$ The DOT PRODUCT, or scalar product, of vectors $\vec{a}$ and $\vec{b}$ [in n dimensions]

$\vec{a} \times \vec{b}$ The CROSS PRODUCT, or vector product, of three-dimensional vectors $\vec{a}$ and $\vec{b}$

$\vec{a}$ Instantaneous vector acceleration

$\vec{a}_{av}$ Average acceleration in n-d throughout a stated or implicit time interval

$\vec{a}_c$ Centripetal [*i.e.,* centre-seeking] acceleration

$\vec{g}$ The [local] acceleration due to gravity; the local gravitational field

$\vec{I}_{if}[\vec{F}]$ Mechanical impulse produced by a force, $\vec{F}$, acting through the time interval from t_i to t_f

$\vec{p}$ Linear momentum of a particle

$\vec{r}$ Vector position in n-d

$\vec{r}_0$ A specific or initial value of vector position

$\vec{u}$ Velocity of a particle expressed in the CofM frame

$\vec{v}$ Instantaneous vector velocity

$\vec{v}_0$ A specific or initial value of vector velocity

$\vec{v}_{av}$ Average velocity in n-d throughout a stated or implicit time interval

$\vec{\mathbb{a}}$ Constant (instantaneous) vector acceleration

a Instantaneous acceleration (in 1-d)

a_{av} Average acceleration in 1-d occurring throughout a stated or implicit time interval

a_t Instantaneous tangential acceleration

$a_{\mathrm{av},t}$ Average tangential acceleration

b Impact parameter; distance of closest approach

b_1, b_2 Coefficients of linear and quadratic drag, respectively

E Total mechanical energy: the sum of the kinetic and potential energies of a particle or system

G The gravitational coupling constant

g $|\vec{g}|$: the magnitude of the local acceleration due to gravity

I Moment of inertia of a system of particles or rigid body about an axis

I_{CofM}, I_c Moment of inertia of a system of particles or a rigid body about an axis through its CofM

K Kinetic energy—some authors use "T"

k Spring [force] constant

K_{CofM} Bulk kinetic energy of a system (arising from the centre of mass motion)

K_{relative} Relative kinetic energy possessed by a system (owing to the internal motions of parts of the system WRT the CofM)

K_{Total} Total kinetic energy of the particles in a system [or constituents of a rigid body]

m, M Mass of a particle

M_{Total} Total mass of a system of particles or a rigid body

P Instantaneous power

P_{av} Average mechanical power

Q A generic (quantified) particle property

R Radial distance from fixed point; radius of circle

s Arc length along the parameterised trajectory of a particle

t Time: a parameter

$U_{[\mathrm{C}]}(\vec{r})$ A potential energy function associated with a conservative force, $\vec{F_{\mathrm{C}}}$

v Instantaneous velocity in 1-d

v_0 A specific [initial] value of the 1-d velocity

v_{av} Average velocity in 1-d occurring throughout a stated or implicit time interval

v_{T} Terminal velocity

v_t Instantaneous tangential velocity

$v_{\mathrm{av},t}$ Average tangential velocity

$V_{\mathrm{G}}(\vec{r})$ Gravitational potential produced by a source at a specified point in space: gravitational potential energy per unit mass

$W_{if}[\vec{F}]_{\mathrm{(path)}}$ Mechanical work performed by a force, $\vec{F}$, acting through a displacement along a prescribed path

$W_{if}[\,\tau\,]$ Mechanical work performed by a torque, τ, acting through an angular displacement [along a specified or implicit angular path]

x, y, z 1-d position; Cartesian component of position in 2- or 3-d

x_0 A specific [initial or equilibrium] value of the 1-d position

$\mathcal{O}$, $\mathcal{P}$, $\mathcal{Q}$ Points in space

d   The [perpendicular] distance from an axis to a particular point in space

$\mathcal{F}$ The condition that the net external force acting on a system must vanish for that system to remain in static equilibrium

K1, K2, K3 Kepler's Empirical Laws of Planetary Motion

$\mathcal{T}$ The condition that the net external torque acting on a system must vanish for that system to remain in static equilibrium

CofM   Centre of mass of a system of particles or a continuous substance

FBD   Free body diagram: a sketch displaying the forces acting on a particle

IRF   Inertial reference frame: a set of coordinates and clock(s) borne by a class of observers for whom N1 is evident

LHS, RHS Left and right hand sides (of a mathematical expression), respectively

N1   Newton's First Law of Motion: Law of Inertia

N3   Newton's Third Law of Motion: Law of Action–Reaction

N2   Newton's Second Law of Motion: $m\,\vec{a} = \vec{F}_{\mathrm{NET\,ext'l}}$

NUG   Newtonian Law of Universal Gravitation

RHR   Right hand rule: a method for assigning an orientation to a vector

WRT   With respect to

WLOG   Without loss of generality

Index